Lecture Notes in Computer Science 5926

Commenced Publication in 1973
Founding and Former Series Editors:
Gerhard Goos, Juris Hartmanis, and Jan van Leeuwen

Asunción Gómez-Pérez   Yong Yu
Ying Ding (Eds.)

The Semantic Web

Fourth Asian Conference, ASWC 2009
Shanghai, China, December 6-9, 2009
Proceedings

 Springer

Volume Editors

Asunción Gómez-Pérez
Universidad Politécnica de Madrid
Facultad de Informática
Dpto. de Inteligencia Artificial
Ontology Engineering Group
Campus de Montegancedo s/n
28660, Boadilla del Monte, Madrid,
E-mail: asun@fi.upm.es

Yong Yu
Shanghai Jiao Tong University
Data and Knowledge Management
Shanghai, 200030, China
E-mail: yyu@apex.sjtu.edu.cn

Ying Ding
Indiana University
School of Library and Information Science
1320 E. 10th St., LI 025
Bloomington, IN 47405, USA
E-mail: dingying@indiana.edu

Library of Congress Control Number: 2009939925

CR Subject Classification (1998): H.2.8, I.2.4, I.2.6, H.3.5, H.1, H.3.3, H.2

LNCS Sublibrary: SL 3 – Information Systems and Application, incl. Internet/Web and HCI

ISSN   0302-9743
ISBN-10   3-642-10870-9 Springer Berlin Heidelberg New York
ISBN-13   978-3-642-10870-9 Springer Berlin Heidelberg New York

springer.com

© Springer-Verlag Berlin Heidelberg 2009
Printed in Germany

Typesetting: Camera-ready by author, data conversion by Scientific Publishing Services, Chennai, India
Printed on acid-free paper   SPIN: 12814473   06/3180   5 4 3 2 1 0

Preface

The Annual Asian Semantic Web Conference is one of the largest regional events in Asia with focused topics related to the Semantic Web. With the decade-round endeavor of Semantic Web believers, researchers and practitioners, the Semantic Web has made remarkable progress recently. It has raised significant attention from US and UK governments, as well as the European Commission who are willing to deploy Semantic Web technologies to enhance the transparency of eGovernment. The Linked Open Data initiative is on its way to convert the current document Web into a data Web and to further enabling various data and service mashups. The fast adoption of Semantic Web technologies in medical and life sciences has created impressive showcases to the world. All these efforts are a crucial step toward enabling the take-off and the success of the Semantic Web.

The First Asian Semantic Web Conference was successfully held in China in 2006. With the following editions in Korea in 2007 and Thailand in 2008, it fostered a regional forum for connecting researchers and triggering innovations. This year, the 4th Asian Semantic Web Conference was held in Shanghai, China. We received 63 submissions from Asia, Europe, and North America, and 25 papers were accepted (the acceptance rate is around 40%). Each submission was reviewed by at least three members of the Program Committee. The Chairs moderated the discussion of conflict reviews or invited external reviewers to reach the final decisions. These submissions cover a broad range of topics including, query languages and optimization, rule and logics, scalable reasoning, semantic content generation, database and semantics, Semantic Web services, eSemantics (e.g., e-Business, e-Science, e-Learning, e-Culture, e-Health), social Web and semantics, semantic graph mining, security for Semantic Web, ontology modeling, ontology management, to name a few.

We, on behalf of the conference Program Committee, are grateful to numerous individuals for their continuous contribution and support. We are especially thankful to Linyun Fu from Shanghai Jiao Tong University and Shanshan Chen from Indiana University for their diligent work to facilitate the organization of this conference.

October, 2009

Asun Gómez-Pérez
Yong Yu
Ying Ding

Organization

Conference Chairs

James A. Hendler   Rensselaer Polytechnic Institute, USA
Ruqian Lu   Chinese Academy of Sciences, China

Program Committee Chairs

Asunción Gómez Pérez   Universidad Politecnica de Madrid, Spain
Yong Yu   Shanghai Jiao Tong University, China
Ying Ding   Indiana University, USA

Workshop Chair

Hong-Gee Kim   Seoul National University, Korea

Tutorial Chair

Jie Bao   Rensselaer Polytechnic Institute, USA

Poster and Demo Chair

Raúl García-Castro   Universidad Politécnica de Madrid, Spain

Local Organizing Chairs

Xuanjing Huang   Fudan University, China
Xiangyang Xue   Fudan University, China

Program Committee Members

Guadalupe Aguado-de-Cea   Universidad Politecnica de Madrid, Spain
Harith Alani   Knowledge Media Institute, Open University, UK
Yuan An   Drexel University, USA
Grigoris Antoniou   FORTH-ICS, Greece
Lora Aroyo   Vrije Universiteit Amsterdam, The Netherlands
Budak Arpinar   University of Georgia, USA
Walter Binder   University of Lugano, Switzerland
Paolo Bouquet   University of Trento, Italy
Dan Brickley   FOAF project

Yutaka Matsuo	University of Tokyo, Japan
Diana Maynard	University of Sheffield, UK
Brian McBride	Hewlett Packard, UK
Dunja Mladenic	J.Stefan Institute, Slovenia
Wolfgang Nejdl	L3S and University of Hannover, Germany
Barry Norton	Knowledge Media Institute, The Open University, UK
Daniel Oberle	SAP, Germany
Leo Obrst	The MITRE Corporation, USA
Daniel Olmedilla	Telefonica R&D, Spain
Jeff Z. Pan	University of Aberdeen, UK
Massimo Paoluccci	DoCoMo Euro labs
Terry Payne	University of Liverpool, UK
Carlos Pedrinaci	Knowledge Media Institute, The Open University, UK
Ruzica Piskac	Swiss Federal Institute of Technology (EPFL), Switzerland
Dimitris Plexousakis	Institute of Computer Science, FORTH, Greece
Guilin Qi	University of Karlsruhe, Germany
Yuzhong Qu	Southeast University, China
Ulrich Reimer	University of Applied Sciences St. Gallen, Switzerland
Marta Sabou	Knowledge Media Institute, The Open University, UK
Twittie Senivongse	Chulalongkorn University, Thailand
Michael Stollberg	SAP Research CEC Dresden, Germany
Umberto Straccia	ISTI-CNR, Italy
Heiner Stuckenschmidt	University of Mannheim, Germany
Rudi Studer	University of Karlsruhe, Germany
Mari Carmen Suárez-Figueroa	Universidad Politécnica de Madrid, Spain
Vojtech Svatek	University of Economics, Prague, Czech Republic
Jie Tang	Tsinghua University, China
Sergio Tessaris	Free University of Bozen - Bolzano, Italy
Robert Tolksdorf	Freie Universität Berlin, Germany
Ioan Toma	STI Innsbruck, Austria
Thanh Tran	AIFB, University Karlsruhe, Germany
Raphael Troncy	EURECOM, France
Victoria Uren	University of Sheffield, UK
Tomas Vitvar	University of Innsbruck, Austria
Holger Wache	University of Applied Science Northwestern Switzerland, Switzerland
Haofen Wang	Shanghai Jiao Tong University, China
Krzysztof Wecel	Poznan University of Economics, Poland
Takahira Yamaguchi	Keio University, Japan
Zhixiong Zhang	Library of Chinese Academy of Science, China
Hai Zhuge	Institute of Computing Technology, Chinese Academy of Sciences, China

Additional Reviewers

Chris van Aart
Sebastian Blohm
Juergen Bock
Stefano Bortoli
Gianluca Correndo
Martin Doerr
Jun Fang
Cristina Feier

Angela Fogarolli
Chiara Ghidini
Rigel Gjomemo
Andreas Harth
Zhixing Huang
Qiu Ji
Beate Krause
Lina Lubyte

Jose Mora
Takeshi Morita
José Ángel Ramos
Yuan Ren
Merwyn Taylor
Alex Villazon

Posters and Demos Program Committee

Yuan An — Drexel University, USA
Walter Binder — University of Lugano, Switzerland
Paolo Bouquet — University of Trento, Italy
Mari Carmen Suarez-Figueroa — Universidad Politecnica de Madrid, Spain
Tom Heath — Talis
Philipp Kaerger — L3S Research Center, Germany
Michel Klein — Vrije University Amsterdam, The Netherlands
Kouji Kozaki — Osaka University, Japan
Reto Krummenacher — STI Innsbruck, Austria
Adrian Mocan — STI Innsbruck, Austria
Malgorzata Mochol — FU Berlin, Germany
Carlos Pedrinaci — The Open University, UK
Ruzica Piskac — EPFL, Switzerland
Chris Poppe — Ghent University, Belgium
Martin Szomszor — University of Southampton, UK
Denny Vrandecic — Universitaet Karlsruhe, Germany

Sponsors

http://research.google.com/

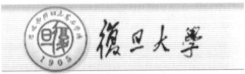

http://www.fudan.edu.cn/

http://www.sti2.org/

Table of Contents

Ontologies

Semantic Web Application

Semantic Web Services

Semantic Web Technology

Demo Papers

Cross-Lingual Ontology Mapping – An Investigation of the Impact of Machine Translation

Bo Fu, Rob Brennan, and Declan O'Sullivan

Centre for Next Generation Localisation & Knowledge and Data Engineering Group,
School of Computer Science and Statistics, Trinity College Dublin, Ireland
{bofu, rob.brennan, declan.osullivan}@cs.tcd.ie

Abstract. Ontologies are at the heart of knowledge management and make use of information that is not only written in English but also in many other natural languages. In order to enable knowledge discovery, sharing and reuse of these multilingual ontologies, it is necessary to support ontology mapping despite natural language barriers. This paper examines the soundness of a generic approach that involves machine translation tools and monolingual ontology matching techniques in cross-lingual ontology mapping scenarios. In particular, experimental results collected from case studies which engage mappings of independent ontologies that are labeled in English and Chinese are presented. Based on findings derived from these studies, limitations of this generic approach are discussed. It is shown with evidence that appropriate translations of conceptual labels in ontologies are of crucial importance when applying monolingual matching techniques in cross-lingual ontology mapping. Finally, to address the identified challenges, a semantic-oriented cross-lingual ontology mapping (SOCOM) framework is proposed and discussed.

Keywords: Cross-lingual Ontology Mapping, Multilingual Ontologies, Ontology Rendering.

1 Introduction

The evolution of the World Wide Web in recent years has brought innovation in technology that encourages information sharing and user collaboration as seen in popular applications during the Web 2.0 era. The future of the Web – the Semantic Web will "allow for integration of data-oriented applications as well as document-oriented applications" [1]. In the process of achieving this goal, ontologies have become a core technology for representing structured knowledge as well as an instrument to enhance the quality of information retrieval [2] [3] and machine translation [4]. Benjamins et al [5] identify multilinguality as one of the six challenges for the Semantic Web, and propose solutions at the ontology level, annotation level and the interface level. At the ontology level, support should be provided for ontology engineers to create knowledge representations in diverse native natural languages. At the annotation level, tools should be developed to aid the users in the annotation of ontologies regardless of the natural languages used in the given ontologies. Finally, at the interface level, users should be able to access information in natural languages of

A. Gómez-Pérez, Y. Yu, and Y. Ding (Eds.): ASWC 2009, LNCS 5926, pp. 1–15, 2009.

their choice. This paper aims to tackle challenges at the annotation level, in particular, it investigates issues involved in cross-lingual ontology mapping and aims to provide the necessary support for ontology mapping in cross-lingual environments. *Cross-lingual ontology mapping (CLOM) refers to the process of establishing relationships among ontological resources from two or more independent ontologies where each ontology is labeled in a different natural language.* The term multilingual ontologies in this paper refers to independent ontologies o and o' where the labels in o are written in a natural language which is different from that of the labels in o'. It must not be confused with representing concepts in one ontology using multilingual labels. In addition, this paper focuses on multilingual ontologies that have not been linguistically enriched, and are specified according to the Resource Description Framework (RDF) schema[1]. Furthermore, this paper presents a first step towards achieving CLOM in generic knowledge domains, which can be improved upon to accommodate more sophisticated CLOM mapping strategies among ontologies in more refined, particular knowledge domains.

A generic approach is investigated in this paper, CLOM is achieved by first translating the labels of a source ontology into the target natural language using freely available machine translation (MT) tools, then applying monolingual ontology matching techniques to the translated source ontology and the target ontology in order to establish matching relationships. In particular, the impact of MT tools is investigated and it is shown with evidence that when using the generic approach in CLOM, the quality of matching results is dependent upon the quality of ontology label translations. Based on this conclusion, a semantic-oriented cross-lingual ontology mapping (SOCOM) framework is proposed which is specifically designed to map multilingual ontologies and to reduce noise introduced by MT tools. The remainder of this paper is organised as follows, section 2 discusses related work. Section 3 details the application of the aforementioned generic approach in CLOM experiments which involve mappings of ontologies labeled in Chinese and English. Findings and conclusions from these experiments are presented and discussed in section 4. The proposed SOCOM framework and its current development are discussed in section 5.

2   Related Work

Considered as light weight ontologies, thesauri often contain large collections of associated words. According to the Global WordNet Association[2], (at the time of this publication) there are over forty WordNet[3]-like thesauri in the world covering nearly 50 different natural languages, and counting. Natural languages used include Arabic (used in ArabicWordNet[4]); Bulgarian (used in BulNet[5]); Chinese (used in HowNet[6]); Dutch, French, German, Italian, Spanish (used in EuroWordNet[7]); Irish (used in

---

[1] http://www.w3.org/TR/rdf-schema
[2] http://www.globalwordnet.org
[3] http://wordnet.princeton.edu
[4] http://www.globalwordnet.org/AWN
[5] http://dcl.bas.bg/BulNet/general_en.html
[6] http://www.keenage.com
[7] http://www.illc.uva.nl/EuroWordNet

LSG[8]) and many others. To make use of such enormous knowledge bases, research has been conducted in the field of thesaurus merging. This is explored when Carpuat et al [6] merged thesauri that were written in English and Chinese into one bilingual thesaurus in order to minimize repetitive work while building ontologies containing multilingual resources. A language-independent, corpus based approach was employed to merge WordNet and HowNet by aligning synsets from the former and definitions of the latter. Similar research was conducted in [7] to match Dutch thesauri to WordNet by using a bilingual dictionary, and concluded a methodology for vocabulary alignment of thesauri written in different natural languages. Automatic bilingual thesaurus construction with an English-Japanese dictionary is presented in [8], where hierarchies of words can be generated based on related words' co-occurrence frequencies. Multilinguality is not only found in thesauri but also evident in RDF/OWL ontologies. For instance, the OntoSelect Ontology Library[9] reports that more than 25% (at the time of this publication) of its indexed 1530 ontologies are labeled in natural languages other than English[10]. To enable knowledge discovery, sharing and reuse, ontology matching must be able to operate across natural language barriers. Although there is already a well-established field of research in monolingual ontology matching tools and techniques [9], as ontology mapping can no longer be limited to monolingual environments, tools and techniques must be developed to assist mappings in cross-lingual scenarios.

One approach of facilitating knowledge sharing among diverse natural languages builds on the notion of enriching ontologies with linguistic resources. A framework is proposed in [10] which aims to support the linguistic enrichment process of ontological concepts during ontology development. A tool – OntoLing[11] is developed as a plug-in for the ontology editor Protégé[12] to realise such a process as discussed in [11]. Similar research aiming to provide multilingual information to ontologies is discussed in [12], where a linguistic information repository is proposed to link ontological concepts with lexical resources. Such enrichment of ontologies provide knowledge engineers with rich linguistic data and can be used in CLOM, however, in order for computer-based applications to make use of these data, standardisation of the enrichment is required. As such requirement is currently not included in the OWL 2 specification[13], it would be difficult to make use of the vast number of monolingual ontology matching techniques that already exist.

Similar to linguistically enriching ontologies, translating the natural language content in ontologies is another approach to enable knowledge sharing and reuse. The translation of the multilingual AGROVOC thesaurus[14] is discussed in [13], which involves a large amount of manual work and proves to be time and human resource consuming. An ontology label translation tool, LabelTranslator is demonstrated in [14]. It is designed to provide end-users with ranked translation suggestions for

---

[8] http://borel.slu.edu/lsg

[9] http://olp.dfki.de/ontoselect

[10]http://olp.dfki.de/ontoselect;jsessionid=3B72F3160F4D7592EE3A5CCF702AAE00?wicket:
   bookmarkablePage=:de.dfki.ontoselect.Statistics

[11] http://art.uniroma2.it/software/OntoLing

[12] http://protege.stanford.edu

[13] http://www.w3.org/TR/owl2-profiles

[14] http://aims.fao.org/en/website/AGROVOC-Thesaurus/sub

ontology labels. The motivation of its design is to ensure that information represented in an ontology using one particular natural language could still achieve the same level of knowledge expressivity if translated into another natural language. Users must select labels to be translated one at a time, LabelTranslator then returns the selected label's suggested translations in one of the three target natural languages, English, Spanish and German. It can be used to provide assistance in the linguistic enrichment process of ontologies as discussed in [15]. LabelTranslator is designed to assist the human to perform semi-automatic ontology label translations and linguistic enrichments, it is not concerned with generations of machine-readable ontologies in the target natural language so that matching tools can manipulate. In contrast to LabelTranslator, the ontology rendering process presented in this paper differs in its input, output and aim. Firstly, the input of our ontology rendering process is ontologies and not ontology labels. Secondly, the output of this rendering process is machine-readable formally defined ontologies that can be manipulated by computer-based systems such as monolingual matching tools. Lastly, such an ontology rendering design aims to facilitate CLOM, it is designed to assist further machine processing whereas the LabelTranslator tool aims to assist humans.

An example of CLOM scenario is illustrated by the Ontology Alignment Evaluation Initiative[15] (OAEI) contest in 2008, where a test case requiring the mapping of web directories written in English and Japanese was defined[16]. Among thirteen participants, only four took part in this test scenario with results submitted from just one contestant. Zhang et al. [16] used a dictionary to translate Japanese words into English (it is unclear whether this translation process is manual or automated) before carrying out the matching process using RiMOM. The generic approach presented in this paper is based on Zhang et al.'s method, instead of using a dictionary, freely available MT tools are used. Montiel-Ponsoda & Peters [17] classify three levels to localizing multilingual ontologies, at the terminological layer, at the conceptual layer and at the pragmatic layer. The translation process presented in the generic CLOM approach concerns translations at the terminological layer, i.e., the terms used to define classes and properties are translated into the target natural language. Pazienza & Stellato propose a linguistically motivated approach to ontology mapping in [18]. The approach urges the usage of linguistically enriched expressions when building ontologies, and envisions systems that can automatically discover the embedded linguistic evidence and establish alignments that support users to generate sound ontology mapping relationships. However, as mentioned previously, the multilingual linguistically enriched ontologies demanded by this approach are hard to come by when such specifications are not currently included in the OWL 2 standardization effort. Trojahn et al. propose a multilingual ontology mapping framework in [19], which consists of smart agents that are responsible for ontology translation and capable of negotiating mapping results. For each ontology label, the translation agent looks up a dictionary and returns a collection of results in the target natural language. The ontology labels are then represented with a group of the returned translation results. Once source and target ontologies are in the same natural language, they are passed to the mapping process which consists of three types of mapping agents,

---

[15] http://oaei.ontologymatching.org
[16] http://ri-www.nii.ac.jp/OAEI/2008

lexical, semantic and structural. These agents each conclude a set of mapping results with an extended value-based argumentation algorithm. Finally, globally accepted results are generated as the final set of mappings [20]. Such an approach is based on the assumption that correct mapping results are and always will be generated by various matching techniques regardless of the algorithms used. However, as stated by Shvaiko & Euzenat [21], "despite the many component matching solutions that have been developed so far, there is no integrated solution that is a clear success", therefore, looking for globally accepted results may limit the scope of correct mapping relationship discovery. In contrast, the proposed SOCOM framework in this paper aims to maximize the performance of individual monolingual ontology matching algorithms in CLOM by providing them with ontology renditions that contain appropriate label translations.

3   A Generic Approach to Cross-Lingual Ontology Mapping

A generic approach to achieve CLOM is presented in this section, as shown in figure 1. Given two ontologies representing knowledge in different natural languages, the ontology rendering process first creates a translated source ontology which is an equivalent of the original source ontology, only labeled in the target natural language. Then monolingual matching tools are applied to generate matching results between the translated source ontology and the target ontology. An integration of the generic approach is discussed in section 3.1. To evaluate the soundness of this approach, two experiments involving the Semantic Web for Research Communities (SWRC) ontology[17] and the ISWC ontology[18] were designed to examine the impact of MT tools in the process of ontology rendering (discussed in section 3.2), also the quality of matching results generated using such an approach (discussed in section 3.3).

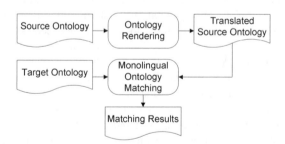

Fig. 1. A Generic Cross-Lingual Ontology Mapping Approach

3.1   Integration of the Generic Approach

The ontology rendering process shown in figure 1 is achieved with a Java application – OntLocalizer, which generates machine-readable, formally defined ontologies in the target natural language by translating labels of the given ontology's concepts using

---

[17] http://ontoware.org/frs/download.php/298/swrc_v0.3.owl
[18] http://annotation.semanticweb.org/ontologies/iswc.owl

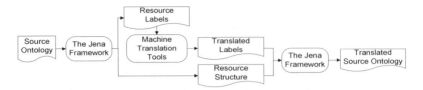

Fig. 2. OntLocalizer Component Overview

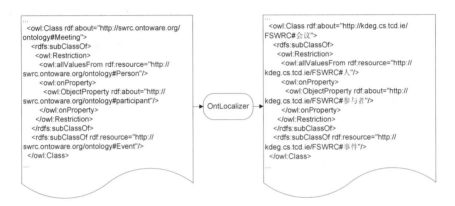

Fig. 3. An Example of Ontology Translation

MT tools, assigning them with new namespaces and structuring these resources – now labeled in the target natural language – using the Jena Framework[19] in the exact same way as the original ontology. Figure 2 shows the components of the OntLocalizer tool. Labels of ontology resources are extracted first by the Jena Framework, which are then passed onto the MT tools to generate translations in the target natural language. Given the original ontology's structure, these translated labels can be structured accordingly to create the translated source ontology. The integrated MT tools include the GoogleTranslate API[20] and the SDL FreeTranslation[21] online translator.

As white spaces are not allowed in the naming of the ontological resources, ontology labels often contain phrases that are made up by two or more words. An example of such labels can be a class named "AssistantProfessor", where the white space between two words has been removed and capital letters are used to indicate the beginning of another word. Another example can be an object property labeled as "is_about", where the white space between two words has been replaced by an underscore. As these labels cannot be translated by the integrated MT tools, the OntLocalizer tool thus breaks up such labels to sequences of constituent words based on the composing pattern, before sending them to the MT tools. In the aforementioned examples, "AssistantProfessor" is transformed to "Assistant Professor", and

---

[19] http://jena.sourceforge.net
[20] http://code.google.com/p/google-api-translate-java
[21] http://www.freetranslation.com

"is_about" is transformed to "is about". Now both in their natural language forms, phrases "Assistant Professor" and "is about" are passed to the MT tools to generate results in the target natural language. Such a procedure is not required when translating labels written in languages such as Chinese, Japanese etc., as phrases written in these languages naturally do not contain white spaces between words and can be processed by the integrated MT tools. Finally, when structuring the translated labels, white spaces are removed to create well-formed resource URIs. *Translation collisions* can happen when a translator returns the same result for several resources in an ontology. For instance, in the SWRC ontology, using the GoogleTranslate API (version 0.4), the class "Conference" and the class "Meeting" are both translated into "会议" (meaning "meeting" in Chinese). To differentiate the two, the OntLocalizer tool checks whether such a resource already exists in the translated source ontology. If so, a number is assigned to the resource label which is under consideration. In the aforementioned example, "Conference" becomes "会议" and "Meeting" becomes "会议0" in the translated ontology. As the integrated MT tools only return one translation result for each intake phrase, it is therefore unnecessary to disambiguate the returned translations in the experiment. A part of the SWRC ontology and its translation in Chinese using the OntLocalizer tool is shown in figure 3.

Once the source ontology is labeled in the target natural language, monolingual ontology matching techniques can be used to generate matching results. Currently, this is achieved by the Alignment API[22] (version 2.5).

3.2 Experiment One Design and Integration

Experiment one is designed to examine the impact of MT tools in the process of ontology rendition, specifically, the quality of machine translated resource labels. In this experiment, labels in the SWRC ontology are translated from English to Chinese through two media, the OntLocalizer tool and a human domain expert – being the lead author. Three translated versions of the SWRC ontology are then created, the GSWRC ontology when using the GoogleTranslate API, the FSWRC ontology when using the FreeTranslation online translator, and the HSWRC ontology which is created manually using the Protégé ontology editor. Each translated version has the original structure of the SWRC ontology with new namespaces assigned to labels in the target natural language. The SWRC ontology is mapped to itself to generate a gold standard of the matching results as M(1), which consists of pairs of matched ontology resources in English. M(A) which contains results of matched resources in Chinese, is then created when the HSWRC ontology is mapped to itself. If exactly the same pairs of resources are matched in M(A) as those found in M(1), then M(A) can be considered as the gold standard in Chinese. The GSWRC ontology and the FSWRC ontology are then each mapped to the HSWRC ontology to create the mappings M(B) and M(C), both containing matched resources in Chinese. Finally, M(B) and M(C) are compared against M(A). This process is shown in figure 4. Eight matching algorithms supported by the Alignment API are used in this experiment.

---

[22] http://alignapi.gforge.inria.fr

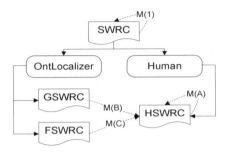

Fig. 4. Experiment One Overview

The hypothesis of this experimental setup is to verify whether the label translation procedure using MT tools would impact on the quality of translated ontologies. If M(B) and M(C) show the same set of results as suggested by M(A), it would mean that MT tools are able to perform like humans and a generic approach using them in CLOM is ideal. If M(B) and M(C) proves to be poorly generated, it would mean that the ontology rendition process is flawed.

3.3   Experiment Two Design and Integration

The second experiment is designed to further investigate the impact of MT tools in CLOM by evaluating the quality of matching results generated using the generic approach. An overview of the experimental steps is shown in figure 5. The English SWRC ontology and the English ISWC ontology are both translated by OntLocalizer to create ontologies labeled in Chinese. The GSWRC ontology and the GISWC ontology are created when using the GoogleTranslate API, and the FSWRC ontology and the FISWC ontology are generated when using the SDL FreeTranslation online translator integrated in OntLocalizer.

The original SWRC ontology is mapped to the original ISWC ontology to generate M(2) as the gold standard which contains matched resources in English. M(B') is generated when the GSWRC ontology is mapped to the GISWC ontology, similarly M(C') is generated when the FSWRC ontology is mapped to the FISWC ontology. Both M(B') and M(C') contain matched resources in Chinese. Again eight matching

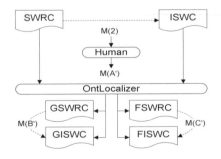

Fig. 5. Experiment Two Overview

algorithms provided by the Alignment API were used in every mapping. To evaluate the quality of M(B') and M(C'), they are compared against the gold standard. Since M(2) contains matched resources written in English, the labels of these resources are translated manually to Chinese by the lead author as M(A'). M(A') is then regarded as the gold standard. Evaluations of M(B') and M(C') are finally conducted based on comparisons to M(A'). The hypothesis of this experiment is, if M(B') and M(C') generated the same sets of matching results as M(A'), it would mean that the generic approach is satisfactory to achieve CLOM. If M(B'), M(C') fail to conclude the same results as found in the gold standard, it would mean that the generic approach would be error-prone when applied to CLOM scenarios.

Precision, recall, fallout and f-measure scores were calculated in both experiments for all the matching algorithms used. Precision measures the correctness of a set of results. Recall measures the completeness of the number of correct results. Fallout measures the number of incorrect matching results based on the gold standard. Finally, f-measure can be considered as a determination for the overall quality of a set of results. If the established gold standard has R number of results and a matching algorithm finds X number of results, among which N number of them are correct according to the gold standard, then precision = N/X; recall = N/R; fallout = (X-N)/X; and f-measure = 2/(1/precision + 1/recall). All scores range between 0 and 1, with 1 being very good and 0 being very poor. An example can be that low fallout score accompanied by high precision and recall scores denote superior matching results.

4 Findings and Conclusions

Findings and conclusions from the two experiments are presented in this section. The results of experiment one is presented and discussed in section 4.1. Section 4.2 shows the results from the second experiment. Finally, data analysis is given in section 4.3.

4.1 Experiment One Findings

Regardless of the matching algorithms used from the Alignment API, the exact same sets of matching results generated in M(1) were found in M(A). Thus, it is with confidence that M(A) can be considered as the gold standard in Chinese. Figure 6 shows an overview of the evaluation results of experiment one. As M(A) equals M(1), its precision, recall and f-measure scores are 1.00 and with 0.00 fallout. The results generated by the eight matching algorithms from the Alignment API are evaluated based on comparisons made to M(A). In M(B) and M(C), a pair of matched resources is considered correct when it is found in the gold standard regardless of its confidence level. Such an evaluation approach aims to measure the maximum precision, recall and f-measure scores that can be achieved in the generated results.

As figure 6 shows, in experiment one, *NameEqAligment* and *StringDistAlignment* algorithm had the highest precision score, however, their low recall scores resulted just above the average f-measure scores. Structure-based matching algorithms had lower recall scores and higher fallout scores comparing to lexicon-based matching algorithms. For each set of results evaluated, the precision score is always higher than its other scores, which suggests that a considerable number of correct matching results

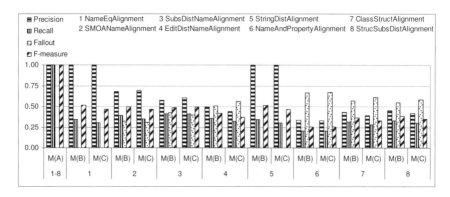

Fig. 6. Experiment One Results

is found, however, they are always incomplete. On average, regardless of the matching algorithms used, f-measure scores are almost always less than 0.50, showing that none of the matching algorithms could meet the standard which is set by the gold standard. Moreover, M(B)'s average f-measure is 0.4272, whereas M(C)'s average f-measure is 0.3992, which suggests that GoogleTranslate API performed slightly better than SDL FreeTranslation online translator in this experiment. Nevertheless, it must be noted that neither of the MT tools was able to generate a translated ontology which, when mapped to itself, could produce a same set of results that are determined by the gold standard. This finding suggests that MT tools had a negative impact on the quality of ontology rendition output.

4.2 Experiment Two Findings

To further validate this finding, the same evaluation approach is used in the second experiment, where a pair of matched result is considered correct as long as it is found in the gold standard, regardless of its confidence level. A series of gold standards were generated for each of the eight matching algorithms in M(2) – written in English, and later manually translated as M(A') – written in Chinese. The evaluation of the results found in M(B') and M(C') is shown in Figure 7.

The *StringDistAlignment* matching algorithm had the highest precision and recall scores in this experiment, thus yielding the highest f-measure score in M(B') and M(C'). Similar to the results found in experiment one, structure-based matching algorithms had lower recall scores comparing to lexicon-based matching algorithms. In experiment two, fallout scores for all the matching algorithms are higher than that of experiment one's, which suggests that the matching procedure was further complicated by the translated ontologies. Also, f-measure scores indicate that structure-based matching algorithms were unable to perform as well as lexicon-based matching algorithms. The average f-measure in M(B') was 0.2927 and 0.3054 in M(C'), which suggests that the FreeTranslation online translator had a slightly better performance than the Google Translate API in this experiment. Nevertheless, from an ontology matching point of view, such low f-measure scores would mean that when used in CLOM, the generic approach would only yield less than fifty percent of the correct matching results. The findings from experiment two show that it is difficult

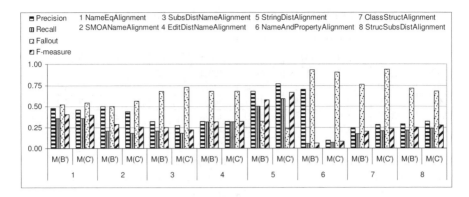

Fig. 7. Experiment Two Results

for matching algorithms to maintain high-quality performance when labels have been translated in isolation using MT tools, and the generic approach in CLOM can only yield poor matching results.

4.3 Result Analysis

So far, the evaluation results that are shown in the previous sections disregard confidence levels. When these confidence levels are taken into account, it is shown that there is a drop in the number of matching results generated with absolute confidence. Table 1 gives an overview of the percentages of matching results with 1.00 confidence levels. In both experiments, all pairs of matched resources generated by the *NameEqAlignment* algorithm and the *NameAndPropertyAlignment* algorithm have 1.00 confidence levels. This is not the case for other algorithms however, where more than half of the results with absolute confidence was not found. For example, every matched pairs of resources by the *EditDistNameAlignment* algorithm from the gold standard in experiment one had 1.00 confidence levels. This was not achieved in M(B) or M(C), where the former contained 47.31% of confident results and only 41.94% for the latter. Averagely, the gold standard in experiment one established a 96.25% of confident results, whereas only 49.53% were found in M(B) and 49.37% in M(C). A similar finding can be concluded for experiment two based on the statistics shown in table 1.

Findings from the experiments suggest that if automated MT tools are to be used in CLOM, more specifically, in the ontology rendering process, the quality of translated ontologies needs to be improved in order for monolingual matching tools to generate high quality matching results. Translation errors introduced by the MT tools in the experiments can be categorized into three main categories. *Inadequate translation* – as mentioned earlier in section 3.1, "Conference" and "Meeting" were both translated into the same words in Chinese. However, since conference is a specified type of meeting, the translated term was not precise enough to capture the intended concept presented in the original ontology. This can be improved if given the context of a resource label to be translated, i.e. the context of a resource can be indicated by a collection of associated property labels, super/sub-class labels. *Synonymic translation* – where

Table 1. Matched Pairs of Results with 1.00 Confidence Levels (%)

	1	2	3	4	5	6	7	8	Avg.
M(A)	100.00	77.34	100.00	100.00	100.00	92.68	100.00	100.00	96.25
M(B)	100.00	33.78	47.83	47.31	100.00	37.25	15.05	15.05	49.53
M(C)	100.00	35.38	44.32	41.94	100.00	34.62	19.35	19.35	49.37
M(A')	100.00	30.89	26.56	48.57	100.00	30.36	0.00	10.94	43.42
M(B')	100.00	16.00	30.86	36.23	100.00	11.63	3.23	3.23	37.65
M(C')	100.00	18.00	30.59	38.24	100.00	13.95	1.30	4.30	38.30

1 = NameEqAlignment	2 = SMOANameAlignment
3 = SubsDistNameAlignment	4 = EditDistNameAlignment
5 = StringDistAlignment	6 = NameAndPropertyAlignment
7 = ClassStructAlignment	8 = StrucSubsDistAlignment

the translation result of a label is correct, however it is different with the one that was used by the target ontology. This can be accounted by algorithms that take structural approaches when establishing matching results, however, it can be very difficult for lexicon-based algorithms to associate them. This can be improved if several candidates are provided in the translation process, and the selection of these candidates gives priority to labels which are used by the target ontology. *Incorrect translation* – where the translation of a term is simply wrong, yielding poor matching results. Similar to inadequate translations, this can be improved if the context of an ontology resource is known to the translation process.

To overcome these challenges and maximise the performance of monolingual matching tools in CLOM, *appropriate translations* of ontology labels must be achieved. A Semantic-Oriented Cross-lingual Ontology Mapping (SOCOM) framework designed to achieve this is proposed and discussed in the next section.

5   The SOCOM Framework and On-Going Research

The semantic-oriented cross-lingual ontology mapping (SOCOM) framework is presented and discussed in this section. The SOCOM framework illustrates a process that is designed specifically to achieve CLOM, it has an extensible architecture that aims to accommodate easy integrations of off-the-shelf software components. To address challenges identified in the experiments and reduce noise introduced by the MT tools, the selection of appropriate translated labels is under the influence of labels used in the target ontology. The SOCOM framework divides the mapping task into three phases – an ontology rendering phase, an ontology matching phase and a matching audit phase. The first phase of the SOCOM framework is concerned with the rendition of an ontology labeled in the target natural language, particularly, appropriate translations of its labels. The second phase concerns the generation of matching results in a monolingual environment. Finally, the third phase of the framework aids ontology engineers in the process of establishing accurate and confident mapping results. Ontology matching is the identification of candidate matches between ontologies, whereas ontology mapping is the establishment of the actual correspondence between ontology resources based on candidate matches [22],

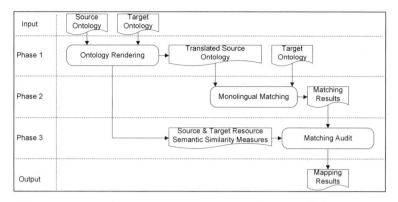

Fig. 8. The SOCOM Framework Process Diagram

this distinction is reflected in the SOCOM framework. Figure 8 shows a process diagram of the proposed framework.

In phase one, the SOCOM framework searches for the most appropriate translation results for ontology labels in the target natural language. To achieve this, the selection of translation candidates is defined by the context a resource is used in, and influenced by the labels that appear in the target ontology. As experimental results show that translating ontology labels in isolation leads to poorly translated ontologies which then yields low-quality matching results, thus, label translations should be conducted within context. As the meaning of a word vary depending on the context it is used in, it is therefore important to capture what a word/phrase signifies as accurately as possible in the target natural language. For instance, the sentence *there is a shift in the tone of today's news broadcasts* and the sentence *research shows that an inevitable side effect of night shifts is weight gain* both use the word *shift*. However, in the first sentence, it is used to express *a change*, whereas in the second sentence, it refers to *a period of work*. In the SOCOM framework, to capture the meaning of a word/phrase in the ontology rendering phase, the context is characterised by the surrounding ontology concepts. As the purpose of translating the source ontology is so that it can be mapped to the target ontology for generations of high quality mapping results (i.e. the translation of the source ontology concepts is within a specific context), the identification of the most appropriate translation results is aided by the labels that appear in the target ontology. Instead of blindly accepting translation results that are returned from a MT tool, for each resource label, a group of translation results are collected and treated as translation candidates. A *translation repository* containing source labels and their translation candidates can be created given a source ontology. On the other hand, a *lexicon repository* can be constructed based on the labels presented in a given target ontology. For each target label, a collection of synonyms can be assembled to maximize knowledge representation with various words and phrases other than those that originally appeared in the target ontology. This can be achieved by querying dictionaries, WordNet, etc., or accessing refined lexicon bases for precise knowledge domains with strict vocabularies such as medicine. Each of the candidates can then be compared to the phrases in the lexicon repository. When matches are found with a target label or a target label's synonym, the target label is chosen as the most appropriate translation result. In addition, when translations are compared to terms in the lexicon repository, similarity measures can be calculated using string comparison techniques, which can then assist the ontology engineers in the final mapping process.

In the second phase, as the source ontology is now labeled in the target natural language, the SOCOM framework can apply existing monolingual ontology matching techniques. It is assumed that prior to CLOM using the SOCOM framework, human experts are involved to establish that it is meaningful to map the concerned ontologies, i.e. they cover the same/similar domain of interest, they are reliable, complete and similar in granularity.

Lastly, in phase three, the matching audit procedure aids ontology engineers in the process of generating the final mapping results. This procedure makes use of the semantic similarity measures that have been concluded in phase one, and displays these findings to the mapping expert providing background information to assist the final mapping. Phase one and two of the SOCOM framework have been integrated, phase three of the proposed framework is currently under development. In the near future, evaluation results of the SOCOM framework and suitability of matching algorithms will become available.

The SOCOM framework is semantic-oriented for two reasons. Firstly, during the ontology rendition phase, the context of an ontological resource is studied in order to determine the most appropriate translation result for its label. This context is defined by the semantics an ontology resource represents, which can be obtained by studying its surrounding concepts, i.e. super/sub-classes and property restrictions. Secondly, the mapping process makes uses of the similarity measures established in the ontology rendition phase in order to generate mapping results. The similarity measures are determined based on the semantics from each pair of ontology resources. An experimental version of the SOCOM framework has been integrated and is currently being evaluated.

Acknowledgement. This research is partially supported by the Science Foundation Ireland (Grant 07/CE/I1142) as part of the Centre for Next Generation Localisation at Trinity College Dublin.

References

1. Powell, S.: Guru Interview: Sir Timothy Berners-Lee. KBE (October 2006)
2. Soergel, D.: Multilingual Thesauri and Ontologies in Cross-Language Retrieval. Presentation at the AAAI Spring Symposium on Cross-Language Text and Speech Retrieval (1997)
3. Simonet, M., Diallo, G.: Multilingual Ontology Enrichment for Semantic Annotation and Retrival of Medical Information. MedNET (October 2006)
4. Shi, C., Wang, H.: Research on Ontology-Driven Chinese-English Machine Translation. In: Proceedings of NLP-KE, pp. 426–430 (2005)
5. Benjamins, R.V., Contreras, J., Corcho, O., Gomez-Perez, A.: Six Challenges for the Semantic Web. SIGSEMIS Bulletin (April 2004)

6. Carpuat, M., Ngai, G., Fung, P., Church, W.K.: Creating a Bilingual Ontology: A Corpus-Based Approach for Aligning WordNet and HowNet. In: Proceedings of the 1st Global WordNet Conference (2002)
7. Malaise, V., Isaac, A., Gazendam, L., Brugman, H.: Anchoring Dutch Cultural Heritage Thesauri to WordNet: Two Case Studies. In: Proceedings of the Workshop on Language Technology for Cultural Heritage Data (2007)
8. Shimoji, Y., Wada, T.: Dynamic Thesaurus Construction from English-Japanese Dictionary. In: Proceedings of International Conference on Complex, Intelligent and Software Intensive Systems, pp. 918–923 (2008)
9. Euzenat, J., Shvaiko, P.: Ontology Matching. Springer, Heidelberg (2007)
10. Pazienza, M.T., Stellato, A.: An Open and Scalable Framework for Enriching Ontologies with Natural Language Content. In: Ali, M., Dapoigny, R. (eds.) IEA/AIE 2006. LNCS (LNAI), vol. 4031, pp. 990–999. Springer, Heidelberg (2006)
11. Pazienza, M.T., Stellato, A.: Exploiting Linguistic Resources for Building Linguistically Motivated Ontologies in the Semantic Web. In: Proceedings of the OntoLex workshop, Interfacing Ontologies and Lexical Resources for Semantic Web Technologies (2006)
12. Peters, W., Montiel-Ponsoda, E., Aguado de Cea, G.: Localizing Ontologies in OWL. In: Proceedings of the OntoLex workshop, From Text to Knowledge: The Lexicon/Ontology Interface (2007)
13. Chang, C., Lu, W.: The Translation of Agricultural Multilingual Thesaurus. In: Proceedings of the 3rd Asian Conference for Information Technology in Agriculture, pp. 526–528 (2002)
14. Espinoza, M., Gómez-Pérez, A., Mena, E.: LabelTranslator – A Tool to Automatically Localize an Ontology. In: Bechhofer, S., Hauswirth, M., Hoffmann, J., Koubarakis, M. (eds.) ESWC 2008. LNCS, vol. 5021, pp. 792–796. Springer, Heidelberg (2008)
15. Espinoza, M., Gómez-Pérez, A., Mena, E.: Enriching An Ontology with Multilingual Information. In: Bechhofer, S., Hauswirth, M., Hoffmann, J., Koubarakis, M. (eds.) ESWC 2008. LNCS, vol. 5021, pp. 333–347. Springer, Heidelberg (2008)
16. Caracciolo, C., Euzenat, J., Hollink, L., Ichise, R., Issac, A., Malaisé, V., Meilicke, C., Pane, J., Shvaiko, P., Stuckenschmidt, H., Sváb-Zamazal, O., Svátek, V.: Results of the Ontology Alignment Evaluation Initiative (2008)
17. Montiel-Ponsoda, E., Peters, W., Aguado de Cea, G., Espinoza, M., Gómez-Pérez, A., Sini, M.: Multilingual and Localization Support for Ontologies. NeOn Deliverable (2008), http://www.neon-project.org/web-content/images/ Publications/neon_2008_d242.pdf
18. Pazienza, T.M., Stellato, A.: Linguistically Motivated Ontology Mapping for the Semantic Web. In: Proceedings of the 2nd Italian Semantic Web Workshop, Semantic Web Applications and Perspectives (2005)
19. Trojahn, C., Quaresma, P., Bieira, R.: A Framework for Multilingual Ontology Mapping. In: Proceedings of the 6th edition of the Language Resources and Evaluation Conference, pp. 1034–1037 (2008)
20. Trojahn, C., Moraes, M., Quaresma, P., Vieira, R.: A Cooperative Approach for Composite Ontology Mapping. In: Spaccapietra, S. (ed.) Journal on Data Semantics X. LNCS, vol. 4900, pp. 237–263. Springer, Heidelberg (2008)
21. Shvaiko, P., Euzenat, J.: Ten Challenges for Ontology Matching. In: Proceedings of the 7th International Conference on Ontologies, Databases and Applications of Semantics (2008)
22. O'Sullivan, D., Wade, V., Lewis, D.: Understanding as We Roam. IEEE Internet Computing 11(2), 26–33 (2007)

Modeling Common Real-Word Relations Using Triples Extracted from n-Grams

Ruben Sipoš, Dunja Mladenić, Marko Grobelnik, and Janez Brank

Jozef Stefan Institute, Jamova 39, 1000 Ljubljana, Slovenia
{ruben.sipos,dunja.mladenic,marko.grobelnik,janez.brank}@ijs.si

Abstract. In this paper, we present an approach providing generalized relations for automatic ontology building based on frequent word n-grams. Using publicly available Google *n*-grams as our data source we can extract relations in form of triples and compute generalized and more abstract models. We propose an algorithm for building abstractions of the extracted triples using WordNet as background knowledge. We also present a novel approach to triple extraction using heuristics, which achieves notably better results than deep parsing applied on *n*-grams. This allows us to represent information gathered from the web as a set of triples modeling the common and frequent relations expressed in natural language. Our results have potential for usage in different settings including providing for a knowledge base for reasoning or simply as statistical data useful in improving understanding of natural languages.

1 Introduction

Solving non-trivial problems using a computer can usually be augmented by including background knowledge. When humans try to solve complex problems they use a large pool of background knowledge attained during their life. However, computers do not have this knowledge nor is it yet available in a computer-friendly format. A possible solution is hiring an expert to construct an ontology that we can use when needed. On the other hand, increasing amounts of publicly available data (especially with the growth of web and conversion of legacy data into digital form) are unlocking new possibilities for automatic knowledge acquisition.

Our motivation is similar to that in [1]. We believe that currently available sources of real-world text are large enough and can be used to attain general world knowledge represented as triples (possibly connected into an ontology). This can then be used for various applications ranging from improvements in automatic text analysis to extending existing ontologies and reasoning over them (e.g., textual entailment [1]). The approach we are proposing in this paper consists of two major components: triple extraction and construction of generalized models of common relations. The advantages of the triple extraction approach described in this paper are efficiency (more than 100 times faster compared to full deep parsing) and quality of extracted triples when applied on text fragments. The second component is used for constructing models which describe a larger set of concrete triple instances with more general terms. This reduces

A. Gómez-Pérez, Y. Yu, and Y. Ding (Eds.): ASWC 2009, LNCS 5926, pp. 16–30, 2009.

the noise and creates informative triples which can be saved in a widely accepted format such as RDF[1] and used as additional knowledge in various applications.

Because our approach works with text fragments we used Google's n-grams[2] dataset for evaluating performance of triple extraction and model construction. This dataset is interesting due to its size (it contains a few billion n-grams) and its origin (it was computed from Google's web index). New challenges posed by this dataset required the development of new approaches that can deal with highly noisy data and can process n-grams instead of complete sentences.

Related Work. There exists a plethora of related work that deals with similar topics. However, in all cases their research focuses on a more specific problem, only deals with some of challenges presented here or presents possible alternatives.

Triple extraction can be approached from various angels. The most common and intuitive is by using full deep parsing and constructing triples from parse trees or similar structures. Such approaches include [2] (which we used in our comparison in section 4.2) and [3, 4, 5]. The main problem with using Stanford Parser, Minipar and similar tools in our case is that we deal with text fragments and not complete sentences as those tools expect. Moreover, those approaches are too slow and can not be used to process large datasets such as Google's n-grams. Some related work [6, 7] uses different sets of handcrafted rules to extract triples. Nevertheless, those rules are different from our patterns because they are more specific and work with different underlying structures. There are also some other methods which take an entirely different approach to triple extraction, for example [8] (using kernel methods, can only extract relations it was trained on), [9] (using statistics about frequently occurring nouns near verbs marked by a POS tagger) and [10] (extracting only subject and predicate and adding possible objects later).

In many of the related papers we can find (similar) approaches to normalization. In [6, 10] they analyze noun phrases to find head words. In [11] they remove attributes (i.e. they keep only head words) and lemmatize words to base forms. We do all that; mark head words, treat attributes separately and lemmatize all words.

We can also find various approaches for generalizing extracted relations. Some of them are graph-based [12, 13, 14] while others define generalization more in the form of textual entailment [1, 11]. Compared to our work, they are meant for different settings than ours and can not be used in our case. One similarity that does exist is the use of WordNet (or some other domain specific ontology) for finding concept abstractions. Another possible approach, described in [17], tries to find generalizations in a way similar to learning selectional restrictions of predicates.

Contributions. The main contributions of this paper are the following:

- We present a scalable approach that can be used even with very large datasets such as Google's n-grams.
- Our method for triple extraction is more than 100 times faster than methods which use deep parsing. Moreover, the proposed method gives vastly better results when used on text fragments.

---

[1] http://www.w3.org/RDF/
[2] http://www.ldc.upenn.edu/Catalog/CatalogEntry.jsp?catalogId=LDC2006T13

- We can construct general models that describe a set of concrete triple instances and thus represent common knowledge implicitly contained in text.
- Evaluation of our results showed that the constructed models are meaningful. We can expect them to be useful for automatic ontology construction and as additional knowledge for various reasoners.

The paper is structured as follows. We first present our novel approach to triple extraction from text segments that does not require deep parsing. Next we describe an approach for constructing models which describe relations represented as triples in a more general way. We experimentally compare our approach to triple extraction to a commonly used existing approach. We also propose a methodology for evaluating generated models and apply it to obtain quality of models constructed from triples. We conclude with a discussion of the results, possibilities for their use and ideas for future work.

2 Triple Extraction

In this section we will present the proposed approach to triple extraction used in our pipeline. Main design goals were scalability and ability to process text fragments. For example, this allows us to process large datasets based on n-grams such as Google's n-grams used in our evaluation.

Text fragments represented as word n-grams differ from sentences in some important properties. We have to consider how they were created if we want to adapt existing approaches for triple extraction to work with them. Text fragments as addressed in this paper never contain words from two different sentences. Our (very basic) filter removes all text fragments containing punctuation marks before the triple extraction step. In this way we ensure alignment of text fragments inside sentence and avoid difficult (and usually less useful) cases which contain words from several sentences or clauses if they are separated with a punctuation mark. Secondly, even short text fragments are enough to cover a lot of possible simple sentences. In the best case scenario text fragments will still resemble sentences although maybe cropped at the beginning and end. Such text fragments can be viewed as simple sentences consisting of a subject, predicate and object. This limits useable text fragments to those consisting of at least three words. In the experiments presented here, text fragments consist of at most five words, therefore we can extract at most one subject-predicate-object triple and two attributes.

The main motivation for developing a new method, instead of using deep parsing, was the time complexity of deep parsing and specificities of text fragments in comparison to complete sentences. Using deep parsing as a basis for triple extraction even on a lot smaller datasets than ours might be unfeasible due to time constraints. We wanted a significantly faster approach than the one using deep parsing even if that meant possibly lowering the quality of extracted triples. Moreover, approaches based on deep parsing usually expect complete sentences. Training sets for parsers such as Stanford Parser[3] and OpenNLP[4] consist of annotated sentences. However, we are dealing with text fragments and therefore might get worse results than expected.

---

[3] http://nlp.stanford.edu/
[4] http://opennlp.sourceforge.net/

Passive smoking increased the risk.
NP   VP   NP

Fig. 1. Text fragment matching pattern noun phrase (NP), verb phrase (VP), noun phrase (NP)

Table 1. Patterns used for triple extraction listed by priority

Patterns represented as sequences of phrase types
NP, VP, NP
NP, VP, PP, NP
NP, ADVP, VP, NP
NP, ADVP, VP, PP, NP
NP, ADJP, VP, NP
NP, ADJP, VP, PP, NP
NP, PP, VP, NP
NP, PP, VP, PP, NP
NP, VP, ADVP, NP
NP, VP, PRT, NP

Our idea for an alternative approach to deep parsing is simple. Full deep parsing is slow; therefore we try to extract triples using only information provided by POS tags and chunking into phrases. This way we can avoid the most time consuming step – building a parse tree. Even though we have incomplete information we can still find the subject, predicate, object and their attributes.

We manually constructed patterns for triple extraction based on a sample of text fragments with added part of speech tags (POS tags) [15]. We incrementally expanded our set of patterns by tagging a set of text fragments with SharpNLP[5], manually noting the patterns, removing already covered examples and repeating this a few times. It turns out that even a very small sample (less than a hundred text fragments) already contains almost all patterns we found.

Our patterns describe the subsequence of phrases. The simplest pattern is: noun phrase (NP), verb phrase (VP), noun phrase (NP). A text fragment matching this pattern is shown in Figure 1. Constructing a triple is trivial: the first NP chunk represents the subject, the VP chunk is the predicate and the second NP chunk is the object. All patterns are listed in Table 1. A pattern does not have to match the whole text fragment, but it has to occur in it as a subsequence. For example, the pattern NP, VP, NP matches the text fragment consisting of NP, NP, VP, NP but not NP, PP, VP, NP. To resolve conflicts if multiple matches are possible we assign each pattern a priority.

After deciding on the subject, predicate and object, we have to choose the main word in each part. Because we did not compute the parse tree we rely on the POS tags. We decided to select the first word with an appropriate POS tag in each part as a main word. For example, word marked with a DT POS tag is certainly not the main word. Set of POS tags used for selecting main words was selected based on the descriptions of those tags.

---

[5] http://www.codeplex.com/sharpnlp

3 Modeling Triples

Our method for modeling triples consists of two phases. During the first phase we use machine learning to determine concept coverage. This is necessary due to some anomalies in our background knowledge. Models can still be constructed even if we skip this phase, but their quality might drop. The second phase uses information from the first and constructs models describing input triples.

In the previous chapter we describe an approach to extracting triples from text fragments. In order to construct more general models we want to abstract triples and find abstractions with the largest set of supporting instances. We used background knowledge which provides us with hypernymy relations and therefore enables concept abstraction. Then we extended this to triples and developed a method for finding the most interesting abstractions.

3.1 Background Knowledge

In order to construct our models we need a way to generalize concepts. One possibility is to use a hypernym. The desired level of abstraction and number of steps we have to traverse in the concept hierarchy depends on the specific goals we want to achieve.

We are using WordNet[6] as our background knowledge. More specifically, we are using hypernymy relations in WordNet for generalization of concepts. However, the tree representing hypernymy relations in WordNet has different levels of detail in different branches. This is mainly due to the fact that different parts of WordNet were constructed by different people. For some topics there were domain experts available, for others maybe not. This resulted in different levels of detail in different parts of WordNet. This means that the same number of steps taken while generalizing some concept might not yield the same amount of generalization if it happens in different parts of WordNet. Therefore, we have to find a way to compensate for this and normalize hypernymy tree to the same amount of generalization per one step in all parts of the tree.

Let us suppose that the words in the input text fragments have a nice and intuitive distribution. If we assign words as instances to possible concepts in WordNet hypernymy tree we expect that less detailed branches will have higher density of assigned instances. To compute the density of instances in different parts of hierarchy we use an unsupervised learning method. Every triple is represented as a bag-of-words. In comparison with using whole text fragments, we use only main words and leave out all additional attributes which might add noise. Additionally, we expand each bag-of-words with related words. Specifically, we add words which are hypernyms of the words from the triple. We limit ourselves to 1-2 steps of generalization using WordNet; alternative solutions are left for future work. This expansion increases the overlap between examples and reduces sparsity. In particular, this introduces some overlap between examples containing related words.

For unsupervised learning we used the k-means clustering method due to its simplicity and reasonable speed. The distance measure was cosine similarity, as commonly used for measuring similarity on text data. Computed centroids are used for

---

[6] http://wordnet.princeton.edu/

determining coverage of concepts in WordNet. We can assume that the majority of cluster mass exists near the centroid and ignore the examples farther off. Next, we can map areas of high density from bag-of-words space onto nodes in WordNet hierarchy.

Instance density information can help us determine the right amount of steps required when generalizing some concept using WordNet. We believe that in areas of lower density we have to make more generalization steps to achieve the same level of abstraction in comparison to more densely covered parts of hierarchy. On the other hand, when dealing with concepts from more densely covered parts of hierarchy we have to be more conservative because those concepts are more widely used and can easily cover larger amount of concrete instances.

An alternative approach to this would be using just the number of instances in a node to determine density in that part. However, we hope to achieve some improvement over that by using clustering and thereby being more robust when dealing with high level of noise (as is the case in our dataset).

3.2 Constructing Models

Some of the findings discussed in Section 3.1 can be used for generalizing words where we strive to select still interesting but more abstract concepts. Now we have to extend this to work with triples consisting of three main words. Naïve approach that tries to generalize all three words independently soon faces too big combinatorial complexity if applied to large datasets.

We can expect high combinatorial complexity due to ambiguity of mapping between words and synsets in WordNet. Sense disambiguation uses context to select the correct synset. However, our context is very limited and probably insufficient for acceptable quality of sense disambiguation. While humans can in most cases still identify the correct sense even when given only a triple with no or few additional attributes, it still is impossible to do using only computers. We concluded that we can not expect to solve this with any simple approach and will have to deal with ambiguities.

The method we are proposing for constructing model can be considered as unsupervised learning. To deal with the large amount of noise present in our data we largely rely on redundancy and reliable statistic information due to large amount of data.

We can not consider each part of a triple separately because of the time and space constraints; we have to find a more efficient solution. Besides, we have to check the majority of possible models with brute-force method because we do not have a good guide for choosing generalizations and appropriate models yet. We define word depth as the number of steps required to reach the tree root from that word in the WordNet hierarchy. We also assume that main word depths in a triple are usually similar because we believe that examples such as "Anna is drawing abstract entity" are rare. Searching for appropriate models is done in multiple passes; each pass at a specific triple depth, which is defined as a sum of main word depths. This defines the order of exploring of model space. To get a triple with a given triple depth we have to generalize main words in this triple to an appropriate word depth using WordNet.

The inability to disambiguate senses is an unavoidable constraint. Our approach uses a trivial solution to this. Every triple is converted to a set of triples where each one represents a possible combination of senses based on words from source triple. However, this drastically increases the amount of data we have to process. One possible solution would be using only the most frequently occurring senses and ignoring others. The current approach does not filter senses in any way because we wanted to construct a baseline first so we can compare results and measure effects of not using all possible word senses. Another possible solution described in [18] would be to use some of the additional information contained in extracted relations to disambiguate senses as some word senses occur more frequently in some relations than other.

WordNet contains only lemmatized word forms. That is why our input data requires lemmatization before we can use it. Simple rule based stemmers, such as Porter Stemmer, are useful for example when using bag-of-words representation of text in text mining [16]. We are using our own lemmatizer to improve the chances of finding a matching concept in WordNet for a given word. This lemmatizer is based on machine learning from correctly annotated examples. Experiments have shown that the use of lemmatizer is essential in achieving acceptable level of matching between words in triples and WordNet concepts.

We remove triples that contain stop-words. This applies only to main words and not additional attributes as we use only main words for constructing models. Also, we did not include some common stop-words in our list. For example, words such as "he" or "is" can be considered as stop-words but we keep them because they are building blocks for many informative triples. Still, we have to remove at least some very frequent words, otherwise we get a lot of useless abstractions such as "it is this" as the models with the highest support.

The algorithm given below describes the whole process of model construction we discussed in this section. The *BuildModels()* function iteratively explores possible candidates for models and keeps only the most promising ones. The *AbstractTriple()*, which is called from *BuildModels()*, is responsible for generalizing triples to a desired triple depth.

```
function AbstractTriple(wordSenses[3], desiredDepth[3])
//generalize all three parts of a given triple
newTriple[3] = new()
for i=0..2
  sense = wordSenses[i]
  depth = desiredDepth[i]
  //look up density information
  //for concepts in WordNet
  //that we previously computed
  density = GetDensity(sense)
  //set the max number of steps we can make
  if density < DENSITY_TRESH
    maxSteps = LOW_DENSITY_MAXSTEPS;
  else
    maxSteps = HIGH_DENSITY_MAXSTEPS;
  //check the constraints
  currentDepth = WordNet.getDepth(sense)
  if |depth - currentDepth| <= maxSteps
    newTriple[i] = WordNet.getHyper(sense, depth)
return newTriple
```

```
function BuildModels()
//we process data in multiple passes
//depending on triple depth
for depthSum=MIN_DEPTH..MAX_DEPTH
  foreach (xd, yd, zd) where xd+yd+zd == depthSum
    //process all triples
    foreach triple in allTriples
      //consider all possible senses
      foreach si in WordNet.getSenses(triple[0])
        foreach sj in WordNet.getSenses(triple[1])
          foreach sk in WordNet.getSenses(triple[2])
            //generalize triple
            model = AbstractTriple({si, sj, sk}, {xd, yd, zd})
            //and add it to candidate list
            mHash[model] += triple;
  //keep only the best model candidates
  ApplyTreshold(mHash, MIN_SUPPORT)
```

The largest part of time complexity is due to inclusion of all possible senses because we can not disambiguate word senses with so limited context. Space requirements of this algorithm depend mostly on hash table which is storing all current model candidates. The possibilities of parallelization are limited because we have to synchronize or distribute hash tables storing model candidates. To reduce the number of model candidates we set a threshold on the lowest acceptable number of supporting instances. The selection of this value depends on the target application in which we will use our models.

In this section we described an algorithm for constructing a set of triples that models concrete input instances. The final output is relations represented as subject, predicate and object which can be used for example as background knowledge in different applications.

4 Evaluation

4.1 Dataset

Evaluation of our approach was done on Google's *n*-grams dataset. This dataset contains word *n*-grams of lengths 1 to 5 and their frequency counts. It was computed from their web search engine's index and was published on 6 DVDs (about 26 GB of compressed textual data). The dataset includes only *n*-grams from the English index (but we can still find a few *n*-grams from other languages due to impreciseness of the methods used for language detection). They captured data in January 2006.

Text fragments represented as word *n*-grams were computed from more than 1 trillion words of source text containing more than 95 billion sentences. The dataset includes more than 1 billion unique 5-grams. Still, only *n*-grams with frequency count over 40 are included. All words that occur less than 200 times are mapped to the special token <UNK>. We believe that this dataset is sufficiently large to contain enough redundancy and useful information so we can build meaningful models on top of it.

Before we extracted triples from those text fragments we applied some filtering. Because the published dataset was not processed in any way (except thresholding frequency counts) it still contained a lot of random junk. We decided to filter away all text fragments containing border between two sentences, non alphanumeric characters and other similar things that are probably useless and would only complicate triple extraction. Our filter was very simple and rule based. For example, about 1/3 of 5-grams remained (8.79 GB of uncompressed binary data) after the filtering step. If later on we found that we filtered away some useful text fragments, we could still add additional rules to filter and keep those text fragments for further processing.

4.2 Triple Extraction Comparison

In this subsection we will present a comparison of triple extraction from text fragments using deep parsing with our approach presented in section 2. The main advantage of our method is efficiency. Processing 393,507,635 text fragments (5-grams) took about 1.2 million seconds (14 days of cumulative CPU time on 2.2 GHz Opteron 875) and created 63,814,809 triples. The time required for a single triple is about 3 ms. Our approach is almost 100 times faster if compared to deep parsing, which needs almost a second for one sentence (0.3 s per sentence according to results in [2] and our own tests). For example, if we used deep parsing we would need one year of CPU time on a 4-core processor and less than a week with our approach.

The second big advantage of our approach is that it was designed to work with text fragments. Tools for constructing parse trees used complete sentences as their training set and perform subpar when used with text fragments. A possible solution would be to artificially create text fragments from a training corpus and use it for training a new parser. Text fragments can always be created from unstructured text (although with some loss of information) if we want to use our method instead of some other approach designed for complete sentences (although doing this would be suboptimal). However, it is almost impossible to use methods designed to work with full sentences if we have only text fragments as they rely on too many assumptions which do not hold when working with text fragments.

The main disadvantage of our approach is the fact that it was designed for simple and short text fragments. If we use it for triple extraction from a complex sentence it will perform worse than approaches using deep parsing because we take into account only the local structure of sentence. However, some preliminary attempts to remedy this showed promising results. With minimal changes to patterns we can achieve roughly the same level of quality and recall as by using deep parsing.

We analyzed a random sample of 100 text fragments (5-grams) and manually reviewed the cases in which neither our approach nor approach using deep parsing returns a triple. In 81 cases neither approach successfully extracted a triple. In most of them (79) the text fragment was not even remotely similar to a sentence and therefore it was impossible to extract a triple from it. Most of those useless text fragments consisted only of nouns (thus we cannot find a predicate) or they started with a verb (and were thus missing a subject). Some other reasons were: the text fragment content was just noise, the text fragment was a fragment of a question, the main word was cropped and only attributes remained etc.

In the second part of comparison we wanted to evaluate the quality of extracted triples. We manually reviewed triples extracted from a random sample of text fragments (5-grams). The results of this evaluation are given in Table 2.

Table 2. Results of comparison between deep parsing and our approach

Method	Triple quality	No. of triples	
Deep parsing	correct	33	52 %
	useful	4	6 %
	forced	18	29 %
	wrong	8	13 %
	total	**63**	
Our approach	correct	89	62 %
	useful	33	23 %
	wrong	22	15 %
	total	**144**	
Both	same	14	40 %
	similar	20	57 %
	different	1	3 %
	total	**35**	
	no extracted triple	828	

First, we evaluated both approaches separately. We defined quality of extracted triples as following:

- **correct:** The extracted triple correctly describes the content of the text fragment. If we had to extract a triple manually we would select the same one. Example: "discussion document | is | a sample".
- **useful:** The extracted triple does not correctly describe the content of the text fragment. However, it still represents a meaningful relation and remains relatively similar in meaning to the original text fragment. Example: "little advice | to help | you".
- **forced:** Sometimes the approach using deep parsing extracts a triple from a text fragment that contains only nouns. It marks one of the nouns as verb although it is obvious to the human that that word represented a noun in the original context. Example: "Pictures Random | Thoughts | Rants Recreation".
- **wrong:** The extracted triple is wrong and useless. The source text fragment in these cases is usually still similar to a sentence but is otherwise meaningless. Example: "order | to understand | these files".

Next, we reviewed the cases in which both approaches extract a triple and compared the returned triples. We defined the following categories:

- **same:** Returned triples are identical.
- **similar:** Triples differ only in attributes; all three main words are the same.
- **different:** Triples differ in one or more main words.

The results we obtained show that our approach is successful. Using it we can extract 122 triples (correct and useful) while deep parsing returns only 37 triples (correct and useful). Furthermore, our approach returns only 15% of wrong triples while the other one returns 41% wrong triples. We can see that almost all triples extracted using deep parsing are included in intersection. Our approach misses and does not extract only 2 of the triples we get by using deep parsing. The new approach we are proposing for triple extraction from text fragments turned out to be better for our purposes as it is faster and extracts more and better triples than approach using deep parsing.

4.3 Models Evaluation

There is no standardized benchmark yet for what we are presenting in this paper. Moreover, it is quite hard to construct synthetic data that simulates real world data with high enough accuracy because we do not know all the properties of our data yet. We decided to manually evaluate models constructed from a subset of text fragments.

We are proposing an evaluation method that is based on very specific criteria for evaluating results. The main guidelines for selecting criteria were possible future applications of our results for automatic text analysis, ontology building, extending knowledge bases and reasoning. The criteria we used for evaluating the constructed models are:

- Appropriate level of abstraction:

  * All three parts of a triple should be on approximately the same level of abstraction.
  * Abstract concepts are useless (e.g. "entity").
  * Instance names are too specific. If we wanted concrete instances we could just skip generalization and directly use extracted triples.
  * Commonly used concepts are preferred over rare ones.

- Usefulness of information:

  * Trivial relations with pronouns are useless (e.g. "it is this").
  * Relation has to be meaningful.

- Ease of interpretation:

  * We prefer non ambiguous concepts.
  * Metaphors etc. are usually too hard to use.

We have to evaluate results using the previously listed criteria but we should take into account all of them simultaneously and not separately. We manually evaluated the most promising models with the following marks:

+1   conforms to most of the criteria
0   only about one half of criteria are satisfied
-1   most of the criteria are not satisfied

Because we decided to compensate the lack of sense disambiguation by taking into account all the possible word senses we wanted to evaluate in what amount do the constructed models include the correct sense. Two human evaluators manually reviewed models and decided if they use the correct word sense when we can discover it from the context (using our human knowledge). Results obtained during evaluation of sample A are presented in Table 3.

Table 3. Correctness of selected concepts in models constructed from sample A

rater	**Number of models**		
	correct	partially correct	Wrong
I	31	28	11
II	35	18	17

We are presenting results of evaluation of models constructed from sample A in the Table 4. We used the criteria described above. Most (9) of the models marked with -1 contained the concept "abstract entity" and were therefore considered useless (but we could easily avoid that by simple filtering or limiting ourselves to models with higher triple depth).

Table 4. Quality of models constructed from sample A

rater	**Number of models**		
	good (+1)	borderline (0)	bad (-1)
I	44	19	3+9
II	47	13	6+9

Table 5. Quality of models constructed from sample B

rater	**Number of models**		
	good (+1)	borderline (0)	bad (-1)
I	11	5	2+3
II	12	6	2+3

We performed our manual evaluation with two independent raters so we could check inter-rater agreement and verify the quality of our instructions for evaluating results. From the results presented in tables 3, 4 and 5 we can see substantial agreement in all choices except the "partially correct" and "wrong" selection of concepts in models. This disagreement is probably due to the vague definition of "partially correct".

To analyze the impact of sample selection on results we evaluated models for three more samples. Table 5 gives results of evaluating sample B where we gave negative mark three times due to too abstract models (they contained "abstract entity"). Samples A' and A" gave a lot worse results than those we got for samples A and B. Almost two thirds of models were marked with -1 mostly because they contained concept "abstract entity".

Samples A, A' and A" included 1000 triples. For a bit more than 50% of triples we could find matching synsets in WordNet for all three main words. We did not use the remaining triples for constructing models because we could not generalize them using WordNet. All three samples consist of small sequential groups of triples randomly selected from input file (which is partially sorted). If we chose entirely random triple we probably would not get any similar triples (within so small sample) and could not find useful generalizations (because they would contain only top level concepts such as "entity"). This does coincide with results (not presented in this paper) that we got from entirely random samples.

Selection of sample B was done in a different way. It contains 5000 triples. We randomly chose a few subjects (from the ones described by triples) and then randomly chose some triples containing that subject. In this way we assured at least some overlap between triples for the same reasons as in samples A, A' and A".

We evaluated results only at triple depths of 5, 10, 15 and 20 to shorten time needed for manual evaluation. This should not have any significant impact on evaluation results as it turned out that there are a lot of same models at different triple depths due to small sizes of samples. Models in samples A and B had about the same quality at different triple depths. Similarly, models constructed from samples A' and A" were bad at all triple depths.

Important bit of information we have to present is how we chose the most promising models from all the candidates. Even when we use really small sample size we get a large amount of candidates in hash table during the run of algorithm. This number is about 20,000 to 4,000,000 for sample B and 10,000 to 1,500,000 for smaller samples A, A' and A". Different numbers of candidates are due to different sizes of model space at different triple depths. Because candidates include all possible models for all possible senses we have to select only the most promising ones at the end. We decided to select the best ones for each triple depth separately because then we can observe the impact of different triple depth on constructed models. For every model candidate we also computed the number of instances that support it and use this number as a guide when choosing the best models. In evaluation we included 5-10 most supported models from selected triple depths.

From the presented results of evaluation we can conclude that the best models adequately satisfy chosen criteria. Also, we expect even better results when we will construct and evaluate models from all input triples (instead of only small sample). Small samples can be biased when randomly chosen from such large dataset depending on which triples we include: meaningful or not, with similar meaning or not, containing very specific or very general words etc. Additionally, we found out that we should add additional weights for some concepts because they have a very large negative impact on quality of results. For example, the word "he" is in a lot of cases represented with the concept of Helium although we know that most of the text contains the word "he" as a pronoun.

5 Conclusions and Future Work

In this paper we presented an approach to constructing generalized models describing triples extracted from text fragments. One of the challenges we tackled was the size of the dataset we used. Google's *n*-grams dataset is one of the largest publicly available datasets describing real world text. A lot of work is required to develop and implement algorithms that are fast enough so we can process such amount of data in a reasonable time. One important property of the dataset we used is that the source data for it was taken from the web. Therefore, we can observe some significant differences when we compare our extracted triples to manually or semi-automatically constructed triples currently available to the public. One major difference it the level of noise present in our dataset and consequently in extracted triples.

We described an approach for efficient triple extraction from text fragments. Compared to full deep parsing we achieved a major speedup (about 100 times) and better results. This will allow us to extract triples from other interesting data sources (e.g. Wikipedia articles) that are too large to process using deep parsing. We also presented an approach for constructing generalized models describing similar triples. Evaluation showed that constructed models were meaningful and could be used in many interesting settings.

Our approach still has to be more thoroughly tested. We hope that in the near future we will be able to test it indirectly by measuring the performance of applications using our data. This will allow us to fine-tune the parameters. The current choice of parameters was mainly based on the limited knowledge we have about the future applications.

In the future we want to extend our approach to work with wider choice of input data. Our design allows for easy extension and modification of various parts. We can use a different ontology (currently we use WordNet) for generalizing concepts and therefore cover other more specific domains. Moreover, we can add new or change existing patterns for triple extraction and by doing so modify our approach to work with longer text fragments or even whole sentences. By changing the POS tagger and chunker we can extract triples and construct models for other natural languages, too.

Acknowledgements

This work was supported by the Slovenian Research Agency and the IST Programme of the EC under PASCAL2 (IST-NoE-216886) and ACTIVE (IST-2008-215040).

References

[1] Clark, P., Harrison, P.: Large-Scale Extraction and Use of Knowledge from Text. In: Proc. Fifth Int. Conf. on Knowledge Capture, KCap 2009 (2009)

[2] Rusu, D., Dali, L., Fortuna, B., Grobelnik, M., Mladenić, D.: Triplet Extraction from Sentences. In: Proceedings of the 10th International Multiconference Information Society - IS 2007, pp. 218–222 (2007)

[3] Specia, L., Baldassarre, C., Motta, E.: Relation Extraction for Semantic Intranet Annotations. Knowledge Media Institute (2006)

[4] Sahay, S., Li, B., Garcia, E.V., Agichtein, E., Ram, A.: Domain Ontology Construction from Biomedical Text, pp. 28–34. CSREA Press (2007)

[5] Fundel, K., Küffner, R., Zimmer, R.: RelEx - Relation extraction using dependency parse trees. Bioinformatics 23, 365–371 (2007)

[6] Etzioni, M., Cafarella, D., Downey, S., Kok, A.-M., Popescu, T., Shaked, S., Soderland, D.S.: Web-scale information extraction in knowitall (preliminary results), pp. 100–110. ACM, New York (2004)

[7] Banko, M., Cafarella, M.J., Soderland, S., Broadhead, M., Etzioni, O.: Open Information Extraction from the Web, pp. 2670–2676 (2007)

[8] Zelenko, D., Aone, C., Richardella, A.: Kernel Methods for Relation Extraction. Journal of Machine Learning Research 3, 1083–1106 (2003)

[9] Kavalec, M., Svatek, V., Buitelaar, P., Cimmiano, P., Magnini, B. (eds.): A Study on Automated Relation Labelling in Ontology Learning. IOS Press, Amsterdam (2005)

[10] Schutz, Buitelaar, P.: RelExt: A Tool for Relation Extraction from Text in Ontology Extension, pp. 593–606 (2005)

[11] Soderland, S., Mandhani, B.: Moving from Textual Relations to Ontologized Relations. In: Proceedings of the 2007 AAAI Spring Symposium on Machine Reading (2007)

[12] Trampuš, M., Mladenić, D.: Constructing Event Templates from Textual News. In: Workshop on: Intelligent Analysis and Processing of Web News Content (2009)

[13] Leskovec, J., Grobelnik, M., Milic-Frayling, N.: Learning Sub-structures of Document Semantic Graphs for Document Summarization. In: Workshop on Link Analysis and Group Detection (LinkKDD), KDD 2004, Seattle, USA, August 22-24 (2004)

[14] Rusu, D., Fortuna, B., Mladenić, D., Grobelnik, M., Sipoš, R.: Document Visualization Based on Semantic Graphs. In: IV 2009 (2009)

[15] Bies, A., Ferguson, M., Katz, K., Mac-Intyre, R.: Bracketing guidelines for Treebank II style Penn Treebank project. Technical report, University of Pennsylvania (1995)

[16] Grobelnik, M., Mladenić, D.: Text Mining Recipes. Springer, Heidelberg (2009), http://www.textmining.net

[17] Ciaramita, M., Gangemi, A., Ratsch, E., Saric, J., Rojas, I.: Unsupervised Learning of Semantic Relations between Concepts of a Molecular Biology Ontology. In: IJCAI 2005, pp. 659–664 (2005)

[18] Pennacchiotti, M., Pantel, P.: Ontologizing Semantic Relations. In: ACL 2006 (2006)

Supporting the Development of Data Wrapping Ontologies

Lina Lubyte and Sergio Tessaris

Free University of Bozen-Bolzano
{lubyte,tessaris}@inf.unibz.it

Abstract. We consider the problem of designing data wrapping ontologies whose purpose is to describe relational data sources and to provide a semantically enriched access to the underlying data. Since such ontologies must be close to the data they wrap, the new terms that they introduce must be "supported" by data from the relational sources; i.e. when queried, they should return nonempty answers. In order to ensure nonemptiness, those wrapping ontologies are usually carefully handcrafted by taking into account the query answering mechanism. In this paper we address the problem of supporting an ontology engineer in this task. We provide an algorithm for verifying emptiness of a term in the data wrapping ontology w.r.t. the data sources. We also show how this algorithm can be used to guide the ontology engineer in fixing potential terms unsupported by the data. Finally, we present an implemented tool and an empirical study showing benefits of our approach.

1 Introduction

The use of a conceptual model or an ontology to wrap and describe relational data sources has been shown to be very effective in several frameworks involving management and access of data. These include federated databases [1], data warehousing [2], information integration through mediated schemata [3], and the Semantic Web [4] (for a survey see [5]). Ontologies provide a conceptual view of the application domain, which is closer to the user perspective. The automated reasoning can be leveraged to provide a better user support in exploring and querying the underlying data sources.

In this paper we focus on the problem of designing ontologies which describe relational data sources, and whose purpose is to provide a semantically enriched access to the underlying data. We use the term *data wrapping ontologies* to distinguish these ontologies from domain ontologies; whose purpose is to model a domain.

In order to illustrate the different roles of the two kinds of ontologies let us consider a scenario in which several independent databases containing data on a given domain (e.g. showbiz) should be accessed through a single web portal driven by an ontology. This ontology is tailored to provide a general view over the domain and enable the users (or software agents) to retrieve information from the portal. For this reason the "portal ontology" doesn't necessarily use the same

A. Gómez-Pérez, Y. Yu, and Y. Ding (Eds.): ASWC 2009, LNCS 5926, pp. 31–45, 2009.

vocabulary as the data sources and we consider it a domain ontology. In order to retrieve data from the sources, the domain ontology should be "connected" to the data sources. This can be done either by defining mappings from the relational tables in the data sources to the terms in the ontology (see e.g [6]), or by creating smaller ontologies wrapping the data sources and aligning those with the domain ontology.

We believe that the second approach has several advantages over the use of direct mappings over the data sources. First of all, it enhances the modularity of the system, since the different components can be designed independently and then integrated. In fact, the domain ontology can be designed almost independently of the details of the data sources; while data wrapping ontologies are smaller and can be automatically bootstrapped from the data sources (see [7]). Moreover, there are well established techniques to support the alignment of ontologies (see [8]).

In order to maximise the benefits of using data wrapping ontologies, these should be rich enough to ease their integration with the domain ontology and, at the same time, precisely characterise the data they wrap. Ontologies extracted automatically from data sources (e.g. by analysing the constraints in the logical schema) are faithful representations of the data sources; however, they are usually shallow and with a limited vocabulary. For this reason, they can be used as bootstrap ontologies, and the task of enriching the extracted ontology is crucial in order to build a truly effective ontology-based information access system. The process of enriching an ontology involves at least the introduction of new axioms and/or new terms. While, from a purely ontological viewpoint, an ontology can be arbitrarily modified, we need to bear in mind that the ultimate purpose of the data wrapper is to access the information available from the data sources. This means that we should be able to use the newly introduced terms in order to retrieve data from the sources.

It is easy to provide examples where such new terms can be completely useless; in the sense that queries over these terms will always return empty answers. This not necessarily because they are unsatisfiable in the usual model theoretic meaning, but because there is no underlying data supporting them. Let us consider a simple example depicted in Figure 1, where the bottom part represents the logical schema, the middle part the data source terms (connected with the relational sources by means of mappings, depicted with dashed arrows) and the top part the enriched fragment of the ontology. It is obvious that any query on Actor would always return empty answer, whatever the data sources may contain; while the concept represented by the same term would be satisfiable. The situation would be different if Actor was also restricted to elements whose range w.r.t. person_role was bound to ActingRole[1]. In this case, there could be instances of the database for which the same query on Actor would return a non empty answer.

---

[1] In Description Logics terminology this corresponds to an inclusion assertion of the form $\exists$person_role.ActingRole $\sqsubseteq$ Actor.

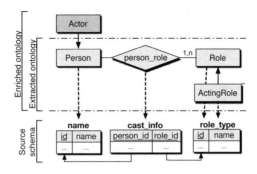

Fig. 1. Example of simple data wrapper

In order to ensure that queries over ontologies wrapping data sources provide sensible answers, these ontologies must be carefully handcrafted by taking into account the query answering algorithm. To the best of our knowledge, little or no research has been devoted to the support of the ontology engineer in such a complex and error prone task. Our research is directed to techniques and tools to support this modelling process.

The foundation of our technique is the problem of verifying the emptiness of a given term w.r.t. a set of data source terms (i.e. terms "connected" to data sources). Given a Description Logic (DL) theory composed by Tbox and Abox over a given vocabulary (see Section 2.1 for details), we define a subset of the concepts and roles as *data source* terms. Given a Tbox, a concept or role term is empty iff the certain answer of the query defined by the term is empty for all possible Aboxes whose assertions are restricted to data source terms. The idea is that data (by means of Abox assertions) can only be associated to data source terms. Clearly the problem is different from classical (un)satisfiability, because we impose a restriction on the kind of allowed Abox assertions. Note that the two problems coincide when all the DL terms are considered as data sources.

In [9] we introduced the above problem and presented some preliminary results; while the contribution of this paper is a generalisation of the results presented in [9], by providing algorithms to verify term emptiness for a more expressive class of ontology languages (see [10]). In particular, a crucial gain in terms of expressive power of the language adopted in this work is the ability to express inclusions among roles. In addition, we describe how this algorithm can be used to support the user in the "repair" of the empty terms (cf. Section 5), and present a Protégé plug-in implementing it (cf. Section 6.1). Finally, we discuss an empirical study showing the benefits of our approach (cf. Section 6.2).

2   Formal Framework

In this section we introduce the formal framework for representing the ontology, queries over the ontology, and show how the actual ontology is linked to relational data sources.

2.1   Ontology Language and Queries

The ontology language is based on the DL $\mathcal{ELHI}$ [10]. An $\mathcal{ELHI}$ *basic concept* B has the form A, $\exists R$, $\exists R^-$, $\exists R.A$, $\exists R^-.A$, or $B_1 \sqcap B_2$, where A is an *atomic concept*, R is an *(atomic) role*. A *TBox* $\mathcal{T}$ is a set of *inclusion assertions* of the form $B_1 \sqsubseteq B_2$ or $R_1 \sqsubseteq R_2$. We consider constants in $\mathcal{ELHI}$ to be defined over the alphabet Γ_O of constant symbols for objects. An *ABox* $\mathcal{A}$ is a set of *membership assertions* of the form $A(a)$ or $R(a,b)$ with a and b constants in Γ_O. An $\mathcal{ELHI}$ *ontology* $\mathcal{K}$ is a pair $\langle \mathcal{T}, \mathcal{A} \rangle$, with $\mathcal{T}$ a TBox and $\mathcal{A}$ an ABox.

We assume the "standard" Description Logics semantics, as defined e.g. in [11], with the *unique name assumption*; i.e., different constants in Γ_O are interpreted as different objects in the interpretation domain. The $\mathcal{ELHI}$ DL is expressive enough to represent the commonly used data modelling constructs.

Example 1. Consider the ER diagram in Figure 1. The constraints expressed in the diagram can be represented in $\mathcal{ELHI}$ ontology using the following assertions:

(1) $\exists$person_role $\sqsubseteq$ Person   (4) ActingRole $\sqsubseteq$ Role
(2) $\exists$person_role$^-$ $\sqsubseteq$ Role   (5) Actor $\sqsubseteq$ Person
(3) Role $\sqsubseteq$ $\exists$person_role$^-$.Person

Assertions (1) and (2) correspond to the role typing of relationship person_role, stating, respectively, that the first component of person_role is of type Person, while the second, i.e., its inverse, is of type Role. Assertion (3) instead states mandatory participation for Role to relationship person_role, at the same time restricting the domain of person_role to the elements of Person. (4) and (5) express is-a relations among the respective terms.

To formulate queries over the $\mathcal{ELHI}$ ontology $\mathcal{K}$ we use *conjunctive queries* (CQs) [12] that are of the form $q(\boldsymbol{x}) \leftarrow body(\boldsymbol{x}, \boldsymbol{y})$, where $body(\boldsymbol{x}, \boldsymbol{y})$ is a conjunction of atoms involving concept and role names in $\mathcal{K}$. Given a CQ q and an $\mathcal{ELHI}$ ontology $\mathcal{K}$, the *answer to q over $\mathcal{K}$* is the set of tuples $q^{\mathcal{I}}$ of constants that substituted to $\boldsymbol{x}$ make the formula $\exists \boldsymbol{y}.body(\boldsymbol{x}, \boldsymbol{y})$ true in $\mathcal{I}$, for $\mathcal{I}$ an interpretation that is a model for $\mathcal{K}$.

2.2   Wrapping Relational Sources

The purpose of data wrapping ontologies is to access data from relational sources. To this end, some of the concepts and roles are mapped to data stored in a database and described by a relational schema. We consider a fixed denumerable alphabet Γ_V of value constants and we consider databases over Γ_V. A *relational schema* $\mathcal{R}$ consists of *relation* symbols, and a *database instance* (or simply a *database*) $\mathcal{D}$ over $\mathcal{R}$ is a set of relations with constants from Γ_V as atomic values.

We adopt an ontology to relational data mapping scheme in the spirit of the framework presented in [6] for linking data to ontologies. We introduce a new alphabet Λ of function symbols in $\mathcal{ELHI}$, where each function symbol has an associated arity. Then, we define the set $\tau(\Gamma_V, \Lambda)$ of all *function terms* of the form $f(v_1, \ldots v_n)$ such that $f \in \Lambda$ with arity $n > 0$, and $v_1, \ldots v_n \in \Gamma_V$. We require that the set Γ_O used to denote objects in $\mathcal{ELHI}$ coincides with $\tau(\Gamma_V, \Lambda)$. Then, we define an $\mathcal{ELHI}$ *data wrapping ontology with mappings* as a triple $\mathcal{K}_m = \langle \mathcal{T}, \mathcal{M}, \mathcal{D} \rangle$, where $\mathcal{T}$ is an $\mathcal{ELHI}$ TBox, $\mathcal{D}$ is a database over the relational schema $\mathcal{R}$, and $\mathcal{M}$ is a set of *mapping assertions* of the form $\phi \rightsquigarrow \psi$, with ψ an arbitrary (SQL) query over $\mathcal{D}$, ϕ an atom of the form $A(f(\boldsymbol{x}))$ or $R(f_1(\boldsymbol{x_1}), f_2(\boldsymbol{x_2})))$, A and R, respectively, an atomic concept and role in $\mathcal{T}$, and $f(\boldsymbol{x})$ a variable term with f in Λ and $\boldsymbol{x}$ tuple of variables.

Given a source relational schema $\mathcal{R}$, the data wrapping ontology with mappings $\mathcal{K}_m = \langle \mathcal{T}, \mathcal{M}, \mathcal{D} \rangle$ is derived from $\mathcal{R}$ in such a way that for each relation $r \in \mathcal{R}$, there are either atomic concepts A_{db} and A or roles R_{db} and R, with $A_{db} \sqsubseteq A$ (respectively, $R_{db} \sqsubseteq R$), such that A_{db} (R_{db}) has an associated mapping specifying how to retrieve the data about A_{db} (R_{db}) from the data sources. By introducing two concepts for a relation, say A_{db} and A, with a corresponding assertion $A_{db} \sqsubseteq A$, we "fix" the db subscripted terms as coming from the data sources, i.e., a syntactic convenience to be used in the emptiness testing algorithm; however, they both mean the same concept. Hence, in the rest of the paper, we will refer to such terms as *data source terms* and denote them by $\Sigma_{\mathcal{DB}}$. For details on the semantics of the mappings we refer to [6].

Example 2. For instance, the following mappings are associated to terms Person, Role and person_role from Figure 1[2].

   Person(pers(id)) $\rightsquigarrow$ SELECT id FROM $name$
   Role(role(id)) $\rightsquigarrow$ SELECT id FROM $role\_type$
   person_role(pers($person\_id$), role($role\_id$)) $\rightsquigarrow$
                         SELECT $person\_id, role\_id$ FROM $cast\_info$

We next introduce the notion of a virtual ABox, whose assertions are generated by "compiling" the mapping assertions starting from the data in $\mathcal{D}$.

Definition 1. *Given a data wrapping ontology with mappings* $\mathcal{K}_m = \langle \mathcal{T}, \mathcal{M}, \mathcal{D} \rangle$ *and a mapping assertion* $M : \phi \rightsquigarrow \psi$ *in* $\mathcal{M}$, *a virtual ABox generated by* M *from* $\mathcal{D}$ *is the set of assertions of the form* $\mathcal{A}(M, \mathcal{D}) = \{\phi[\boldsymbol{f}(\boldsymbol{x})/\boldsymbol{f}(\boldsymbol{v})] \mid \boldsymbol{v} \in ans(\psi, \mathcal{D})\}$, *where* $\phi[\boldsymbol{f}(\boldsymbol{x})/\boldsymbol{f}(\boldsymbol{v})]$ *is a ground atom, obtained from* ϕ *by substituting the n-tuple of variable terms* $\boldsymbol{f}(\boldsymbol{x})$ *with the n-tuple of constant terms* $\boldsymbol{f}(\boldsymbol{v})$. *Then, the* virtual ABox *for* $\mathcal{K}_m$ *is the set of assertions* $\mathcal{A}(\mathcal{M}, \mathcal{D}) = \{\mathcal{A}(M, \mathcal{D}) \mid M \in \mathcal{M}\}$.

Observe that by construction, all concept and role names appearing in $\mathcal{A}(\mathcal{M}, \mathcal{D})$ are data source terms from $\Sigma_{\mathcal{DB}}$. Thus, we will consider $\mathcal{A}(\mathcal{M}, \mathcal{D})$ as an *incomplete database*.

It follows, from the semantics of the mappings [6] and the construction of the virtual ABox, that the models of $\mathcal{ELHI}$ data wrapping ontology with

---

[2] For the sake of simplicity, we don't explicitly display in Figure 1 the corresponding db terms. As mentioned, all entities and relationships in the middle layer are fixed to be database terms.

mappings $\mathcal{K}_m = \langle \mathcal{T}, \mathcal{M}, \mathcal{D} \rangle$ and those of $\mathcal{ELHI}$ ontology with virtual ABox $\mathcal{K} = \langle \mathcal{T}, \mathcal{A}(\mathcal{M}, \mathcal{D}) \rangle$ coincide, i.e., we can express the semantics of $\mathcal{K}_m$ in terms of $\mathcal{ELHI}$ ontology with a virtual Abox. This immediately implies an algorithm to answer queries over an $\mathcal{ELHI}$ ontology with mappings: *(i)* compute $\mathcal{A}(\mathcal{M}, \mathcal{D})$, and *(ii)* apply query answering technique of [13] for $\mathcal{ELHI}$ ontology $\mathcal{K} = \langle \mathcal{T}, \mathcal{A}(\mathcal{M}, \mathcal{D}) \rangle$. Given this, in the rest of the paper, we will only consider an $\mathcal{ELHI}$ ontology and a virtual ABox as a source (incomplete) database. Notice however that for the actual emptiness testing algorithm we will not explicitly build the virtual ABox; nevertheless, we will use this notion for our technical development described in the next section.

3   Emptiness of Ontology Terms

From now on, we will consider a scenario where an $\mathcal{ELHI}$ ontology $\mathcal{K}$ derived form the data sources, i.e., defined over the signature $\Sigma_{\mathcal{DB}}$, has been enriched by adding to $\mathcal{T}$ new terms and/or assertions.

Given a term η in $\mathcal{T}$ of an ontology $\mathcal{K}$, we call a *query for* η a CQ of the form $q(x) \leftarrow \eta(x)$ (resp., $q(x, y) \leftarrow \eta(x, y)$), for η an atomic concept (resp., role) in $\mathcal{T}$. Our goal is to test whether η is empty w.r.t. the data at the sources, i.e., w.r.t. $\Sigma_{\mathcal{DB}}$. Clearly, such a test should involve the query answering process; thus we now define the notion of query answering in the presence of an incomplete database.

Definition 2. *Given an $\mathcal{ELHI}$ ontology $\mathcal{K} = \langle \mathcal{T}, \mathcal{A}(\mathcal{M}, \mathcal{D}) \rangle$ and a query q over $\mathcal{K}$, the* certain answers *to q w.r.t. $\mathcal{K}$, denoted $cert(q, \mathcal{K})$, are the set of tuples $\boldsymbol{t}$ such that $\boldsymbol{t} \in q^{\mathcal{I}}$ for every interpretation $\mathcal{I}$ that is a model for $\mathcal{K}$.*

Then, we say that a given term is empty, if the certain answers to its corresponding query are empty for *every* incomplete database $\mathcal{A}(\mathcal{M}, \mathcal{D})$.

Definition 3. *Let $\mathcal{K} = \langle \mathcal{T}, \mathcal{A}(\mathcal{M}, \mathcal{D}) \rangle$ be an $\mathcal{ELHI}$ ontology and η a term in $\mathcal{T}$ with query q for η. Then, η is empty w.r.t. $\Sigma_{\mathcal{DB}}$ iff $cert(q, \mathcal{K}) = \emptyset$ for every database $\mathcal{A}(\mathcal{M}, \mathcal{D})$.*

This defines the problem studied in this paper: given a term $\eta \in \mathcal{T}$ with a CQ q for η, test whether $cert(q, \mathcal{K}) = \emptyset$ for every $\mathcal{A}(\mathcal{M}, \mathcal{D})$. Note however that this does not imply that we will be necessarily computing $cert(q, \mathcal{K})$.

It is well known that the problem of computing answers in the presence of an incomplete database is often solved via *query rewriting* under constraints. Specifically, based on [13], we have that given a query q over an $\mathcal{ELHI}$ ontology $\mathcal{K} = \langle \mathcal{T}, \mathcal{A}(\mathcal{M}, \mathcal{D}) \rangle$, we can compute another query, a rewriting of q denoted by $rew(q, \mathcal{T})$, such that the answers of q over $\mathcal{K}$ and the answers of $rew(q, \mathcal{T})$ over $\mathcal{A}(\mathcal{M}, \mathcal{D})$ only coincide, i.e., $cert(q, \mathcal{K}) = rew(q, \mathcal{T})^{\mathcal{A}(\mathcal{M}, \mathcal{D})}$. Thus, we have the following:

Theorem 1. *Let $\mathcal{K} = \langle \mathcal{T}, \mathcal{A}(\mathcal{M}, \mathcal{D}) \rangle$ be an $\mathcal{ELHI}$ ontology and η a term in $\mathcal{T}$ with query q for η. Then, η is empty w.r.t. $\Sigma_{\mathcal{DB}}$ iff $rew(q, \mathcal{T})^{\mathcal{A}(\mathcal{M}, \mathcal{D})} = \emptyset$ for every database $\mathcal{A}(\mathcal{M}, \mathcal{D})$.*

4   Testing Emptiness

As follows from the previous section, to test emptiness of a given term we have to rewrite its corresponding query and check whether the obtained rewriting results in being empty. Recent work on rewriting conjunctive queries over $\mathcal{ELHI}$ ontologies [13] shows that for a CQ q over $\mathcal{K} = \langle \mathcal{T}, \mathcal{A}(\mathcal{M}, \mathcal{D}) \rangle$, $rew(q, \mathcal{T})$ is a Datalog program. Therefore, according to Theorem 1, our problem now comes down to testing emptiness of a query predicate q in the rewritten Datalog program.

We first define some notions that will be needed throughout this section.

4.1   Preliminary Notions

A *Datalog program* Π over an $\mathcal{ELHI}$ ontology $\mathcal{K}$ consists of *(i)* a set of rules of the form $head(\boldsymbol{x}) \leftarrow body(\boldsymbol{x}, \boldsymbol{y})$, where $body(\boldsymbol{x}, \boldsymbol{y})$ is a conjunction of atoms involving concept and role names in $\mathcal{K}$; *(ii)* a *special rule* that is a CQ with a *query predicate* q in the head.

The *extensional database (EDB) predicates* of a Datalog program Π are those that do not appear in the head of any rule in Π, all other predicates are called *intentional database (IDB) predicates*. Given a (incomplete) database $\mathcal{A}(\mathcal{M}, \mathcal{D})$, the evaluation $\Pi(\mathcal{A}(\mathcal{M}, \mathcal{D}))$ of Π over an EDB $\mathcal{A}(\mathcal{M}, \mathcal{D})$, is the evaluation of the special rule, denoted by $q_\Pi(\mathcal{A}(\mathcal{M}, \mathcal{D}))$, taken as a CQ, over the minimum Herbrand model of $\Pi \cup \mathcal{A}(\mathcal{M}, \mathcal{D})$ [12].

Given a Datalog program Π and an IDB predicate q in Π, the associated *AND-OR tree for q in Π* is a set of labelled nodes such that *(i)* the root of the tree is a (and-)node labelled by q; *(ii)* for every and-node labelled by g_i, and for every rule r_i of Π having g_i in the head, there exists an or-node, child of g_i, labelled by r_i; *(iii)* for every or-node labelled with a rule r_i in Π, and for every atom name g_{i_j} in the body of r_i, there exists an and-node labelled with g_{i_j}.

An *or-branch* of an AND-OR tree is a set of and-nodes $\mathcal{G}$ that are children of a unique combination of or-nodes in the tree, i.e., when several sibling or-nodes are present, only children of one of the or-nodes are contained in $\mathcal{G}$.

4.2   Emptiness Testing Algorithm

The problem of verifying emptiness of Datalog predicates has been addressed by Vardi [14] in the setting of Datalog optimisation. The author shows that deciding emptiness of IDB predicates in Datalog programs can be done in polynomial time. The key idea underlying this result is the observation that a Datalog program can be viewed as an infinite union of CQs that, in turn, can be described by means of *expansion trees*. Importantly, [14] shows that we can get rid of variables when building expansion trees, obtaining *skeletons of expansion trees*.

We build our approach on the results of [14], and in particular on the possibility of building finitely labelled trees for IDB predicates. Note that while [14] presents the problem as decision problem on emptiness of tree automata ([14] is specifically tailored for the exposition of the use of tree automata), we prefer to present direct algorithms (that do not involve tree automata) because working with skeleton trees, as we will see, is conceptually much simpler.

Given a term η with a CQ q for η in an $\mathcal{ELHI}$ ontology $\mathcal{K}$, we devise our emptiness testing algorithm in four steps: *(i)* rewrite q using procedure of [13], obtaining a Datalog program Π, *(ii)* add to Π auxiliary rules for making IDB predicates explicit, *(iii)* for the resulting Datalog program with a query predicate q, build an AND-OR tree for q, and *(iv)* visit the obtained tree and mark its nodes as empty/nonempty corresponding to empty/nonempty predicates, which, in turn, correspond to empty/nonempty concepts and roles in $\mathcal{K}$. In the following we will elaborate on steps *(ii)-(iv)*; for details on the rewriting algorithm we refer to [13].

Adding Auxiliary Rules to Π. Consider a Datalog program Π resulting from rewriting a CQ for a given term over an ontology $\mathcal{K} = \langle \mathcal{T}, \mathcal{A}(\mathcal{M}, \mathcal{D}) \rangle$. Then, for each term η that is not among data source terms in $\Sigma_{\mathcal{DB}}$ and does not appear in any head of the rules of Π, add to Π an auxiliary rule $\eta(x) \leftarrow \eta(x)$ (resp. $\eta(x, y) \leftarrow \eta(x, y)$) for η corresponding to an atomic concept (resp. role) in $\mathcal{T}$ of $\mathcal{K}$. We denote the resulting Datalog program by Π^*.

The intuition here is that, since, by construction of the data wrapping ontology in Section 2.2, all terms subscripted with *db* appear only on the left-hand sides of inclusion assertions, then, by virtue of the rewriting algorithm [13], the corresponding predicates in Π won't be saturated and thus are guaranteed to occur only in the bodies of the rules of Π (i.e., as EDB predicates). However, it is not the case that the rest of the terms of $\mathcal{T}$ (i.e., non database terms) occur only in the heads of the rules of the rewritten program Π. Therefore, using auxiliary rules above, we explicitly make all non data source terms as IDB predicates. Note that an auxiliary rule $\eta(x) \leftarrow \eta(x)$ is equivalent to a tautology $\eta(x) \vee \neg \eta(x)$. Thus, from a logical point of view, we do not change the semantics of the program Π. That is, an IDB predicate q is empty in Π iff q is empty in Π^*, with $\mathcal{A}(\mathcal{M}, \mathcal{D})$ considered as EDB, i.e., $q_{\Pi}(\mathcal{A}(\mathcal{M}, \mathcal{D})) = \emptyset$ iff $q_{\Pi^*}(\mathcal{A}(\mathcal{M}, \mathcal{D})) = \emptyset$.

Building Skeleton Tree for q in Π^*. The *skeleton tree* for a query predicate q in Π^*, $skel(q, \Pi^*)$, is an AND-OR tree for q in Π^* (we assume all rules are named beforehand), with a condition that an and-node is not expanded (i.e., is a leaf), if either

- it is labelled with an EDB predicate,
- it has an isomorphic and-node (i.e., a node labelled with the same predicate symbol) that has already been expanded, or
- it is marked as empty/nonempty i.e., has already been processed before (in another skeleton tree).

Example 3. Consider the data wrapping ontology in Figure 2(a) and consider the (part of a) Datalog program with rules r_1 to r_7 below obtained by rewriting a CQ $q(x) \leftarrow \mathsf{TVListing}(x)$, i.e., we want to test emptiness of $\mathsf{TVListing}$. We

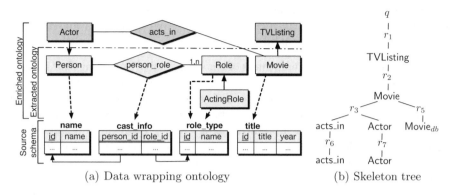

Fig. 2. Data wrapping ontology and skeleton tree for the Datalog program of Example 3

additionally have auxiliary rules r_6 and r_7 that make acts_in and Actor to be IDB predicates in the program.

$r_1 : q(x) \leftarrow \mathsf{TVListing}(x)$

$r_2 : \mathsf{TVListing}(x) \leftarrow \mathsf{Movie}(x)$

$r_3 : \mathsf{Movie}(x) \leftarrow \mathsf{acts\_in}(y, x), \mathsf{Actor}(y)$

$r_4 : \mathsf{Person}(x) \leftarrow \mathsf{Actor}(x)$

$r_5 : \mathsf{Movie}(x) \leftarrow \mathsf{Movie}_{db}(x)$

$r_6 : \mathsf{acts\_in}(x, y) \leftarrow \mathsf{acts\_in}(x, y)$

$r_7 : \mathsf{Actor}(x) \leftarrow \mathsf{Actor}(x).$

The skeleton tree for this Datalog program is shown in Figure 2(b). Note that the children of r_6 and r_7, the and-nodes acts_in and Actor are not expanded anymore, since they have isomorphic nodes that have already been expanded.

Observe that the size of the skeleton tree, as well as the time needed to build it, is linear in the number of the rules in Π^*, since, by construction of the tree, each rule in Π^* is expanded *only once*. Moreover, it is immediate to see that an AND-OR skeleton tree obtained in this way represents all the skeletons of expansion trees as defined in [14]. So, a predicate q is empty, iff all or-branches of $skel(q, \Pi^*)$ are empty. We next define emptiness of nodes in $skel(q, \Pi^*)$.

Visiting Skeleton Tree. Once the tree is built, our algorithm examines it bottom-up (depth-first) and marks the respective nodes as empty or nonempty. Specifically, starting from an and-leaf labelled with g_{i_j}, if g_{i_j} is an EDB predicate in Π^*, then it is marked as nonempty. The algorithm then visits next sibling of g_{i_j} and checks its emptiness/nonemptiness, so that a parent of g_{i_j}, an or-node labelled by a rule r_i, is marked as nonempty iff *all* its children are marked as nonempty. The algorithm then proceeds to an and-node g_i, parent of r_i, marking it as nonempty iff *at least one* of its children or-nodes are nonempty.

Example 4 (Example 3 continued). We start with acts_in leaf and mark it as empty (it is not an EDB predicate). This makes also its parent r_6 and, in turn, acts_in empty. To decide for r_3, we have to check the or-branch starting with Actor, which results in being empty. Therefore, r_3 is empty. Again, to decide for Movie, we look at the or-branch on the right-hand side. Movie_{db} is an EDB predicate, so it is nonempty. Consequently, we mark r_5 and Movie as nonempty, which determines non-emptiness for r_2 and then TVListing, r_1 and finally q. Indeed, we can construct a CQ $q(x) \leftarrow \text{Movie}_{db}(x)$ that guarantees non-emptiness when evaluated over the actual data.

It is important to note that emptiness of a node is "global", meaning that if a node is empty, it will be empty in every skeleton tree it appears in (and the same for non-emptiness). This is due to the rewriting algorithm [13] which "compiles" in the Datalog program all the knowledge about a given term. For this reason, as we have already mentioned, each predicate in a Datalog program is expanded only once. Finally, notice that the technique proposed in this section is applicable to ontology languages in the full spectrum of DLs from $\mathcal{ELHI}$ to *DL-Lite*$_{core}$ [15]. This again because of the rewriting technique [13] being able to deal with this range of languages. The rewriting of a CQ over a *DL-Lite* KB is a union of CQs [13], however, a Datalog program can always be viewed as a union of CQs or a single CQ.

5   Repairing Empty Terms

So far, we have devised a procedure for verifying whether a given term in a data wrapping ontology is empty w.r.t. the database terms at the sources. We next present a method for supporting the repair of empty concepts and roles, consisting of a set of *guidelines* for ontology engineers.

To suggest a repair for an empty term, we naturally resort to the Datalog program Π^* and the skeleton tree generated from Π^* by our emptiness testing algorithm. Indeed, the skeleton tree for a term η, by virtue of its construction, contains as nodes all and only *relevant* terms for η: those that contribute or *could* contribute to its non-emptiness. So an intuitive way to fix an empty term is to *focus* on the relevant nodes (in one of the or-branches) of its corresponding skeleton tree and to possibly *expand* those nodes by rendering them nonempty. The possible expansion should obviously be in correspondence with an addition or refinement of a term or/and assertion in the actual ontology. We elaborate on this idea in the rest of this section.

Given a skeleton tree constructed by the algorithm with an and-node g in the tree, let $\mathcal{G}^* = [G_1, \ldots, G_n]$ denote the sequence of sets of its and-nodes, such that, intuitively, each G_i contains, in a bottom-up fashion, distinct groupings of and-nodes in *one* of the or-branches of the tree that should be marked as nonempty in order for g to be marked as nonempty. Moreover, $\mathcal{G}^*$ is such that, in order for and-nodes in G_i to be marked as nonempty in the tree, all the G_ks, $k = \{1, \ldots, i - 1\}$, have to be marked as nonempty.

Example 5. Suppose rule r_5 was not present in the tree of Figure 2(b). Then, for the and-node TVListing, there would be only one or-branch in the tree, $\mathcal{G}^* = [\{\text{acts\_in}, \text{Actor}\}, \{\text{Movie}\}]$. The intuition here is that both, acts_in and Actor, and Movie have to be repaired in order for TVListing to become nonempty. And similarly, to repair Movie, both acts_in and Actor must be rendered to be no longer empty.

Thus, for each and-node g_{i_j} in every G_i, of every or-branch of the skeleton tree, our strategy is to consider g_{i_j} as a leaf in the tree, and to examine its possible expansions, whereas to expand a leaf we mainly need a new rule with its corresponding atom in the head. Given such a rule, we can track down the needed terms and assertions in the ontology and provide those repairs as guidelines to the user. For this purpose, we distinguish two cases: *(i)* g_{i_j} corresponds to an atomic concept A, and *(ii)* g_{i_j} corresponds to a role R.

For case *(i)*, our repair service provides two guidelines. First, it suggests to add an inclusion assertion with A on the right-hand side. This, from the modelling point of view, results in either defining participation constraints for A to a relationship R, if R appears in any of G_ks, $k = \{1, \ldots, i-1\}$, of $\mathcal{G}^*$, or asserting A as a superclass of some class B, verifying beforehand that B is nonempty. Second, if $B(x) \leftarrow A(x)$ is present in the program Π^* and B is nonempty, the user is warned with misplaced is-a relationship, i.e., maybe $B \sqsubseteq A$ should be added instead of $A \sqsubseteq B$.

For case *(ii)* we have again two possible guidelines. The first one hints to add an inclusion assertion between roles with R on the right-hand side. The second, if a concept A appears in any of G_ks as above, and A is nonempty, the service suggests to add an assertion $A \sqsubseteq \exists R$, i.e., mandatory participation for A in the relationship R.

Example 6 (Example 4 continued). We have seen that Actor and acts_in were empty. For Actor, our repair service suggests to either assert it as a superclass to some (possibly still to be added) nonempty class, or to replace Actor $\sqsubseteq$ Person with Person $\sqsubseteq$ Actor which, evidently, in this case is not appropriate. To repair acts_in, the user will be suggested to either assert mandatory participation for Movie to acts_in: Movie $\sqsubseteq$ $\exists$acts_in$^-$, or to make acts_in more general than some (possibly still to be added) nonempty role.

Finally, if none of the above mentioned repairs are possible, we suggest to explicitly map to the sources either the actual empty term or any set of its relevant terms (i.e. any set from $\mathcal{G}^*$).

6   Implementation and Evaluation

We discuss in this section a Protégé plug-in implementing techniques described in Sections 4 and 5, and present an empirical study showing usage and benefits of our approach.

6.1   Tool for Emptiness Testing and Repair

We have implemented our approach as a (preliminary) plug-in for Protégé[3] that enhances the OBDA plug-in[4] [16]. The OBDA plug-in provides facilities to design

---

[3] http://protege.stanford.edu
[4] http://obda.inf.unibz.it/protege-plugin

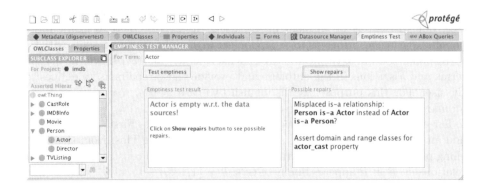

Fig. 3. Emptiness testing and repair in Protégé

Ontology Based Data Access (OBDA) system components (i.e., data sources and mappings). It supports the definition of relational data sources and GAV-like mappings to link concepts and roles of the *DL-Lite$_A$* ontology [6] to data in the defined sources. It also allows for conjunctive query answering (using SPARQL syntax).

The key goal of our plug-in is to provide a support for verifying emptiness of a selected term in an ontology w.r.t. the data at the defined sources, and for repairing empty terms by allowing the user to explore different repair solutions. Figure 3 shows the screenshot of our Protégé plug-in when using the wrapping ontology to query the data (using OBDA plug-in's *ABox Queries* tab) at the underlying database (specified together with mappings in *Datasource Manager* tab). As can be seen, for a selected class or property at the left segment of the plug-in's window, the user can verify, by clicking on *Test emptiness* button, whether that term will return any answer when queried against the source database (without computing the actual evaluation!). When a term results in being empty, the user can ask for guidelines to repair that term by clicking on *Show repairs* button. The repairs are then devised by the tool and shown in a text pane in natural language using standard ontology modelling terminology.

6.2   Usability Study

We conducted a small usability study involving ten subjects, all with homogeneous understanding of description logics reasoning, ontology-based data access, and experience using Protégé – most subjects were graduate students that have attended courses on the mentioned topics and have used Protégé editor for their practical projects.

We used showbiz domain for the study. In particular, for the sources, we used *IMDB* movie database, retrieved using IMDbPY[5]. The wrapping ontology, that we call showbiz, was obtained by first automatically extracting the bootstrap ontology from *IMDB* database together with mappings [7] (21 in total), and

---

[5] http://imdbpy.sourceforge.net/

then by manually enriching it with terms and assertions to (partly) describe TV programmes. The obtained ontology contained 24 classes and 14 properties; we refer to the technical report [17] for the actual terms and assertions.

The subjects were randomly divided into two groups: five subjects without the support for testing emptiness of ontology terms and repairing them (group 1), and five subjects with the support of the above described plug-in (group 2). Then, each subject was given four simple queries over showbiz ontology (even though most subjects were obviously familiar with IMDB domain, none of the subjects had seen before neither the ontology, nor the queries) but having empty answers: e.g. asking for all movies that have a genre, all TV listings and their kinds, etc (see [17]). Given that, the subjects were asked to add to the ontology new assertions so that the given queries were no longer empty. This involved identifying atoms responsible for query emptiness and repairing the corresponding terms. The subjects in group 2 were additionally asked to fill in a questionnaire concerning their experience using the tool. The goal of this study was to compare the time taken and effort needed to complete the task between the two groups, and to evaluate user experience in using the plug-in.

The results of the study are promising. While the assertions added to an ontology in order to arrive to a solution were mostly correct and alike in both groups, the time taken to do it in group 2 was between 2-3 times less than in group 1. Specifically, the average time taken for group 1 was 39 minutes, and 20 minutes for group 2. The average number of changes made to the ontology in order to repair given queries, which we consider to be as key sub-task, for group 1 was 11, and 6 for group 2. The total number of changes needed for all queries was 5. This means that, in average, each subject in group 1 made 5 *erroneous changes* to repair the given queries, while in group 2 – 1 erroneous change.

As mentioned, we have also collected user reactions to the tool. The questionnaire used for this purpose was composed of 10 short statements (e.g., "I found repair guidelines to be adequate"), each accompanied by a 5-point scale of "strongly disagree" (1 point) to "strongly agree" (5 points). Thus, given 5 subjects in group 2, each statement scores to maximum of 25 points. The key aspects, from the usability point of view, are that subjects in group 2 felt that they could effectively identify the reason for query emptiness using the tool (rated a total score of 19) and effectively repair empty terms using the tool (21 points), and strongly agreed that they could identify empty classes/properties and fix them using the tool faster than without it (25 points). Finally, the overall satisfaction of using the plug-in scores to 25.

For more details on the results of the study, the reader is referred to [17].

7   Related Work

To our knowledge, little or no research has been devoted for supporting the problem addressed in this paper. There are several ontology engineering methodologies in the literature (see [18] for a survey); but they are mostly focused on the design of what we call domain ontologies rather than accessing (relational) data sources. Note that the techniques proposed in this paper can be used with most of the well established methodologies.

Recently there has appeared a contribution [19], carried out independently and in parallel with our work, that tackles a very similar problem but comes up in a different setting. The authors define the notion describing when a term is relevant for ABoxes formulated over a given signature, with a given TBox in place, and study its computational complexity for several DLs. Their approach to decide relevance takes also quite a different approach from ours. While our algorithm is via translation to emptiness of IDB predicates in Datalog, [19] instead uses reduction to standard ABox reasoning.

As described in Section 4, our core algorithm is strictly related to the problem of emptiness of intensional predicates in Datalog programs (see e.g. [20]). However, those techniques cannot be applied directly because of the fact that we adopt the DL $\mathcal{ELHI}$ instead of Datalog. The former is better suited for characterising the kind of axioms required for capturing common ER/UML constructs, and part of the upcoming OWL2 W3C recommendation.[6]

Finally, we refer to the work in [21] as related, where, for queries having answers solely determined by the database predicates (the so-called DBox predicates with closed semantics, as apposed to the ABox), the authors show how to find a rewriting over such predicates. The restriction to determinacy may be however in some cases too strong, as for instance **TVListing** in Figure 2(a) is not *determined* by the data source terms but can be (in the classical setting) *rewritten* to a data source term.

8   Conclusions

This paper presents a technique for supporting ontology engineers in the development of ontologies for accessing relational data sources. We introduced the problem of deciding emptiness of a given query w.r.t. a DL theory where data can be accessed only through a subset of the concepts and roles (analogously to the EDB/IDB predicates distinction in Datalog programs). Moreover, we have shown how the algorithm to decide the above problem can be exploited in order to support the engineer in repairing the ontology.

We enhanced the OBDA Protégé plug-in in order to support our technique and we evaluated its effectiveness and usability with an experiment involving external users.

The algorithm presented can be applied in other scenarios, e.g. for optimising the rewriting. Indeed, rules with empty predicates in the rewriting will not contribute to an answer, and thus can be eliminated. For instance, rule r_3 in Example 3 can be removed from the program: when evaluated against the actual data, it won't return any answer.

---

[6] http://www.w3.org/TR/2009/CR-owl2-profiles-20090611/

References

1. Sheth, A.P., Larson, J.A.: Federated database systems for managing distributed, heterogeneous and autonomous databases. ACM Computing Surveys 22(3), 183–236 (1990)
2. Calvanese, D., Giacomo, G.D., Lenzerini, M., et al.: Data integration in data warehousing. Int. J. of Cooperative Information Systems 10(3), 237–271 (2001)
3. Lenzerini, M.: Data integration: A theoretical perspective. In: Proc. of PODS 2002, pp. 233–346. ACM, New York (2002)
4. Berners-Lee, T., Hendler, J., Lassila, O.: The semantic web. Scientific American (2001)
5. Wache, H., Vogele, T., Visser, U., et al.: Ontology-based integration of information - a survey of existing approaches. In: Proc. of the Workshop on Ontologies and Information Sharing, pp. 108–117 (2001)
6. Poggi, A., Lembo, D., Calvanese, D., De Giacomo, G., Lenzerini, M., Rosati, R.: Linking data to ontologies. In: Spaccapietra, S. (ed.) Journal on Data Semantics X. LNCS, vol. 4900, pp. 133–173. Springer, Heidelberg (2008)
7. Lubyte, L., Tessaris, S.: Automatic extraction of ontologies wrapping relational data sources. In: Bhowmick, S., Küng, J., Wagner, R. (eds.) DEXA 2009. LNCS, vol. 5690, pp. 128–142. Springer, Heidelberg (2009)
8. Euzenat, J., Shvaiko, P.: Ontology Matching. Springer, Heidelberg (2007)
9. Lubyte, L., Tessaris, S.: Supporting the design of ontologies for data access. In: Workshop Notes of DL 2008 (2008)
10. Baader, F., Brandt, S., Lutz, C.: Pushing the $\mathcal{EL}$ envelope. In: Proc. of IJCAI 2005, pp. 364–369 (2005)
11. Baader, F., Calvanese, D., McGuinness, D., et al. (eds.): The Description Logic Handbook. Cambridge University Press, Cambridge (2003)
12. Abiteboul, S., Hull, R., Vianu, V.: Foundations of Databases. Addison-Wesley, Reading (1995)
13. Pérez-Urbina, H., Motik, B., Horrocks, I.: Rewriting conjunctive queries under description logic constraints. In: Proc. of LID 2008 (2008)
14. Vardi, M.Y.: Automata theory for database theoreticians. In: Proc. of PODS 1989, pp. 83–92. ACM, New York (1989)
15. Calvanese, D., Giacomo, G.D., Lembo, D., et al.: Tractable reasoning and efficient query answering in description logics: The dl-lite family. J. of Automated Reasoning 39(3), 385–429 (2007)
16. Rodriguez-Muro, M., Lubyte, L., Calvanese, D.: Realizing ontology based data access: A plug-in for protégé. In: Proc. of IIMAS 2008, pp. 286–289 (2008)
17. Lubyte, L., Tessaris, S.: Supporting the design of ontologies for semantic data access. Technical report, KRDB group – Free University of Bozen-Bolzano (2009), http://www.inf.unibz.it/krdb/pub/TR/KRDB09-3.pdf
18. Corcho, Ó., Fernández-López, M., Gómez-Pérez, A.: Methodologies, tools and languages for building ontologies: Where is their meeting point? Data Knowl. Eng. 46(1), 41–64 (2003)
19. Baader, F., Bienvenu, M., Lutz, C., Wolter, F.: Query answering over DL ABoxes: How to pick the relevant symbols. In: Workshop Notes of DL 2009 (2009)
20. Levy, A.Y.: Irrelevance Reasoning in Knowledge Based Systems. PhD thesis, Stanford University (1993)
21. Seylan, I., Franconi, E., de Bruijn, J.: Effective query rewriting with ontologies over DBoxes. In: Proc. of IJCAI 2009, pp. 923–930 (2009)

A Conceptual Model for a Web-Scale Entity Name System*

Paolo Bouquet, Themis Palpanas, Heiko Stoermer,
and Massimiliano Vignolo

University of Trento, Trento, Italy
{bouquet, themis, stoermer, vignolo}@disi.unitn.it

Abstract. The problem of identity and reference is receiving increasing attention in the (semantic) web community and is emerging as one of the key features which distinguish traditional knowledge representation from knowledge representation *on the web* with respect to data interlinking and knowledge integration on a large scale. As part of this debate, the OKKAM project proposed the creation of an Entity Name System which provides **rigid identifiers**, named OKKAMids, for any type of concrete and particular entities, and links OKKAMids to existing identifiers which have been created elsewhere for the same entity. The introduction of these identifiers raises some practical and conceptual concerns. In this paper we address them by extending two proposed ontologies (IRE and IRW) to accomodate the notion of OKKAMid, describe their formal properties, illustrate why they may play an important role in the construction of the Semantic Web and discuss how they can be integrated with other approaches for mapping URIs onto each others.

1   Introduction

One of the most ambitious visions of the Semantic Web is to create an open, decentralized space for sharing and combining knowledge, like the web did for hypertexts. In a note from 1998, Tim Berners-Lee described this vision as follows:

> Knowledge representation is a field which currently seems to have the reputation of being initially interesting, but which did not seem to shake the world to the extent that some of its proponents hoped. It made sense but was of limited use on a small scale, but never made it to the large scale. This is exactly the state which the hypertext field was in before the Web [...]. The Semantic Web is what we will get if we perform the same globalization process to Knowledge Representation that the Web initially did to Hypertext.

---

* This work is partially supported by the FP7 EU Large-scale Integrating Project **OKKAM – Enabling a Web of Entities** (contract no. 215032). For more details, visit http://www.okkam.org/. We are grateful to Stefano Bocconi for his helpful comments on an early draft of this paper.

A. Gómez-Pérez, Y. Yu, and Y. Ding (Eds.): ASWC 2009, LNCS 5926, pp. 46–60, 2009.

As a contribution to this vision, the EU-funded OKKAM project[1] has started the design and development of a so-called *Entity Name System* (ENS) [2], a web-scale, open service which supports users and applications in the systematic reuse of global and stable identifiers for entities which are named and described in distributed collections of data and content. The main goal of the project is to consolidate the information space of the web of data by reducing the number of URIs which are used for referring to the same entity in different datasets, making the integration and fusion of RDF data and content much easier and faster.

However, the very idea of an ENS has raised some theoretical and practical concerns which need to be addressed before such a service may be adopted by the community at large. This paper aims at solving some of these issues by proposing a conceptual model in which we draw a clear distinction between the meaning and the role of the HTTP URIs which are used as identifiers by the ENS (OKKAMids) and the standard RDF URIs which are used in RDF datasets. The underlying intuition is that OKKAMids provide a form of *direct reference* to real-world entities, whereas RDF URIs provide a description-based reference to entities (which means that different RDF URIs may be needed to publish different representations of the "same" entity). We will discuss why, in our view, both types of URIs are needed for building the Semantic Web, and show how they can easily cohexist in practical methods for publishing RDF data.

The OKKAM conceptual model (OCM) distinguishes these two views clearly at a foundational level by providing a formal definition of OKKAMids and RDF URIs. The underlying intuition is that these two types of identifiers need not be perceived as mutually exclusive. On the contrary, they serve different (complementary) purposes, and therefore they should be used together to bring knowledge representation on the web to its full potential. In section 2 we clarify some important conceptual issues concerning the fundamental relations between URIs, the Semantic Web and the real world. Section 3 introduces the OCM model formalized in first order logic. In section 4 we give justification to the distinction between the linguistic function of OKKAMids as rigid and direct identifiers and the linguistic function of RDF URIs as identifiers based on descriptions. In section 5 we discuss some practical consequences of the difference between the two linguistic functions.

2   Identity and Reference in the Semantic Web

Similarly to what happened for the hypertext web with URLs and HREF references, one of the key factors for realizing the vision we quoted above is to enable a global and uniform naming space for the "entities" which are named in a piece of data and content, so that people and machines can always refer unambiguously to whatever entity they need to name. The proposal is to exploit the Web architecture and use URIs (more precisely HTTP URIs[2]) as such naming

---

[1] http://www.okkam.org/

[2] http://www.w3.org/DesignIssues/LinkedData.html

mechanism. Indeed, an important feature of URIs is that the same URI is always dereferenced in the same way, no matter where it appears.

The question is whether the idea works in the other direction as well: does it make sense to request that **the same entity is always referred to by the same (HTTP) URI**? The answer to this question is not so straightforward. The most important objection is conceptual, and has to do with the following issue: does a URI make direct reference to an entity, or is it "equivalent" to a *description* of that entity? The current trend is to view URIs as basically equivalent to descriptions, namely the sets of RDf statements which we obtain when the URIs are dereferenced. In this view, the two identifiers http://www.w3.org/People/Berners-Lee/card#i and http://dblp.l3s.de/d2r/resource/authors/Tim_Berners-Lee provide different (and potentially inconsistent) information about a person (Tim Berners-Lee), and since these two descriptions should not be confused, it makes sense to have two different URIs for the same person. From a different perspective, however, there are researchers who stress that the above issue can be understood only if one assumes that the two URIs are indeed about the same entity. Therefore it must be the case that a name is somehow "attached" directly to an entity without the mediation of any particular description. In short, in the first view *reference is essentially mediated by description*, and the latter is more fundamental than the former; in the second view, reference is not necessarily mediated by any particular description, as *reference is a primitive and direct relation* between a real world entity and its identifier.[3]

The discussion is far from being a mere academic debate on the theory of reference. Indeed, it has a deep practical impact on how people are developing the Semantic Web, and in particular the so-called Web of Data. The first view offers a very elegant DNS-based method for publishing and accessing sets of statements about an entity (and to evaluate their level of trust). In addition, it provides the technical basis for enabling the web-style exploration of semantic data via RDF browsers, like the Tabulator[4], Disco[5] or the OpenLink RDF browser.[6] The second enables very powerful forms of URI-based data retrieval (e.g. through semantic search engines, like Sigma[7], Sindice[8] or Falcons[9]), makes semantic mashups very easy and straightforward, and enables web-scale distributed reasoning. In order to unleash the full power of the technologies that have been developed by the community, we propose a formal model that distinguishes both views at a foundational level.

---

[3] See [8] for a philosophical discussion of this thesis. In [7], the concept of direct reference is presented in a slightly different way as the idea that "on the Web, the resource identified by a URI is whatever was intended by the owner". We'll explain later on why the two definitions have different consequences on our argument.

[4] http://dig.csail.mit.edu/2007/tab/

[5] http://www4.wiwiss.fu-berlin.de/bizer/ng4j/disco/

[6] http://demo.openlinksw.com/DAV/JS/rdfbrowser/index.html

[7] http://sig.ma/

[8] http://sindice.com/

[9] http://iws.seu.edu.cn/services/falcons/objectsearch/index.jsp

3 The OKKAM Conceptual Model

3.1 Basic Concepts

The architecture of the web forces a subdivision of the universe into things that exist or might exist on the web and things that cannot exist on the web. Given that existence on the web amounts to accessibility on the web through dereferencing URIs, most real world entities, indeed all those that are not computational objects, are things that cannot exist on the web because we cannot access them directly but only their representations. Accordingly, the OCM draws the distinction between computational objects and non-web resources.

A computational object is defined as (i) the physical realization of an information object and (ii) something that can participate in a computational process that ensures the resolution of a URI (see [5]). All digital documents, databases, electronic services, files, applications, are computational objects. Once a computational object is assigned a URI that gives it a location on the web, and thereby makes it a web-accessible entity, the computational object becomes a web resource. Non-web resources are all those entities that are not computational objects. The class of OKKAM entity is a subclass of the class of non-web resources, more precisely the class of all particular and concrete entities (events included). This means that classes, properties and abstract concepts do not count as OKKAM entities. It is worth noting that the class of non-web resources is not the complement of the class of web resources. Indeed, a computational object that does not possess a location on the web is neither a web resource nor a non-web resource. Figure 1 shows the relation of inclusion between the above classes of entities. One objective of the Semantic Web is to allow people to talk about things that do not exist on the web and nevertheless to talk about them by using URIs. There is a tension between the objective of the Semantic Web and the idea of using URIs for talking about entities that do not exist on the web. Indeed, from a linguistic point of view, URIs work as descriptions. A URI describes a certain entity as that entity that can be accessed at a certain location on the web. For example, the URI http://www.dit.unitn.it/~bouquet/ denotes by description a web resource, i.e. Paolo Bouquet's homepage. The linguistic function of that

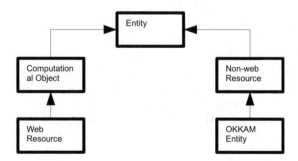

Fig. 1. Relations between classes of entities

URI is the same as the linguistic function of the definite description *the web resource accessible by resolving the URI* http://www.dit.unitn.it/~bouquet/. The idea that the linguistic function of URIs is that of definite descriptions denoting web resources gives rise to two main difficulties. The first is that only computational objects can be accessed on the web. Moreover, while web *locations* persist identical over time, the web resources located there can change. For example, Paolo Bouquet's homepage might change over time due to the updating of his publications, teaching activities, academic appointments etc. The second difficulty, then, is that URIs, as descriptions denoting web resources, are not *rigid designators*,[10] because there is no guarantee that by employing the same URI we will always be talking about the same entity, while in talking about non-web resources we would like to use rigid designators that are guaranteed not to change their referents. Therefore, the very idea of employing URIs as names for entities seems to require the distinction between two separate linguistic functions of URIs. One is the function of denoting by description and the other is the function of naming by reference [10]. The Semantic Web needs URIs that can be used as names that *denote* web resources by description and URIs that can be used as names that *refer* to non-web resources.

There are two ways of solving such linguistic ambiguity. One is to use different URIs according to the linguistic function performed, the other is to use the context of use for disambiguating the two linguistic functions. According to the first solution, RDF URIs are used to make reference to non-web resources. RDF URIs are so configured that when they are dereferenced the web server returns a 303 redirection code redirecting to another URI that might resolve into a web resource or start a further process of redirection. This view makes the distinction between the RDF URIs' linguistic function of making reference to non-web resources and the URIs' linguistic function of denoting web resources by description. It tells that being denoted by description consists in being referenced by a URI, while being referred to consists in being named by a RDF URI which redirects to another URI. Thus, RDF URIs working as names referring to non-web resources are distinguished from URIs working as definite descriptions denoting web resources.

The aim of OCM presented in this paper is to enlarge that way of solving the linguistic ambiguity of URIs to a global scale – though restricted to concrete and particular non-web resources. The main idea underlying OKKAM is that OKKAMids are fundamental tools for enlarging and integrating the use of RDF URIs as identifiers of non-web resources. As is represented in IRW, some RDF URIs can be treated as identifiers of non-web resources. In fact, the irw:identifies property, i.e. the linguistic function of reference defined in IRW over RDF URIs and non-web resources, is functional. The problem with this picture is that the meachanism by which RDF URIs make reference to entities is still based on description. RDF URIs identify their referents as those entities that satisfy the information conveyed in the web resources that are referenced

---

[10] An identifier is a so-called rigid designator if in all possible worlds it denotes the same object. See e.g. [8] for Kripke's introduction to this notion.

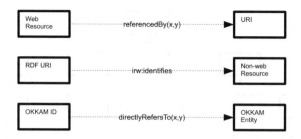

Fig. 2. URI's linguistic functions

by the URIs to which the RDF URIs redirect. It follows that we might have – and indeed this is already the case with many RDF URIs – different RDF URIs expressing different descriptions of the same entity in different contexts. OKKAMids allow us to make it explicit in the web community that two or more RDF URIs identify one and the same entity, though from different points of view. Figure 2 shows the URIs' linguistic function – defined in OCM – of denoting web resources, the RDF URIs' linguistic function – defined in IRW – of designating non-web resources by description and the OKKAMids' linguistic function – defined in OCM – of referring *rigidly* and *directly* to OKKAM entities.

3.2 Formalization

The conceptual model we present is built on top of the ontologies for identity and reference on the Web (IRE, IRW) which have been presented in [6, 5, 10, 7]. OCM adds new concepts which are specifically related to OKKAM in order to model the relations between OKKAM, the real-world and the web. IRE specializes the DOLCE ontology and some of its modular extensions, namely Spatial Relations, DnS with Information Objects, and Knowledge Content Objects (KCO) and Ontology Design Ontology (ODO) modules. Figure 3 shows the relation of inclusion between the concepts defined below and the relations of directlyRefersTo(x, y) defined over OKKAMids and OKKAM entities, referencedBy(x, y) defined over web resources and URIs, and irw:identifies(x, y) defined over RDF URIs and non-web resources. In the following we present the definitions and axioms that form the OKKAM Conceptual Model[11]:

$$OkkamEntity(x) =_{def} Entity(x) \wedge$$
$$Particular(x) \wedge dol : Concrete(x) \wedge \neg od : ComputationalObject(x) \quad (1)$$

Definition 1 states that only concrete and particular entities are OKKAM entities, i.e. entities apt for being assigned OKKAMids. Note that Entity(x) and Particular(x) are implicit in DOLCE. Here we use them for the sake of exposition.

---

[11] The OWL specification of the OCM can be found at http://models.okkam.org/ OKKAM-conceptual_model.owl. The name space for the OCM objects is http:// models.okkam.org/OKKAM-conceptual_model.owl#

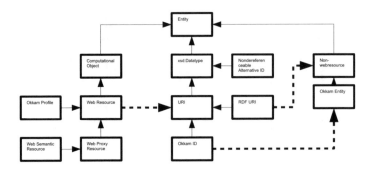

Fig. 3. Map of OCM concepts and relations

$$\texttt{ire}:\texttt{URI}(x) \rightarrow \texttt{xsd}:\texttt{Datatype}(x) \tag{2}$$

Axiom 2 gives a characterization of the concept of URI in terms of XSD datatype as is usual practice.

$$\texttt{OkkamID}(x) =_{def} \texttt{ire}:\texttt{URI}(x)\wedge$$
$$(\texttt{Pattern}(x) =' http://www.okkam.org/ens/id < UUID >') \tag{3}$$

$$\texttt{URIokkamProfile}(x) =_{def} \texttt{ire}:\texttt{URI}(\texttt{x})\wedge$$
$$(\texttt{Pattern}(x) =' http://www.okkam.org/ens/id < UUID > /about.rdf')(4)$$

Definitions 3 and 4 define the concepts of `OKKAMid` and URI for OKKAM profiles by specifying their patterns.[12] The OKKAM web server returns a 303 redirection code response for a request of an `OKKAMid` and gives the URI for the OKKAM profile as the new location of the document.

$$\texttt{ire}:\texttt{hasIdentifier}(x,y) \rightarrow \texttt{dol}:\texttt{Region}(x) \wedge \texttt{xsd}:\texttt{Datatype}(y) \tag{5}$$

Axiom 5 characterizes the relation of having an identifier between regions and datatype identifiers.

$$\texttt{ire}:\texttt{AbstractWebLocation}(x) =_{def}$$
$$\texttt{dol}:\texttt{AbstractRegion}(x) \wedge \exists y(\texttt{ire}:\texttt{URI}(y) \wedge \texttt{ire}:\texttt{hasIdentifier}(x,y)\wedge$$
$$\neg\exists z(\texttt{ire}:\texttt{URI}(z) \wedge y \neq z \wedge \texttt{ire}:\texttt{hasIdentifier}(x,z))) \tag{6}$$

$$\texttt{ire}:\texttt{AbstractWebLocation}(x) \rightarrow$$
$$\neg\exists y,z(\texttt{ire}:\texttt{URI}(y) \wedge \texttt{ire}:\texttt{AbstractWebLocation}(z) \wedge x \neq z$$
$$\wedge\texttt{ire}:\texttt{hasIdentifier}(x,y) \wedge \texttt{ire}:\texttt{hasIdentifier}(z,y)) \tag{7}$$

---

[12] $< UUID >$ as defined in `http://java.sun.com/j2se/1.5.0/docs/api/java/util/` `UUID.html`

Definition 6. and axiom 7. state that an abstract web location is a point in the combinatorial regions identified by the URI metric such that it is identified by at most one URI and cannot be identified by any other URI already employed to identify another abstract web location.

$$\text{ire}: \texttt{webLocationOf}(x,y,t) =_{def}$$
$$\text{dol}: \texttt{eAbstractLocationOf}(x,y,t) \wedge \text{ire}: \texttt{AbstractWebLocation}(x)$$
$$\wedge \text{od}: \texttt{ComputationalObject}(y) \wedge \text{dol}: \texttt{Time}(t) \tag{8}$$

Definition 8 defines a relation between abstract web locations, computational objects and times and specializes the relation $\texttt{dol:eAbstractLocationOf}(x,y)$ imported from the Spatial Relations module.

$$\text{ire}: \texttt{ResolutionMethod}(x) \rightarrow \texttt{edns}: \texttt{Method}(x) \tag{9}$$

$$\texttt{RedirectionMethod}(x) \rightarrow \texttt{edns}: \texttt{Method}(x) \tag{10}$$

$$\text{ire}: \texttt{WebResource}(x) =_{def}$$
$$\exists m(\text{ire}: \texttt{ResolutionMethod}(m) \wedge \texttt{edns}: \texttt{involves}(m,x))$$
$$\wedge \exists y,t(\text{ire}: \texttt{webLocationOf}(y,x,t)) \tag{11}$$

Axioms 9 and 10 and definition 11 state that web resources are computational objects accessible on the web by dereferencing a URI. We add $\texttt{time}(x)$ – which might be an interval of time – because there can be a computational object that lacks an abstract web location at time t, but gets one at time t' or viceversa a computational object that has an abstract web location at t and then it loses it at t'.

$$\texttt{NonWebResource}(x) =_{def} \texttt{Entity}(x) \wedge \neg \text{od}: \texttt{ComputationalObject}(x) \tag{12}$$

Definition 12 defines the concept of non-web resource. $\texttt{OkkamEntity}(x)$ is a subclass of $\texttt{NonWebResources}(x)$.

$$\texttt{OkkamProfile}(x) =_{def}$$
$$\exists u,y,t(\texttt{URIokkamProfile}(y) \wedge \text{ire}: \texttt{hasIdentifier}(u,y) \wedge$$
$$\text{ire}: \texttt{webLocationOf}(u,x,t)) \tag{13}$$

Definition 13 defines the notion of OKKAM profile. An OKKAM profile is a web resource that is accessible on the web by dereferencing an $\texttt{OKKAMid}$.

$$\texttt{referencedBy}(x,y) =_{def}$$
$$\text{ire}: \texttt{WebResource}(x) \wedge \text{ire}: \texttt{URI}(y) \wedge$$
$$\exists m(\text{ire}: \texttt{ResolutionMethod}(m) \wedge \texttt{edns}: \texttt{involves}(m,x)) \wedge$$
$$\exists z,t(\text{ire}: \texttt{webLocationOf}(z,x,t) \wedge \text{ire}: \texttt{hasIdentifier}(z,y)) \tag{14}$$

Definition 14 defines the relation of being referenced over web resources and URIs. The property referencedBy(x,y) in OCM is different from the property irw:isReferencedBy(x, y) and equivalent to the inverse of the irw:accesses (x, y) property.

$$\texttt{redirectsTo}(x, y) =_{def}$$
$$\texttt{ire}: \texttt{URI}(x) \land \texttt{ire}: \texttt{URI}(y) \land \exists m(\texttt{RedirectionMethod}(m) \land$$
$$\texttt{edns}: \texttt{involves}(m, x) \land \texttt{edns}: \texttt{involves}(m, y)) \land$$
$$\exists z(\texttt{referencedBy}(z, y)) \tag{15}$$

Definition 15 defines the relation of redirection over URIs.

$$\texttt{assignsTo}(x, y, z) =_{def}$$
$$\texttt{OkkamProfile}(x) \land \texttt{OkkamID}(y) \land \texttt{OkkamEntity}(z) \land \exists t, s(\texttt{dol}: \texttt{Time}(t)$$
$$\land \texttt{edns}: \texttt{InformationObject}(s) \land \texttt{edns}: \texttt{realizes}(x, s, t) \land$$
$$\texttt{edns}: \texttt{about}(s, z, t) \land \exists u, w(\texttt{ire}: \texttt{URIokkamProfile}(w) \land$$
$$\texttt{ire}: \texttt{hasIdentifier}(u, w) \land \texttt{ire}: \texttt{webLocationOf}(u, x, t) \land$$
$$\texttt{RedirectsTo}(y, w))) \tag{16}$$

Definition 16 captures the idea that an OKKAM profile does not describe an entity but is used to perform a baptism of an entity with an OKKAMid. (More on the idea of baptism in the following section).

$$\exists x, y(\texttt{assignsTo}(x, y, z) \land \texttt{assignsTo}(x, y, w)) \rightarrow z = w \tag{17}$$

Axiom 17 states that two different OKKAM entities cannot have the same OKKAMid.

$$\texttt{directlyRefersTo}(x, y) =_{def}$$
$$\texttt{OkkamID}(x) \land \texttt{OkkamEntity}(y) \land \exists z(\texttt{OkkamProfile}(z)$$
$$\land \texttt{assignsTo}(z, x, y)) \tag{18}$$

Definition 18 defines the relation of direct reference over OKKAMids and OKKAM entities. The directlyRefersTo(x, y) property in OCM is a functional property and aims to capture the Berners-Lee's *direct reference position*. The property directlyRefersTo(x, y) is distinct from the irw: refersTo(x, y) property, since the latter is not a functional property. The irw:identifies property, too, is a functional property defined over RDF URIs and non-web resources. However, OKKAMids make *rigid* and *direct* reference to entities, whereas RDF URIs refer to entities through the mediation of a description.

$$\texttt{Okkamised}(x) =_{def}$$
$$\texttt{OkkamEntity}(x) \land \exists y(\texttt{OkkamID}(y) \land \texttt{directlyRefersTo}(y, x)) \tag{19}$$

Definition 19 defines the concept of having an OKKAMid and states that only Okkam entities can be assigned OKKAMids.

$$\texttt{webProxyFor}(x, y, t) =_{def}$$

$$\texttt{ire : WebResource}(x) \land \neg\texttt{OkkamProfile}(x) \land \texttt{Entity}(y) \land \texttt{dol : Time}(t)$$
$$\land \exists z(\texttt{edns : InformationObject}(z) \land \texttt{edns : realizes}(x, z, t) \land$$
$$\texttt{edns : about}(z, y, t)) \tag{20}$$

Definition 20 states that in order for x to bear the $\texttt{webProxyFor}$(x, y, t) relation, x must be a resource that realizes an information object about the entity y at t and cannot be an OKKAM profile.

$$\texttt{WebProxyResource}(x) =_{def} \exists y, t(\texttt{webProxyFor}(x, y, t)) \tag{21}$$

Definition 21 states that a web proxy resource x is a web resource that stands in the webProxyFor relation to an entity y at time t.

$$\texttt{WebSemanticResource}(x) =_{def}$$
$$\exists y, t, z, w(\texttt{webProxyFor}(x, y, t) \land \texttt{edns : InfomationObject}(z) \land$$
$$\texttt{edns : FormalLanguage}(w) \land \texttt{edns : realizes}(x, z, t) \land \texttt{edns : about}(z, y, t) \land$$
$$\texttt{edns : orderedBy}(z, w)) \tag{22}$$

Definition 22 states that a web semantic resource is a web proxy resource that realizes an information object about an entity by a codification in a formal language for the web. Example: `http://dbpedia.org/page/Eiffel_Tour`

$$\texttt{RDFURI}(x) =_{def}$$
$$\texttt{ire : URI}(x) \land \exists w, y, u, t(\texttt{WebSemanticResource}(y) \land \texttt{ire : URI}(w) \land$$
$$\texttt{ire : hasIdentifier}(u, w) \land \texttt{ire : webLocationOf}(u, y, t) \land$$
$$\texttt{redirectsTo}(x, w)) \tag{23}$$

Definition 23 states that a RDF URI is a URI that redirects to the URI of a web semantic resource.[13]

$$\texttt{dereferenceableAlternativeIDOf}(x, y) =_{def}$$
$$\texttt{RDFURI}(x) \land \texttt{okkamised}(y) \land \exists w, z, u, t(\texttt{ire : URI}(w) \land$$
$$\texttt{ire : hasIdentifier}(u, w) \land \texttt{ire : webLocationOf}(u, z, t) \land$$
$$\texttt{webProxyFor}(z, y, t) \land \texttt{redirectsTo}(x, w)) \tag{24}$$

Definition 24 defines the relation of being a dereferenceable alternative ID of an OKKAM entity that has an $\texttt{OKKAMid}$. Dereferenceable alternative IDs of OKKAM entities are RDF URIs.

$$\texttt{corefer}(x, y) =_{def}$$
$$\texttt{RDFURI}(x) \land \texttt{OkkamID}(y) \land \exists z(\texttt{Okkamized}(z) \land$$
$$\texttt{directlyRefersTo}(y, z) \land \texttt{dereferenceableAlternativeIDOf}(x, z)) \tag{25}$$

---

[13] It must be noted that the content negotiation might ask for "text/html", so what is here presented as a RDF URI might redirect to a URI that retrieves a HTML page. Therefore, replacing the term "RDFURI" with another term like, say, "Linked-DataURI" might seem more appropriate. We leave such terminological question aside. Nothing conceptually important follows if one uses "LinkeDataURI", since the redirection is to only one representation per media type.

Definition 25 defines the relation of coreference between RDF URIs and `OKKAMids` that holds when the entity identified by an RDF URI is the same entity directly referred to by an `OKKAMid`. Of course, more than one RDF URI can bear the `corefer`(x,y) relation to the same `OKKAMid`.

$$\texttt{coidentify}(x, y) =_{def}$$
$$\texttt{RDFURI}(x) \land \texttt{RDFURI}(y) \land \exists z(\texttt{OkkamID}(z) \land$$
$$\texttt{corefer}(x, z) \land \texttt{corefer}(y, z)) \tag{26}$$

Definition 26 defines the relation of coidentification between RDF URIs. The raltion can be inferred from the fact that two RDF URIs are deferenceable alternative IDs of the same OKKAM entities.

4   OKKAM IDs and RDF URIs

One aspect of OCM above others deserves special clarification. The fact is that one might consider the objection that `OKKAMids` and RDF URIs are not really distinguished. Indeed, URIs of both types can be dereferenced and the act of dereferencing them triggers a process of redirection to URIs for web resources. `OKKAMids` redirect to URIs for OKKAM profiles, whereas RDF URIs redirect to URIs for other web resources. OKKAM profiles give information about the OKKAM entities referred to. Therefore, the mechanism of reference of `OKKAMids`, too, seems to be mediated by description. The objection, then, is that there is no structural and web architectural difference between `OKKAMids` and RDF URIs to the effect that the linguistic distinction between them looks arbitrary and unjustified. Why do `OKKAMids` make *rigid* and *direct* reference to OKKAM entities, whereas RDF URIs identify non-web resources by descriptions?

We reply to this objection by granting the indiscernibility of `OKKAMids` and RDF URIs from the structural and web architectural point of view. However, the ground of the distinction can be found elsewhere, namely in the purpose of using such URIs. Our reply is that the linguistic distinction between `OKKAMids` and RDF URIs has a pragmatic ground. The purpose of creating and using `OKKAMids` is to give the start to a linguistic practice by an act of baptism and by following acts of subscription to that linguistic practice. Such a practice is not assessed in terms of truth and falsity, which amounts to saying that the information contained in an OKKAM profile need not be true of the entity being assigned the `OKKAMid`. It is sufficient that the web community converges on that information in order to fix the referent of that `OKKAMid`.

To make the point clear it might be helpful to adapt a famous example (from [3]) in philosophy of language to our case. Imagine that Jane and John are enjoying a party. They give a look at a man holding a martini glass. For some reasons they are willing to assign a proper name to that man and agree on the following convention: *lets us call the man drinking martini "Jack"*. In order for their convention to be successful it is not necessary that the liquid in the glass be martini. However, even if the description *the man drinking martini* does not denote the man standing in front of Jane and John – because, say, the liquid in the glass is water – the act of baptism is successful provided that both Jane and John share the belief that that man is drinking martini, no matter how false that belief is. In fact, the description *the man drinking martini* is not used to express information about the man standing in front of Jane and John, but to fix the referent of the newly introduced name "Jack". Jane and John are not so much concerned as to whether that man is drinking martini or not, as to the fact that they both share that belief and use it to fix the referent of the name "Jack".

The use of the information conveyed in an OKKAM profile is the same as the use of the description *the man drinking martini* in the above scenario: it is not used to express information about an entity but to fix the referent of a name, i.e. an OKKAMid. Very likely, most of the information conveyed in an OKKAM profile will be true *de facto* of the entity to which that profile assigns the OKKAMid. That circumstance does not alter the fact that users need not endorse such information as true. To make the point clear, consider the above scenario again. Imagine Clark, too, is at the party and comes to know Jane and John's convention of naming the man standing in front of them "Jack" and that they believe that that man is drinking martini. Clark, however, knows that the liquid in the glass is water and not martini. Nevertheless, Clark can appeal to Jane and John's false belief to disambiguate utterances of the name "Jack", although Clark does not endorse such belief as true. For example, if Clark says "Jack is a computer scientist" and Jane or John replies "Jack who?", Clark might answer "the man who is drinking martini" to fix the referent of his utterance of the name "Jack" and to say of the man named "Jack" by Jane and John that he is a computer scientist.

The idea underlying our view is that OKKAM profiles are not about OKKAM entities in the same way as the information conveyed in other web resources is about non-web resources. More precisely, an OKKAM profile is not a description of an entity but constitutes the virtual context for the assignment of an OKKAMid to an entity. One should think of the assignment of an OKKAMid to an entity as a baptism that dubs that entity with a name. A baptism is a performative speech act. Performatives, unlike constatives, which are assessed in terms of truth or falsity, can only be assessed as felicitous or infelicitous (see Austin's felicity conditions [1]). A baptism is a speech act with its own felicity conditions. One of them is the existence and the salience of the entity being dubbed. No baptism can take place if there is no entity to be dubbed and if that entity is not cognitively available as the most salient to the persons who have the authority to make the baptism. An OKKAM profile serves exactly to make the entity to be dubbed salient, and its purpose is not to provide a description of that entity. On the other hand, accessing an OKKAM profile by dereferencing an OKKAMid amounts to the speech act of subscribing to the linguistic practice of using that OKKAMid as a name for a certain non-web resource. The creator of the OKKAM profile for an entity is the producer of that linguistic practice, whereas the users who access that OKKAM profile since its creation are the consumers of that linguistic practice(see [4] Ch. 11). It is not necessary that the information in the OKKAM profile be true of a non-web resource in order for the baptism to be successful. For the accomplishment of the baptism it is sufficient that the web community shares or converges on that (mis)information. Consider the following example, borrowed from [4]. Take the poet known to his contemporaries as "Homer" (or known by some name from which "Homer" descends); we think of the claim "Homer wrote the *Iliad*" as a substantial hypothesis about the authorship of the poem. But suppose the hypothesis is false. We might still use that piece of (mis)information to create an OKKAM profile assigning an OKKAMid to the *Iliad*, and saying that the *Iliad* was written by Homer. So long as a community converges on that piece of (mis)information, the baptism is felicitous.

OCM mirrors the semantic distinction between OKKAMids and RDF URIs by the stipulation that the former redirect to URIs for OKKAM profiles and the latter to URIs for web proxy resources, and that OKKAM profiles are not web proxy resources. The justification of that distinction is not fully expressed by the definitions and the axioms in OCM. Indeed, the axioms by themselves simply stipulate that there is a linguistic difference between OKKAMids and RDF URIs. Nevertheless, the distinction can be justified from pragmatic reflections on the purpose of using OKKAMids and RDF URIs. RDF URIs are used to express and endorse information about entities, whereas OKKAMids are used to fix the referent of RDF URIs within the whole web community and eventually to make it explicit that two or more different RDF URIs are different names of the same entity independently of the information retrievable by dereferencing those RDF URIs.

5   Conclusions

Practice shows that RDF URIs are commonly used for three different things:

1. Redirecting to a set of assertions *about* a non-web resource. As mentioned in the first part of this paper, dereferencing a RDF URI usually results in the retrieval of RDF triples describing non-web resources.
2. Linking from one set of assertions to another. Employing the owl:sameAs construct, a link can be established between one RDF URI and another. The semantics of this will be further addressed in this section.
3. Providing a surrogate/substitute/proxy for non-web resources. This is the typical case for the notion of "identifier for an individual" in a Description Logics knowledge base.

Cases (1) and (2) form a vital mechanism of the Linked Data approach. From our point of view, case (2) implies some very important semantics that have to be respected. First of all, there is a certain mismatch of the use of the owl:sameAs property in the Linked Data approach, and its intended semantics in the OWL specification [9]: collapsing all equivalent RDF nodes into a single one and thus joining the set of all axioms about these equivalent nodes onto the collapsed node, thus losing the ability to distinguish which nodes the statements were about in the first place, is the defined semantics of owl:sameAs. The actual use of this construct today however is one of linkage, i.e. the author of such a statement rather intends semantics of pointing to, or even endorsing, more axioms about the same real-world entity provided by another source. And indeed, losing the provenance of the axioms is not only undesirable, but also not commonly practiced. In Semantic Web applications, the `owl:sameAs` property is often directly translated to a hypertext link which the user can click to navigate to another set of assertions, as for example in the Tabulator application. Or the assertions are retrieved following case (1) and presented in an aggregated view, but preserving provenance, in order for example to gather feedback from users which is fed into a trust model about data sources, as practiced e.g. in the sig.ma application.

Case (3) has been the cause for lengthy discussions especially within W3C, which started from the opinion that a URI cannot identify a non-web resource and a web resource at the same time. The agreed recommendation [11] on how to solve this conflict is to use status codes of the underlying HTTP protocol to inform an agent whether a URI is identifying a web resource or not, and use a redirection mechanism that provides a web resource.

While this approach solves the problem of knowing *what kind of resource* a URI identifies, it does not address the question of *which* non-web resource such a URI identifies, and does not guarantee that it identifies always the same one. This fact makes mere RDF URIs problematic, even if they are well-implemented (i.e. providing the right status codes and redirection mechanism).

In OCM we are devising a way to add precision to the management and interpretation of identifiers on the Semantic Web. While RDF URIs satisfy cases (1) and (2), and can be implemented technically to conform to W3C recommendations, `OKKAMids` add the possibility to refer *rigidly* and *directly* to a non-web resource. This means that to become a "cool URI", we recommend one of the following solutions: (i) using directly `OKKAMids` for non-web resources, whenever possible; or (ii) adding a `corefer` statement for each non-web resource named in the dataset (in OKKAM, this is called "OKKAMization"); or (iii) makes sure that applications aiming at aggregating different RDF datasets make a runtime call the ENS for retrieving the `OKKAMids` of the non-web resources named in a dataset.

The adoption of such an approach has three important benefits; first, RDF URIs will maintain their intended use of being interpreted into a set of triples. Secondly, as a consequence, RDF URIs are perfectly suited to implement *Linked Data*, preserving provenance and context. Finally, *ad-hoc* solutions for calculating the transitive closure over `owl:sameAs` statements can be often avoided because identity is syntactically evident. If transitive closure is required, the ENS accomodates for the notion of *dereferenceable alternative ID* and provides the community with the practical solution of maintaining these closures in a defined location.

To sum up, our recommendation is that the `owl:sameAs` statements are reserved to cases in which one intends to express a strong semantic link of compatibility between two different *descriptions* of the same non-web resource. Whereas coidentification statements, i.e. statements about the fact that two or more RDF

URIs identify the same non-web resource, should be inferred from the fact that different RDF URIs are mapped onto the same `OKKAMid` in the ENS through the `corefer`(x,y) relation defined in 25. In this picture, one can think of an OKKAM profile as a **gateway** to information about a non-web resource existing on the Web. It turns out that the relation of redirection should be thought of as performing two distinct functions for RDF URIs and `OKKAMids`. The processes of redirection and resolution that connect RDF URIs to pieces of information should be thought of as functional in the following sense: a RDF URI – via redirection and the `owl:sameAs` relation – should be connected to one and only one *coherent* piece of information about a non-web resource and keep tracks of its sources. On the contrary, the process of redirection and resolution that connect `OKKAMids` to pieces of information in general is not functional, as the OKKAM profile of an entity should not be interepreted nor used as an additional piece of information about the entity, but only as information which a community agrees to use in order to fix the referent of that OKKAMid.

References

[1] Austin, J.L.: How to Do Things with Words: The William James Lecture. Oxford University Press, Oxford (1962)
[2] Bouquet, P., Stoermer, H., Niederee, C., Mana, A.: Entity Name System: The Backbone of an Open and Scalable Web of Data. In: Proceedings of the IEEE International Conference on Semantic Computing, ICSC 2008, pp. 554–561. IEEE Computer Society, Los Alamitos (2008); CSS-ICSC 2008-4-28-25
[3] Donnellan, K.S.: Reference and definite descriptions. The Philosophical Review 77, 281–304 (1966)
[4] Evans, G.: The Varieties of Reference. Oxford University Press, Oxford (1982)
[5] Gangemi, A., Presutti, V.: Towards an OWL Ontology for Identity on the Web. In: Semantic Web Applications and Perspectives, SWAP 2006 (2006)
[6] Gangemi, A., Presutti, V.: A grounded ontology for identity and reference of web resources. In: i3: Identity, Identifiers, Identification. Proceedings of the WWW 2007 Workshop on Entity-Centric Approaches to Information and Knowledge Management on the Web, Banff, Canada, May 8 (2007)
[7] Halpin, H., Presutti, V.: An ontology of resources: Solving the identity crisis. In: Aroyo, L., et al. (eds.) Proceedings of ESWC 2009. Studies in Logic and Computation, pp. 121–140. Research Studies Press/Wiley (2009)
[8] Kripke, S.: Naming and necessity. Harvard University Press (1972)
[9] Patel-Schneider, P.F., Hayes, P., Horrocks, I.: Web Ontology Language (OWL) Abstract Syntax and Semantics. Technical report, W3C (February 2003), http://www.w3.org/TR/owl-semantics/
[10] Presutti, V., Gangemi, A.: Identity of resources and entities on the web. International Journal on Semantic Web and Information Systems 4(2) (2008)
[11] W3C. Cool URIs for the Semantic Web. W3C Interest Group Note 03 December (December 2008), http://www.w3.org/TR/cooluris/

What Makes a Good Ontology?
A Case-Study in Fine-Grained Knowledge Reuse

Miriam Fernández, Chwhynny Overbeeke, Marta Sabou, and Enrico Motta

Knowledge Media Institute,
The Open University, Milton Keynes, United Kingdom
{M.Fernandez, C.Overbeeke, R.M.Sabou, E.Motta}@open.ac.uk

Abstract. Understanding which ontology characteristics can predict a "good" quality ontology, is a core and ongoing task in the Semantic Web. In this paper, we provide our findings on which structural ontology characteristics are usually observed in high-quality ontologies. We obtain these findings through a task-based evaluation, where the task is the assessment of the correctness of semantic relations. This task is of increasing importance for a set of novel Semantic Web tools, which perform fine-grained knowledge reuse (i.e., they reuse only appropriate parts of a given ontology instead of the entire ontology). We conclude that, while structural ontology characteristics do not provide statistically significant information to ensure that an ontology is reliable ("good"), in general, richly populated ontologies, with higher depth and breadth variance are more likely to provide reliable semantic content.

Keywords: semantic relations, knowledge reuse, Semantic Web.

1 Introduction

Ontologies are fundamental Semantic Web (SW) technologies, and as such, the problem of their evaluation has received much attention from areas such as ontology ranking [8], selection [16][21], evaluation [11] and reuse [22]. Various approaches have been proposed in these fields, ranging from manual evaluation to (semi-) automatic evaluation of a single ontology to benchmark evaluation of the entire Semantic Web, and, finally, to task-based evaluations of a single ontology or a collection of ontologies. These studies have explored a variety of ontology characteristics that could predict ontology quality, including characteristics such as the modeling style of the ontologies, their vocabulary, structure, or performance within a given task. In this paper we continue the investigation of what makes a "good" ontology by using a *task-based approach* to evaluate the *collection of ontologies* available on the SW in terms of measures relating to their *structure*.

The context of our work is that of fine-grained knowledge reuse, i.e., the reuse of ontology parts rather than the ontology as a whole. This kind of knowledge reuse is increasingly frequent, particularly for the new family of applications that take advantage of the large scale of the Semantic Web and the set of mature technologies

A. Gómez-Pérez, Y. Yu, and Y. Ding (Eds.): ASWC 2009, LNCS 5926, pp. 61–75, 2009.

for accessing its content[1] in order to reuse online knowledge. In the case of these applications, knowledge reuse happens at run-time, and therefore it primarily focuses on the reuse of small parts of ontologies, typically at the level of a semantic relation [17]. This is why it is essential to automatically detect the quality of such relations.

The task we focus on in this paper is the evaluation of a single semantic relation (and not that of an entire ontology). We have built an algorithm that explores online ontologies in order to perform this task [18]. The performance of the task depends on the selection of these ontologies. We experiment with a set of structure-based ontology characteristics to select appropriate ontologies and decide which characteristics are more important by measuring their influence on the performance achieved when predicting the quality of relations. The correlation between structure-based ontology characteristics and ontology correctness arises from our own experience in previous works [10][18], and other ontology evaluation studies where this distinction seems to be natural, useful and recurrent (see e.g. [15]).

Our findings show that while structural ontology characteristics do not provide statistically significant information to identify a correct ontology, some of them point to valuable information that can help enhance ontology selection techniques. In particular, we conclude that richly populated ontologies with a high breadth and depth variance are more likely to be correct, and should be ranked higher by ontology selection algorithms.

The contribution of our paper is two-fold. On the one hand, we further advance work on automatic relation evaluation by providing our findings on the ontology characteristics which could predict which ontologies are most likely to provide correct relations. On the other hand, a side-effect of this work is a large-scale investigation of what are the core structural characteristics that can predict a good-quality ontology.

The rest of the paper is structured as follows. We present related work in Section 2 and describe some motivating scenarios in the context of fine-grained knowledge reuse in Section 3. Section 4 introduces the task we focus on, the evaluation of a single semantic relation, and its implementation. We present the evaluation setup in Section 5 and detail experimental results in Section 6. We conclude in Section 7.

2   Related Work

As the number of ontologies on the Web increases, the need arises to determine which ontologies are of the highest quality or are the most appropriate for a certain task. There are several conceptions of what makes a "good" ontology, which will be discussed in this section.

Significant work has been done in the area of ontology quality assessment [6][14]. Most of these attempts try to define a generic quality evaluation framework. As a result, specific applications of ontologies are not taken into account, and the ontology is considered as a whole during its quality evaluation.

Existing evaluation methods rely on rather simple ways of specifying an information need, such as (sets of) keywords or a corpus from which sets of keywords are abstracted and output their results as a ranked list of ontologies [21].

---

[1] http://esw.w3.org/topic/TaskForces/CommunityProjects/LinkingOpenData/
SemanticWebSearchEngines

Table 1. Summary of existing approaches to ontology evaluation and the evaluation criteria they explore (adapted from [22])

Quality Framework	Syntax Evaluation	Domain cohesion	Structural evaluation	Population of classes	Usage statistics
AKTiveRank [2]		X	X		
OntoClean [11]			X		
OntoKhoj [16]		X			X
Ontometric [14]	X				
OntoQA [23]			X	X	
OntoSelect [5]			X		
Semiotic metrics [6]	X	X			X
Swoogle [8]					X

There are three major categories of ontology evaluation approaches:

- *Manual approaches* are those based on human interaction to measure ontology features not recognizable by machines [14].
- *Automatic approaches* are those that evaluate an ontology by comparing it to a Golden Standard, which may itself be an ontology [15] or some other kind of representation of the problem domain [4].
- *Task-based approaches* are those that evaluate the ontologies by plugging them in an application, and measuring the quality of the results that the application returns [19].

The different existing methods of evaluation also vary with regard to their selection criteria and evaluation metrics. Aspects that are generally considered to be useful for the evaluation of the quality of an online ontology, shown in Table 1 are:

- Evaluation of *syntax* checks if an ontology is syntactically correct. This is most important for ontology-based applications as the correctness reflects on the application [14].
- *Cohesion to domain and vocabulary* measures the congruence between an ontology and a domain [4][6][16].
- *Structural* evaluation deals with the assessment of taxonomical relations versus other semantic relations, i.e., the ratio of Is-A relationships and other semantic relationships in an ontology is evaluated [5].
- *Population of classes* measures instance-related metrics such as how instances are distributed across classes or average population [23].
- *Usage statistics and metadata* evaluate those aspects that focus on the level of annotation of ontologies, i.e., the metadata of an ontology and its elements [6][9][8].

In this work we report on a task-based evaluation of online available ontologies, where we investigate which structural and popularity characteristics of these ontologies are good indicators to measure their quality.

3 Use Cases in the Context of Fine-Grained Knowledge Reuse

In this section, we describe two motivating scenarios where fine-grained knowledge reuse is performed rather than reuse of ontologies as a whole.

Embedded in the NeOn Toolkit's ontology editor, the Watson plugin[2] allows the user to reuse a set of relevant ontology statements (equivalent to semantic relations) drawn from online ontologies in order to construct new knowledge. Concretely, for a given concept selected by the user, the plug-in retrieves all the relations in online ontologies that contain this concept (i.e., concepts that have the same label). The user can then integrate any of these relations into his ontology through a mouse click. For example, for the concept Book the plugin would suggest relations such as: Book $\subseteq$ Publication, Chapter $\subseteq$ Book or Book -containsChapter- Chapter. These semantic statements are presented in an arbitrary order. Because of the typically large number of retrieved semantic statements it would be desirable to rank them according to their correctness.

Our second scenario is provided by PowerAqua [13], an ontology-based Question Answering (QA) system which receives questions in natural language and is capable of deriving an answer by combining knowledge gathered from multiple online ontologies. In a nutshell, the system breaks up the user query in several triple-like structures, which are then matched to appropriate triples (or relations) within online ontologies. PowerAqua derives the final answer by combining these ontology triples. As in the case of the Watson plug-in, PowerAqua does not evaluate the quality of these relations. Our work on establishing a correlation between certain ontology characteristics and the quality of the relations they provide would improve PowerAqua's ability to discard noise or irrelevant semantic information.

4 The Task: Evaluating the Quality of Semantic Statements

The task we use as a means to get an insight into the quality of online ontologies is that of evaluating the quality of a semantic relation. We define a semantic relation $<s, R, t>$ as a triple where s represents the source term, t represents the target term, and R represents the relation between those terms, e.g., $<Helicopter, \subseteq, Aircraft>$. R can represent a wide range of relation types, such as hyponymy, disjointness, or simply any associative relation.

In our work, for any given relation we want to evaluate, we are capable to identify all online ontologies that directly or indirectly link s and t. Fig. 1 shows the example of three ontologies (O_1, O_2, O_3) that can lead to a relation between Aircraft and Helicopter. O_1[3] contains a direct subclass relation while O_2[4] contains a direct disjoint relation between Aircraft and Helicopter. O_3[5] provides an implicit subclass relation between these two concepts, which can be inferred from the following derivation path: *Helicopter $\subseteq$ Rotorcraft $\subseteq$ HeavierThanAirCraft $\subseteq$ Aircraft*

---

[2] http://watson.kmi.open.ac.uk/WatsonWUI/

[3] http://reliant.teknowledge.com/DAML/Transportation.owl

[4] http://reliant.teknowledge.com/DAML/Mid-level-ontology.owl

[5] http://www.interq.or.jp/japan/koi_san/trash/aircraft3.rdf

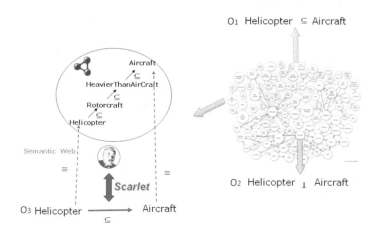

Fig. 1. Example of finding relations between Helicopter and Aircraft on the SW

In this example we can see that different ontologies provide information of a different quality. While O_1 and O_3 provide a correct relation in terms of domain modeling, this is not the case for O_2. Further, note that even if they agree on the relation between Helicopter and Aircraft, ontologies O_1 and O_3 have different ways to declare this relation: explicitly (derivation path length = 1) or implicitly (derivation path length = 3). In this work we make use of the fact that different ontologies provide relations between the same terms in order to investigate which ontology characteristics can predict the ontology that is most likely to provide a correct relation.

To perform our task we use a software package that, given two terms, can identify all online ontologies that lead to a relation between these terms, as well as the actual relation and its derivation path. We implemented this package using the services of the Watson[6] SW gateway. Watson crawls and indexes a large number of online ontologies[7] and provides a comprehensive API which allows us to explore these ontologies.

The relation extraction algorithm is highly parameterized[8]. For the purposes of this study we have configured it such that for each pair (A,B) of terms it identifies all ontologies containing the concepts A' and B' corresponding to A and B from which a semantic relation can be derived between these terms. Correspondence is established if the labels of the concepts are lexical variations of the same term. For a given ontology (O_i) the following derivation rules are used:

- If $A'_i \equiv B'_i$ then derive $A \equiv B$.
- If $A'_i \subseteq B'_i$ then derive $A \subseteq B$.
- If $A'_i \supseteq B'_i$ then derive $A \supseteq B$.
- If $A'_i \perp B'_i$ then derive $A \perp B$.

---

[6] http://watson.kmi.open.ac.uk
[7] Estimated to 250,000 during the writing of this paper.
[8] A demo of some of these parameters and an earlier version of the algorithm are available at http://scarlet.open.ac.uk/

- If R(A$'_i$, B$'_i$) then derive A $_R$ B.
- If P$_i$ such that A$'_i$ ⊆ P$_i$ and B$'_i$ ⊆ P$_i$ then derive A sibling B.

Note that in the above rules the relations between A$'_i$ and B$'_i$ represent both explicit and implicit relations (i.e., relations inherited through reasoning) in O$_i$. For example, in the case of two concepts labeled *DrinkingWater* and *tap_water*, the algorithm deduces the relation *DrinkingWater* ⊆ *tap_water* through the following subsumption chain in the TAP[9] ontology: *DrinkingWater* ⊆ *FlatDrinkingWater* ⊆ *TapWater*.

5   Evaluation Setup

This section describes the evaluation setup. Here we explain the set of measures and datasets that we have selected to perform the evaluation.

5.1   Measures

Twelve different measures have been considered to evaluate the quality of the ontologies. Because these measures are investigated in the context of applications that need to select semantic knowledge at runtime, they must accomplish two main requirements: *generality* and *performance*. *Generality* refers to the applicability of the measures to any potential ontology available in the Web, independent of its language, size, or any other characteristic. *Performance* refers to the availability of these measures at runtime. This requirement generally implies that the measures are either lightweight in terms of computational requirements or pre-computed. The list of selected measures has been divided in two main groups:

a) Knowledge coverage and popularity measures

- Number of classes: number of classes in a given ontology.
- Number of properties: number of properties in a given ontology.
- Number of individuals: number of individuals in a given ontology.
- Direct popularity: number of ontologies importing a given ontology.

b) Structural ontology measures

- Maximum depth: size of the longest branch in the given ontology.
- Minimum depth: size of the shortest branch in the given ontology.
- Average depth: average size of the branches of the given ontology.
- Depth variance: variance of the size of the branches in the ontology.
- Maximum breadth: size of the largest level of the ontology.
- Minimum breadth: size of the narrowest level of the ontology.
- Average breadth: average size of the levels of the ontology.
- Breadth variance: variance of the size of the levels in the ontology.

5.2   Datasets

As experimental data we used datasets from the domain of ontology matching, in the form of alignments obtained in two different test cases put forward by the Ontology

---

[9] http://139.91.183.30:9090/RDF/VRP/Examples/tap.rdf

Table 2. Overview of the experimental datasets and their characteristics

Data Set	Nr. Of Relations	Type of Relations	Domain
AGROVOC/NALT	380	$\subseteq$, $\supseteq$, $\perp$	Agriculture
OAEI'08 301	112	$\subseteq$, $\supseteq$, $\perp$, named relations	Academia
OAEI'08 302	116	$\subseteq$, $\supseteq$, $\perp$, named relations	Academia
OAEI'08 303	458	$\subseteq$, $\supseteq$, $\perp$, named relations	Academia
OAEI'08 304	386	$\subseteq$, $\supseteq$, $\perp$, named relations	Academia
Total	**1452**		

Alignment Evaluation Initiative[10] (OAEI), an international body that coordinates evaluation campaigns for this task.

The AGROVOC/NALT dataset has been obtained by performing an alignment between the United Nations' Food and Agriculture Organization (FAO)'s AGROVOC ontology and its US equivalent NALT. The relations established between the concepts of the two ontologies are of three types: $\subseteq$, $\supseteq$, and $\perp$. Each relation has been evaluated by experts, as described in more detail in [17].

The OAEI'08 dataset represents the alignments obtained by the Spider system on the 3** benchmark datasets and their evaluation [20]. This dataset contains four distinct datasets representing the alignment between the benchmark ontology and the MIT (301), UMBC(302), KARLSRUHE(303) and INRIA(304) ontologies respectively. Besides the $\subseteq$, $\supseteq$, and $\perp$ relation types, this dataset also contains named relations, e.g. *<Article, inJournal, Journal>*. Table 2 provides a summary of these datasets and their characteristics.

6   Evaluation Results

In this section we describe the study we conducted to evaluate the discriminative effect of the proposed measures when selecting the ontologies that are most likely to provide correct relations. For this purpose we have used the datasets presented in Section 5.2 and the implementation described in Section 4.

6.1   Evaluating the Quality of Semantic Statements: Types of SW Matches

As we can see in Section 5.2, the datasets selected for the study contain four different types of relations R: $\subseteq$, $\supseteq$, $\perp$ and named. For each individual triple $<s, R, t>$ in the dataset a user evaluation is available, stating whether the relation R between s and t is correct.

Each triple $<s, R, t>$ is then searched in the SW using the methodology described in Section 4. As a result, all online ontologies that directly or indirectly link s and t are identified. For each relation R to be evaluated we consider five different potential matches within online ontologies: $\subseteq$, $\supseteq$, $\perp$, named and sibling.

For example, for the semantic relation $<fog, \subseteq, weather>$, which users have evaluated as a correct relation, we found two different matches in the SW: The

---

[10] http://oaei.ontologymatching.org/

ontology *http://morpheus.cs.umbc.edu/aks1/ontosem.owl* with the match *<fog, ⊆, weather>* and the ontology http://sweet.jpl.nasa.gov/ontology/phenomena.owl with the match *<fog, hasAssociatedPhenomena, weather>*.

Considering the semantic relation, its corresponding user evaluation and the relation matched in the ontology, we distinguish three different types of matches:

- *Correct matches*: they provide exactly same relation that the users are considering true.
- *Incorrect matches*: they provide exactly the same relation that the users are considering false or a different relation to the one the users are considering true.
- *Unknown matches*: the rest of the cases in which we cannot determine if the ontologies are providing correct or incorrect information without a manual evaluation.

Table 3 summarizes the rules that we use to automatically judge the correctness of a match in online ontologies based on the value of the original relation (column 1) and the user evaluation of the original relation (column 2).

Table 3. Quality of identified matches

Original relation	User evaluation	Match quality		
		Correct	Unknown	Incorrect
⊆	True	⊆	Named, sibling	⊇, ⊥
⊇	True	⊇	Named, sibling	⊆, ⊥
⊥	True	⊥	Named	⊆, ⊇, sibling
named	True		named, sibling, ⊆, ⊇, ⊥	
⊆	False		⊇, ⊥, named, sibling	⊆
⊇	False		⊆, ⊥, named, sibling	⊇
⊥	False		⊆, ⊇, sibling, named	⊥
named	False		⊆, ⊇, ⊥, named, sibling	

For the 1452 semantic relations described in the five different datasets we have found 53726 matches in 283 online ontologies using the services provided by Watson. Following the classification mechanism described above, we have extracted 1498 correct matches from 140 different ontologies (O_{cm}), 2279 incorrect matches from 148 different ontologies (O_{im}) and 49949 unknown matches from 275 different ontologies(O_{um}). Note that the same ontology can fall within the three different subsets if it provides correct, incorrect and unknown mappings for the various semantic relations of the dataset.

6.2 Selecting Correct and Incorrect Ontologies

The identified correct and incorrect matches will help us distinguish between two different subsets of ontologies: O_r, reliable ontologies when assessing the quality of a semantic relation and $O_{nr,}$ unreliable ontologies. In order to select these subsets of ontologies we try to maximize two different criteria: a) the number of matches generated by the ontology and b) over those matches, the percentage of correct ones in the case of O_r and incorrect ones in the case of O_{nr}. Fig. 2 and 3 show the distribution of the ontologies meeting these two criteria. Note that in these figures the percentages are expressed on a per unit basis.

As we can see in both figures, the percentage of correct and incorrect matches decreases in correlation with the increase in the number of matches. This is due to the fact that, for those ontologies that are able to provide a higher number of matches, the majority of identified matches have an unknown quality, i.e., we cannot determine if they are correct without a manual evaluation. This effect partially invalidates the criterion of maximizing the number of matches in order to select O_r and O_{nr}. To avoid this effect we consider that: a) those ontologies that provide a number of matches greater than or equal to the average obtain the maximum score for this criterion and b) the criterion of maximizing the percentage of correct and incorrect matches should have slightly more relevance than the criterion of maximizing the number of matches. Considering these constraints we define O_r and O_{nr} as:

$$O_r = \{o_i \in O_{cm}, where, \alpha * \min(1, \frac{m_{oi}}{Avg_n(m_{oi})}) + (1-\alpha) * \frac{mc_{oi}}{m_{oi}} > \lambda\}$$

$$O_{nr} = \{o_i \in O_{im}, where, \alpha * \min(1, \frac{m_{oi}}{Avg_n(m_{oi})}) + (1-\alpha) * \frac{mi_{oi}}{m_{oi}} > \lambda\}$$

Where: m_{oi} is the set of matches found for the ontology o_i, mc_{oi} is the subset of correct matches found for the ontology o_i, mi_{oi} is the subset of incorrect matches found for the ontology o_i, n is the total number of ontologies that provided matches for the relations in our dataset (283), α is a constant parameter that determines the relevance for each criterion and λ is a certain threshold that discriminates the final subset of ontologies.

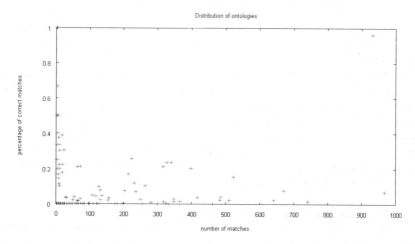

Fig. 2. Distribution of ontologies according to the number of matches and percentage of correct matches

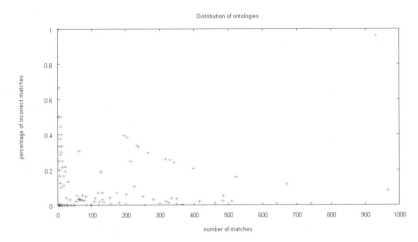

Fig. 3. Distribution of ontologies according to the number of matches and percentage of incorrect matches

For our experiments α has been empirically set to 0.4, providing less relevance to the criterion of maximizing the number of matches. λ has been empirically set to 0.5 in the selection of O_r and to 0.6 in the selection of O_{nr} in order to obtain the top 40 ontologies for each subset ($|O_r| = |O_{nr}| = 40$).

A relevant aspect to consider in the selection of O_r is that we have discarded all the ontologies potentially involved in the generation of the experimental dataset in order to avoid biased information. An example of these ontologies is: http://oaei.ontologymatching.org/2004/Contest/228/onto.rdf.

6.3 Studying the Discriminative Effect of Ontology Quality Measures

O_r and O_{nr} represent respectively reliable an unreliable online semantic content in the context of assessing the quality of semantic relations, i.e., O_r represents the subset of correct ontologies and O_{nr} the subset of incorrect ontologies. In this section we study how well the previously introduced measures (Section 5.1) are able to discriminate between these two types of ontologies. We therefore compute the measures for the 80 ontologies selected (40 belonging to O_r and 40 belonging to O_{nr}).

The analysis has been performed using the preprocessing tools of the Weka[11] data mining software. For each measure we present a figure that contains the ranges of values for the measure on the x-axis and the number of reliable versus unreliable ontologies that fall in each of these ranges on the y-axis. Reliable ontologies are presented in blue, and unreliable ones are presented in red.

6.3.1 Knowledge Coverage and Popularity Measures

Fig. 4 shows the results obtained for the knowledge coverage and the population measures. As we can see in the figure, the number of classes in the ontologies varies

---

[11] http://www.cs.waikato.ac.nz/ml/weka/

Number of classes

Number of properties

Number of instances

Ontology Direct Popularity

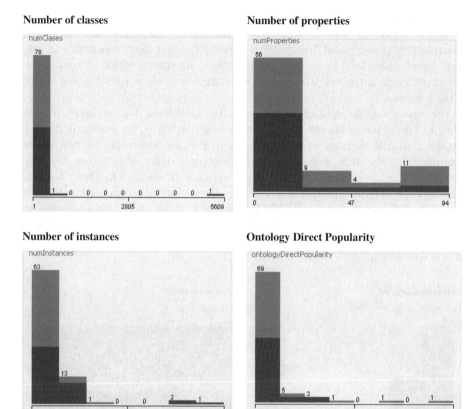

Fig. 4. Discriminative effect of the knowledge coverage and popularity measures

between 1 and 5609. The higher percentage of ontologies contains between 1 and 1000 classes and this includes reliable and unreliable ones. Only two reliable ontologies present a number of classes higher than 1000 but this number of ontologies is not statistically significant to claim that ontologies with a higher number of classes provide more reliable semantic relations.

The number of properties varies from 0 to 94 in the selected subset of ontologies. We can see that reliable ontologies tend to have fewer properties than the unreliable ones on average. However, this measure does not draw a clear line between the two subsets of ontologies either.

The number of individuals varies between 0 and 287. While there is a small subset of reliable ontologies able to provide a higher number of individuals, again this is not discriminative enough to consider that, in general, more populated ontologies provide better semantic relations.

The popularity measure varies between 0 and 14 imports per ontology. All ontologies with a popularity value higher than 6 are considered unreliable. However, there are only three ontologies in the dataset showing this effect, and therefore this measure cannot be considered discriminative either.

6.3.2 Structural Ontology Measures

Structural measures aim to study the topology of the ontologies, and more concretely their *depth* and *breadth*. We hypothesize that these measures can help us to better understand how conceptual relations are spread within the ontologies and therefore to determine which ontologies are better when assessing the quality of the relations.

The first group of measures that we have considered for this evaluation is related to the depth of the ontology. As we can see in Fig. 5, the minimum depth is always 2 so this measure is not discriminative at all. However, the rest of the measures slightly show that in general, those ontologies with higher levels of maximum depth, average depth and depth variance belong to O_r. Over the three measures we should highlight the ontology depth variance, since all ontologies with values higher than 0.9 are considered reliable. Even though these results are not statistically significant, there is a tendency that shows that those ontologies with higher depth variance can be considered "better" when assessing the quality of semantic relations.

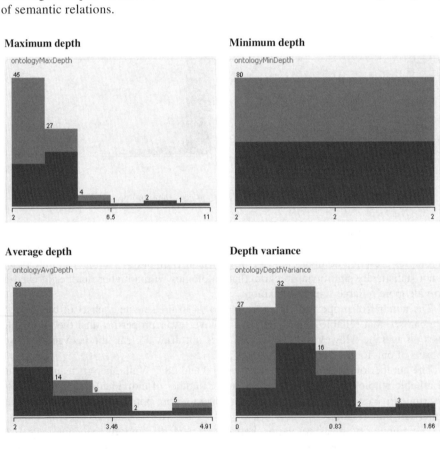

Fig. 5. Discriminative effect of the ontology depth measures

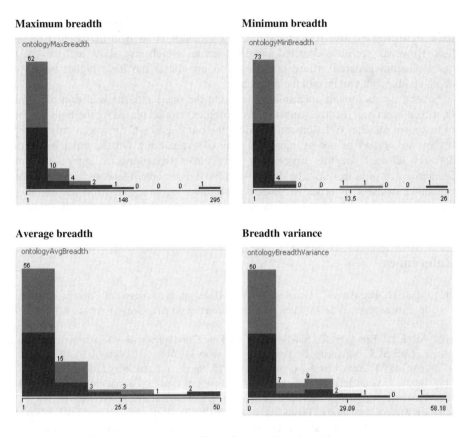

Fig. 6. Discriminative effect of the ontology breadth measures

The second group of measures that we have considered in this study is related to the breadth of the ontology. Here the tendency is not as visible as in the case of the depth measures, but we can also see that all ontologies with maximum breadth values higher than 100 and breadth variance values higher than 20 always belong to O_r.

In summary we can conclude that, even though there is no statistically significant information to affirm that the topology characteristics of the ontologies are discriminative measures to distinguish reliable versus unreliable ontologies, there is a tendency showing that those ontologies that present higher values of depth and breadth variance are able to provide better semantic relations.

7   Conclusions

Understanding which ontology characteristics can predict "good quality ontologies" is a core and ongoing task in the SW. In this paper we studied the effect of several structural ontology measures to discriminate a "good ontology" in the context of a task-based evaluation, the assessment of correct semantic relations.

Our study shows that there is no statistically significant information to assure that these measures are able to identify the best semantic content in the context of this task. However, we have detected some tendencies which may show that the "best" ontologies are generally those that are more populated and have higher values of depth and breadth variance in their structure.

Several issues remain open nonetheless. On the one hand, the selection of O_r and O_{nr} (the correct and incorrect subsets of ontologies) can be biased by the high number of unknown matches (relations provided by the ontologies where we can only be sure if they are correct or not by means of manual evaluation). On the other hand, the datasets selected for this experiment only cover two domains: agriculture and academia. It would therefore be desirable to have more heterogeneous and completed evaluated datasets in order to discern with more accuracy if structural ontology measures can identify the best ontologies to assess the correctness of semantic relations.

References

[1] Alani, H., Brewster, C.: Ontology Ranking Based on the Analysis of Concept Structures. In: Proceedings of the 3rd International Conference on Knowledge Capture, K-CAP 2005 (2005)

[2] Alani, H., Brewster, C., Shadbolt, N.: Ranking Ontologies with AKTiveRank. In: Cruz, I., Decker, S., Allemang, D., Preist, C., Schwabe, D., Mika, P., Uschold, M., Aroyo, L.M. (eds.) ISWC 2006. LNCS, vol. 4273, pp. 1–15. Springer, Heidelberg (2006)

[3] Brank, J., Grobelnik, M., Mladenić, D.: A Survey of Ontology Evaluation Techniques. In: Proceedings of the Conference on Data Mining and Data Warehouses (2005)

[4] Brewster, C., Alani, H., Dasmahapatra, S., Wilks, Y.: Data Driven Ontology Evaluation. In: Proceedings of the 4th International Conference on Language Resources and Evaluation, LREC 2004 (2004)

[5] Buitelaar, P., Eigner, T., Declerck, T.: OntoSelect: A Dynamic Ontology Library with Support for Ontology Selection. In: McIlraith, S.A., Plexousakis, D., van Harmelen, F. (eds.) ISWC 2004. LNCS, vol. 3298. Springer, Heidelberg (2004)

[6] Burton-Jones, A., Storey, V., Sugumaran, V., Ahluwalia, P.: A Semiotic Metrics Suite for Assessing the Quality of Ontologies. Data and Knowledge Engineering 55(1), 84–102 (2005)

[7] d'Aquin, M., Baldassarre, C., Gridinoc, L., Angeletou, S., Sabou, M., Motta, E.: Characterizing Knowledge on the Semantic Web with Watson. In: Aberer, K., Choi, K.-S., Noy, N., Allemang, D., Lee, K.-I., Nixon, L.J.B., Golbeck, J., Mika, P., Maynard, D., Mizoguchi, R., Schreiber, G., Cudré-Mauroux, P. (eds.) ASWC 2007 and ISWC 2007. LNCS, vol. 4825. Springer, Heidelberg (2007)

[8] Ding, L., Finin, T., Joshi, A., Pan, R., Cost, R.S., Peng, Y., Reddivari, P., Doshi, V.C., Sachs, J.: Swoogle: A Semantic Web Search and Metadata Engine. In: Proceedings of the 13th ACM Conference on Information and Knowledge Management, CIKM 2004 (2004)

[9] Ding, L., Pan, R., Finin, T., Joshi, A., Peng, Y., Kolari, P.: Finding and Ranking Knowledge on the Semantic Web. In: Gil, Y., Motta, E., Benjamins, V.R., Musen, M.A. (eds.) ISWC 2005. LNCS, vol. 3729, pp. 156–170. Springer, Heidelberg (2005)

[10] Fernández, M., Cantador, I., Castells, P.: CORE: A Tool for Collaborative Ontology Reuse and Evaluation. In: Proceedings of the 4th International EON Workshop (EON 2006) at the 15th International World Wide Web Conference, WWW 2006 (2006)

[11] Guarino, N., Welty, C.: An Overview of OntoClean. In: Staab, S., Studer, R. (eds.) Handbook on Ontologies, pp. 151–172. Springer, Heidelberg (2004)
[12] Hartmann, J., Sure, Y., Giboin, A., Maynard, D., Suarez-Figueroa, M.C., Cuel, R.: Methods for Ontology Evaluation. Knowledge Web Deliverable D1.2.3 (2005)
[13] Lopez, V., Motta, E., Uren, V.: PowerAqua: Fishing the Semantic Web. In: Sure, Y., Domingue, J. (eds.) ESWC 2006. LNCS, vol. 4011, pp. 393–410. Springer, Heidelberg (2006)
[14] Lozano-Tello, A., Gómez-Pérez, A.: ONTOMETRIC: A Method to Choose the Appropriate Ontology. Journal of Database Management 15(2), 1–18 (2004)
[15] Maedche, A., Staab, S.: Measuring Similarity between Ontologies. In: Gómez-Pérez, A., Benjamins, V.R. (eds.) EKAW 2002. LNCS (LNAI), vol. 2473, pp. 251–263. Springer, Heidelberg (2002)
[16] Patel, C., Supekar, K., Lee, Y., Park, E.: OntoKhoj: A Semantic Web Portal for Ontology Searching, Ranking, and Classification. In: Chiang, R.H.L., Laender, A.H.F., Lim, E.P. (eds.) Proceedings of the 5th ACM CIKM International Workshop on Web Information and Data Management (WIDM 2003), pp. 58–61. ACM Press, New York (2003)
[17] Sabou, M., d'Aquin, M., Motta, E.: Exploring the Semantic Web as Background Knowledge for Ontology Matching. In: Spaccapietra, S., Pan, J.Z., Thiran, P., Halpin, T., Staab, S., Svatek, V., Shvaiko, P., Roddick, J. (eds.) Journal on Data Semantics XI. LNCS, vol. 5383, pp. 156–190. Springer, Heidelberg (2008)
[18] Sabou, M., Fernández, M., Motta, E.: Evaluating Semantic Relations by Exploring Ontologies on the Semantic Web. In: Proceedings of the 14th International Conference on Applications of Natural Language to Information Systems, NLDB 2009 (2009)
[19] Sabou, M., Gracia, J.L., Angeletou, S., d'Aquin, M., Motta, E.: Evaluating the Semantic Web: A Task-Based Approach. In: Aberer, K., Choi, K.-S., Noy, N., Allemang, D., Lee, K.-I., Nixon, L.J.B., Golbeck, J., Mika, P., Maynard, D., Mizoguchi, R., Schreiber, G., Cudré-Mauroux, P. (eds.) ASWC 2007 and ISWC 2007. LNCS, vol. 4825, pp. 423–437. Springer, Heidelberg (2007)
[20] Sabou, M., Gracia, Spider, J.: Bringing Non-Equivalence Mappings to OAEI. In: Proceedings of the 3rd International Workshop on Ontology Matching (OM-2008) at the 7th International Semantic Web Conference, ISWC 2008 (2008)
[21] Sabou, M., Lopez, V., Motta, E., Uren, V.: Ontology Selection: Ontology Evaluation on the Real Semantic Web. In: Proceedings of the 4th International EON Workshop (EON 2006) at the 15th International World Wide Web Conference, WWW 2006 (2006)
[22] Strasunskas, D., Tomassen, S.: Empirical Insights on a Value of Ontology Quality in Ontology-Driven Web Search. In: Meersman, R., Tari, Z. (eds.) OTM 2008, Part II. LNCS, vol. 5332, pp. 1319–1337. Springer, Heidelberg (2008)
[23] Tartir, S., Arpinar, I., Moore, M., Sheth, A., Aleman-Meza, B.: OntoQA: Metric-Based Ontology Quality Analysis. In: IEEE Workshop on Knowledge Acquisition from Distributed, Autonomous, Semantically Heterogeneous Data and Knowledge Sources, pp. 45–53. IEEE Computer Society, Los Alamitos (2005)

Repairing the Missing is-a Structure of Ontologies

Patrick Lambrix, Qiang Liu, and He Tan

Department of Computer and Information Science
Linköpings universitet
581 83 Linköping, Sweden

Abstract. Developing ontologies is not an easy task and often the resulting on-tologies are not consistent or complete. Such ontologies, although often useful, also lead to problems when used in semantically-enabled applications. Wrong conclusions may be derived or valid conclusions may be missed. To deal with this problem we may want to repair the ontologies. Up to date most work has been performed on finding and repairing the semantic defects such as unsatisfi-able concepts and inconsistent ontologies. In this paper we tackle the problem of repairing modeling defects and in particular, the repairing of structural relations (is-a hierarchy) in the ontologies. We study the case where missing is-a relations are given. We define the notion of a structural repair and develop algorithms to compute repairing actions that would allow deriving the missing is-a relations in the repaired ontology. Further, we define preferences between repairs. We also look at how we can use external knowledge to recommend repairing actions to a domain expert. Further, we discuss an implemented prototype and its use as well as an experiment using the ontologies of the Anatomy track of the Ontology Alignment Evaluation Initiative.

1 Introduction

Developing ontologies is not an easy task and often the resulting ontologies are not consistent or complete. Such ontologies, although often useful, also lead to problems when used in semantically-enabled applications. Wrong conclusions may be derived or valid conclusions may be missed. Defects in ontologies can take different forms (e.g. [7]). Syntactic defects are usually easy to find and to resolve. Defects regarding style include such things as unintended redundancy. More interesting and severe defects are the modeling defects which require domain knowledge to detect and resolve, and semantic defects such as unsatisfiable concepts and inconsistent ontologies. Most work up to date has focused on finding and repairing the semantic defects in an ontology (e.g. [5,6,7,13]). Recent work has also started looking at repairing semantic defects in a set of mapped ontologies [4] or the mappings between ontologies themselves [11].

In this paper we tackle the other difficult problem, i.e. the repairing modeling de-fects. In particular, we focus on the repairing of structural relations (is-a hierarchy) in the ontologies. In this setting it is known that a number of intended is-a relations are not present in the source ontology. The missing is-a relations can be discovered by in-spection of the ontologies by experts or they can be generated by automated tools. For instance, in the case of task 4 in the Anatomy track in the 2008 Ontology Alignment Evaluation Initiative (OAEI) [10], two ontologies, Adult Mouse Anatomy Dictionary

A. Gómez-Pérez, Y. Yu, and Y. Ding (Eds.): ASWC 2009, LNCS 5926, pp. 76–90, 2009.

[1] (MA, 2744 concepts) and the NCI Thesaurus - anatomy [12] (NCI-A, 3304 concepts), and 988 mappings between the two ontologies are given. Based on the structure of the source ontologies and the given mappings, it can be derived that 178 is-a relations in MA and 146 in NCI-A are missing.[1]

Once missing is-a relations are found, the problem is to add is-a relations (or subsumption axioms) to the ontology such that the missing is-a relations can be derived. Although the easiest way to do this, is to just add the missing is-a relations, this may not be the most interesting solution for the domain expert. For instance, in MA an is-a relation between wrist joint and joint is missing and could be added to the ontology. However, knowing that there is an is-a relation between wrist joint and limb joint in MA, a domain expert may want to add an is-a relation between limb joint and joint. This is more informative and would lead to the fact that the missing is-a relation can be derived. In general, such a decision is preferably made by a domain expert. Therefore, in this work, we develop algorithms to generate and recommend possible ways to repair the structure of the ontology and develop a tool that allows a domain expert to repair the structure of an ontology in a semi-automatic way.

In section 2 we formally define the notion of structural repair. As not all possible ways to repair an ontology are equally useful, we also define a number of preference relations between repairs. Section 3 describes our algorithms for generating, recommending and executing repairing actions. Our prototype system and its use are described in section 4. Further, we discuss experiments on repairing MA and NCI-A in section 5. Related work is presented in section 6 and the paper concludes in section 7.

2   Theory

The setting that we study is the case where the ontology is defined using named concepts and subsumption axioms[2]. Most ontologies contain this case and many of the most well-known and used ontologies, e.g. in the life sciences, are covered by this setting. We therefore use the following definition.

Definition 1. *Let $\mathcal{O} = (\mathcal{C}, \mathcal{I})$ be an ontology with $\mathcal{C}$ its set of named concepts and $\mathcal{I} \subseteq \mathcal{C} \times \mathcal{C}$ a representation of its is-a structure. Let $\mathcal{M} \subseteq \mathcal{C} \times \mathcal{C}$ be a set of missing is-a relations (i.e. $\mathcal{M}$ represents a set of missing subsumptiom axioms). A* **structural repair** *for the ontology $\mathcal{O}$ with respect to $\mathcal{M}$ is a set of pairs of concepts $\mathcal{R} \subseteq \mathcal{C} \times \mathcal{C}$ such that for each $(A_i, B_i) \in \mathcal{M}$: $(\mathcal{C}, \mathcal{I} \cup \mathcal{R}) \models A_i \to B_i$.*

The definition states that a structural repair of an ontology given a set of missing is-a relations, is a set of is-a relations such that when these is-a relations are added to the ontology, then all missing is-a relations can be derived from the extended ontology. The elements in a structural repair we call *repairing actions*.

---

[1] A number of these are actually redundant. For instance, it may be that when repairing one missing is-a relation, others are repaired as well. Using this property we can remove 57 missing is-a relations from MA (with 121 remaining) and 63 from NCI-A (with 83 remaining).

[2] In this paper we denote subsumption axioms often using $\to$. A $\to$ B means that A is-a B.

An immediate consequence is that the set of missing is-a relations is itself a structural repair. Another consequence is that adding is-a relations to a structural repair also constitutes a structural repair.

Not all structural repairs are equally useful or interesting for a domain expert. To deal with this issue we introduce a number of preference relations.

In some structural repairs there may be is-a relations that do not contribute to the derivation of the missing is-a relations. In other structural repairs some of the is-a relations may be derivable from the other is-a relations in the structural repair and therefore redundant. For example, the missing is-a relation between wrist joint and joint can be repaired in MA by adding the is-a relation between limb joint and joint. In that case, the missing is-a relation between elbow joint and joint is also repaired since there is a is-a relation between elbow joint and limb joint. Therefore, an is-a relation between elbow joint and joint in the structural repair is redundant. The first preference relation prefers not to use these redundant or non-contributing is-a relations for repairing.

Definition 2. *Let $\mathcal{R}_1$ and $\mathcal{R}_2$ be structural repairs for the ontology $\mathcal{O}$ with respect to $\mathcal{M}$, then $\mathcal{R}_1$ is axiom-preferred to $\mathcal{R}_2$ (notation $\mathcal{R}_1 \ll_A \mathcal{R}_2$) iff $\mathcal{R}_1 \subseteq \mathcal{R}_2$.*

As discussed in the introduction, just adding the missing is-a relations, is not always the most interesting solution for the domain expert. For instance, repairing the missing is-a relation wrist joint and joint by adding an is-a relation between limb joint and joint may be more informative. When one is-a relation can be derived from another in the context of the ontology, we say that the second is-a relation is more informative than the first. The second preference relation prefers to use as informative is-a relations as possible for repairing.

Definition 3. *We say that (X_1, Y_1) is more informative than (X_2, Y_2) iff $X_2 \rightarrow X_1$ and $Y_1 \rightarrow Y_2$. Let $\mathcal{R}_1$ and $\mathcal{R}_2$ be structural repairs for the ontology $\mathcal{O}$ with respect to $\mathcal{M}$. Then $\mathcal{R}_1$ is information-preferred to $\mathcal{R}_2$ (notation $\mathcal{R}_1 \ll_I \mathcal{R}_2$) iff $\exists (X_1, Y_1) \in \mathcal{R}_1, (X_2, Y_2) \in \mathcal{R}_2$: (X_1, Y_1) is more informative than (X_2, Y_2).*

Further, some structural repairs may introduce equivalence relations for concepts which were only connected by an is-a relation in the original ontology. Although such a structural repair may result in a consistent ontology, this is usually not desired from a modeling perspective. For example, in MA we have is-a relations between posterior communicating artery and artery, and between communicating artery and artery. However, there is a missing is-a relation between posterior communicating artery and communicating artery. This could be repaired by adding an is-a relation between artery and communicating artery. However, this also introduces an equivalence between communicating artery and artery. The third preference relation prefers not to change is-a relations in the original ontology into equivalence relations.

Definition 4. *Let $\mathcal{R}_1$ and $\mathcal{R}_2$ be structural repairs for the ontology $\mathcal{O} = (\mathcal{C}, \mathcal{I})$ with respect to $\mathcal{M}$. Then $\mathcal{R}_1$ is strict-hierarchy-preferred to $\mathcal{R}_2$ (notation $\mathcal{R}_1 \ll_{SH} \mathcal{R}_2$) iff $\exists A, B \in \mathcal{C}: (\mathcal{C}, \mathcal{I}) \models A \rightarrow B$ and $(\mathcal{C}, \mathcal{I}) \not\models B \rightarrow A$ and $(\mathcal{C}, \mathcal{I} \cup \mathcal{R}_1) \not\models B \rightarrow A$ and $(\mathcal{C}, \mathcal{I} \cup \mathcal{R}_2) \models B \rightarrow A$.*

In general, we would want structural repairs that are maximally preferred.

> *Input*:
> Source ontology, missing is-a relations.
> *Output*
> Repairing actions.
> **Algorithm**
> 1. Initialize KB with ontology;
> 2. For every missing is-a relation (A_i, B_i): add the axiom $A_i \rightarrow B_i$ to the KB;
> 3. For each (A_i, B_i):
>   3.1 Source(A_i, B_i) := super-concepts(A_i) - super-concepts(B_i);
>   3.2 Target(A_i, B_i) := sub-concepts(B_i) - sub-concepts(A_i);
> 4. Missing is-a relation (A_i, B_i) can be repaired by choosing an element
>   from Source(A_i, B_i) x Target(A_i, B_i).

Fig. 1. Algorithm for generating repairing actions - 1

Definition 5. *A structural repair $\mathcal{R}$ for the ontology $\mathcal{O}$ with respect to $\mathcal{M}$ is maximally preferred with respect to the preference relation $\ll$ iff for all structural repairs $\mathcal{R}_1$ for $\mathcal{O}$ with respect to $\mathcal{M}$ it holds that if $\mathcal{R}_1 \ll \mathcal{R}$ then $\mathcal{R} \ll \mathcal{R}_1$.*

3   Repairing the Structure of an Ontology

A naive way to compute all possible structural repairs would be to take all sub-sets of $\mathcal{C}$ x $\mathcal{C}$ and for each sub-set, add its elements as is-a relations to the ontology and check whether the missing is-a relations can be derived. This is in practice infeasible as it requires checking too many cases. Even for small ontologies, it is not practical as domain experts usually deal with one or a few missing is-a relations at a time, rather than choosing between large sets of possible repairs including all missing is-a relations. Therefore, we develop algorithms that generate possible repairing actions for the missing is-a relations, taking into account the preferences defined in section 2. We also provide an algorithm that recommends repairing actions. The user can then select a missing is-a relation to repair (and we rank these in terms of the number of possible repairing actions). Further, we developed an algorithm that, upon the repairing of a missing is-a relation, detects for which missing is-a relations the set of repairing actions needs to be updated, and updates these.

3.1   Generating Repairing Actions

Algorithm 1. In our first algorithm (see figure 1), when generating repairing actions for a missing is-a relation, we take into consideration that all missing is-a relations will be repaired, but we do not take into account the actual repairing actions that could be performed for other missing is-a relations.

In the algorithm we store the ontology in a knowledge base and add the missing is-a relations to the ontology. As we know that these should be derivable in the repaired ontology, adding them introduces the desired new connections. Then, we generate correct ways to introduce more informative is-a relations that would allow us to derive the missing is-a relations. Therefore, for a repairing action (S_i, T_i) regarding missing is-a

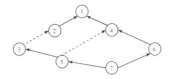

Fig. 2. Example 1

relation (A_i, B_i) we require that $A_i \rightarrow S_i$ and $T_i \rightarrow B_i$ (preference $\ll_I$ in definition 3). This also ensures that we only compute repairing actions that are relevant for repairing the missing is-a relations (preference $\ll_A$ in definition 2.) At the same time we do not want to introduce new equivalence relations, where in the source ontology we have only is-a relations (preference $\ll_{SH}$ in definition 4). This is realized by the selection of the elements in the Source and Target sets.

The proposed repairing actions for a missing is-a relation (A_i, B_i) all lead to the derivation of (A_i, B_i) in the extended ontology. In general, a user may repair the ontology by choosing for each missing is-a relation (A_i, B_i) an element from Source(A_i, B_i) and an element from Target(A_i, B_i). However, as we have not taken into account all influences of possible repairing actions for other missing is-a relations, a better strategy is to repair one missing is-a relation and recompute repairing actions for the other missing is-a relations in the partially repaired ontology.

As an example, consider the case presented in figure 2, where $\mathcal{O}_1 = (\mathcal{C}_1, \mathcal{I}_1)$ is an ontology with concepts $\mathcal{C}_1 = \{1, 2, 3, 4, 5, 6, 7\}$ and is-a relations (shown in full lines in figure 2) $\mathcal{I}_1 = \{(7,5), (7,6), (5,3), (2,1), (6,4), (4,1)\}$. ($\mathcal{I}_1$ represents the is-a hierarchy and thus also all is-a relations derived from the elements in $\mathcal{I}_1$.) The set of missing is-a relations (shown in dashed lines in figure 2) is $\mathcal{M}_1 = \{(5,4), (3,2)\}$. The algorithm will then generate the following Source and Target sets: Source(5,4) = $\{5, 3, 2, 1, 4\}$ - $\{4, 1\} = \{5, 3, 2\}$; Target(5,4) = $\{4, 6, 7, 5\}$ - $\{5, 7\} = \{4, 6\}$; Source(3,2) = $\{3, 2, 1\}$ - $\{2, 1\} = \{3\}$; Target(3,2) = $\{2, 3, 5, 7\}$ - $\{3, 5, 7\} = \{2\}$. For missing is-a relation (3,2) the only generated repairing action is (3,2). For missing is-a relation (5,4) any of the repairing actions (5,4), (5,6), (3,4), (3,6), (2,4), (2,6) together with (any of) the generated repairing action(s) for (3,2) leads to the derivation of the missing is-a relation (5,4) in the extended ontology. The example also shows the importance of initially adding the missing is-a relations to the knowledge base. The possible repairing action (2,4) for missing is-a relation (5,4) would not be generated when we do not take into account that missing is-a relation (3,2) will be repaired.[3] Further, the example also shows that we do not introduce repairing actions that would turn is-a relations in the original ontology into equivalence relations. For instance, adding (1,4) would lead to the fact that missing is-a relation (5,4) would be derivable in the extended ontology, but also leads to making 1 and 4 equivalent.

---

[3] So this means that repairing one is-a relation may influence the repairing actions for other missing is-a relations. However, when *generating* repairing actions in algorithm 1 the only influence that is taken into consideration is the fact that missing is-a relations are or will be repaired (least informative repairing action), but not the actual (possibly more informative) repairing actions that could be performed.

Input:
Source ontology, missing is-a relations.
Output
Repairing actions.
Algorithm
1. Initialize KB with ontology;
2. For every missing is-a relation $A_i \to B_i$:
    2.1 create new concepts X_i and Y_i in the KB;
    2.2 add the axioms $A_i \to X_i$, $X_i \to Y_i$, $Y_i \to B_i$ to the KB;
3. For each (A_i, B_i):
    3.1 Source-ext(A_i, B_i) := super-concepts(A_i) - super-concepts(X_i);
    3.2 Target-ext(A_i, B_i) := sub-concepts(B_i) - sub-concepts(Y_i);
4. Missing is-a relation (A_i, B_i) can be repaired by choosing an original ontology element
    from Source-ext(A_i, B_i) and an original ontology element from Target-ext(A_i, B_i).

Fig. 3. Algorithm for generating repairing actions - 2

Algorithm 2. Our second algorithm for finding repairing actions for a particular missing
is-a relation (see figure 3) takes into account influences of other missing is-a relations
that are valid for all possible choices for repairing actions for the other missing is-a relations. The difference between the basic algorithm and our extended algorithm occurs
mainly in steps 2 and 3. Instead of adding the missing is-a relations to the knowledge
base, in the extended algorithm we introduce for each missing is-a relation (A_i, B_i) two
new concepts X_i and Y_i in the knowledge base as well as the axioms $A_i \to X_i$, $X_i \to$
Y_i, $Y_i \to B_i$. (X_i, Y_i) satisfies the requirements that each possible repairing action for
(A_i, B_i) should satisfy. As they are new concepts in the knowledge base, the properties
and relations of X_i, respectively Y_i, to other concepts in the knowledge base represent
the properties and relations that are common to the source concepts, respectively target
concepts, of the possible repairing actions for (A_i, B_i). The Source and Target sets are
now computed relative to the X_i and Y_i.

As an example, consider the case presented in figure 4, where $\mathcal{O}_2 = (\mathcal{C}_2, \mathcal{I}_2)$ is an
ontology with concepts $\mathcal{C}_2 = \{1, 2, 3, 4, 5, 6, 7, 8, 9, 10\}$ and is-a relations (shown in
full lines in figure 4) $\mathcal{I}_2 = \{(7,6), (6,5), (5,2), (2,1), (7,4), (10,4), (10,9), (9,8), (8,3),$
$(3,1), (4,1)\}$. (As before, $\mathcal{I}_2$ represents the is-a hierarchy and thus also all is-a relations
derived from the elements in $\mathcal{I}_2$.) The set of missing is-a relations (shown in dashed

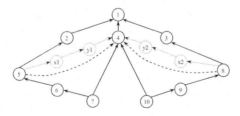

Fig. 4. Example 2

lines in figure 4) is $\mathcal{M}_2 = \{(5,4), (8,4)\}$. The algorithm in figure 1 will then generate the following Source and Target sets: Source(5,4) = $\{5, 4, 1, 2\}$ - $\{4, 1\}$ = $\{5, 2\}$; Target(5,4) = $\{4, 8, 9, 10, 5, 6, 7\}$ - $\{5, 6, 7\}$ = $\{4, 8, 9, 10\}$; Source(8,4) = $\{8, 4, 1, 3\}$ - $\{4, 1\}$ = $\{8, 3\}$; Target(8,4) = $\{4, 8, 9, 10, 5, 6, 7\}$ - $\{8, 9, 10\}$ = $\{4, 5, 6, 7\}$.

The extended algorithm in figure 3 will add the nodes x1, y1, x2, y2 and the is-a relations $5 \rightarrow x1, x1 \rightarrow y1, y1 \rightarrow 4, 8 \rightarrow x2, x2 \rightarrow y2$, and $y2 \rightarrow 4$ (shown in dotted lines in figure 4). It then generates the following Source and Target sets: Source-ext(5,4) = $\{5, 4, 1, 2, x1, y1\}$ - $\{4, 1, x1, y1\}$ = $\{5, 2\}$; Target-ext(5,4) = $\{4, 8, 9, 10, 5, 6, 7, x1, y1, x2, y2\}$ - $\{5, 6, 7, x1, y1\}$ = $\{4, 8, 9, 10, x2, y2\}$; Source-ext(8,4) = $\{8, 4, 1, 3, x2, y2\}$ - $\{4, 1, x2, y2\}$ = $\{8, 3\}$; Target-ext(8,4) = $\{4, 8, 9, 10, 5, 6, 7, x1, y1, x2, y2\}$ - $\{8, 9, 10, x2, y2\}$ = $\{4, 5, 6, 7, x1, y1\}$. The sets generated by the extended algorithm indicate that there is an influence between the two missing is-a relations. Indeed, when a choice is made for repairing the first missing is-a relation, we have essentially added equivalence relations between x1, respectively y1, and concepts in the ontology. The appearance of x1 and y1 in the Target-ext set for the second missing is-a relation indicates that the concept chosen to be equivalent to x1 (and all concepts between this concept and 5) are now also candidates for the Target for the second missing is-a relation. For example, when choosing (2,4) as a repairing action for missing is-a relation (5,4) then (3,2) is a possible repairing action for missing is-a relation (8,4).

Similarly to the basic algorithm, the proposed repairing actions for a missing is-a relation (A_i, B_i) all lead to the derivation of (A_i, B_i) in the extended ontology. In general, a user may repair the ontology by choosing for each missing is-a relation (A_i, B_i) an element from Source(A_i, B_i) and an element from Target(A_i, B_i). However, as the algorithm only takes into account influences that are common to all possible choices for repairing actions, a user may want to repair one missing is-a relation and recompute repairing actions for the other missing is-a relations.

3.2 Recommending Repairing Actions

As there may be many possible repairing actions, we develop a method for recommending repairing actions based on domain knowledge. We assume that we can query the domain knowledge regarding subsumption of concepts. There are several such sources such as general thesauri (e.g. WordNet) or specialized domain-specific sources (e.g the Unified Medical Language System). In our algorithm (see figure 5) we generate recommended repairing actions for a missing is-a relation starting from the Source and Target sets generated by the algorithm in figure 1[4]. The algorithm selects the most informative repairing actions that are supported by evidence in the domain knowledge. The variable *visited* in the algorithm in figure 5 keeps track of already processed repairing actions. The variable *recommended* stores recommended repairing actions at each step and its final value is returned as output. It is initialized with the missing is-a relation itself. This is the least informative repairing action that can be performed for repairing the missing is-a relation. Steps 3 and 4 compute the set X_e of maximal elements with respect to the is-a relation in the Source set and the set Y_e of minimal elements with respect to the is-a relation in the Target set. The elements from X_e x Y_e are then the

---

[4] We have also extended the algorithm in figure 5 to deal with Source and Target sets derived by the algorithm in figure 3.

Input:
domain knowledge, source ontology, missing is-a relation (A_i, B_i),
Source and Target for the missing is-a relation as computed by algorithm in figure 1.
Output
Recommended repairing actions.
Algorithm
Global Variable $visited$: stores already processed repairing actions.
Global Variable $recommended$: stores recommended repairing actions.
1. Set $visited = \{(A_i, B_i)\}$;
2. Set $recommended = \{(A_i, B_i)\}$;
3. Set $X_e = \{x_e : x_e \in \text{Source}(A_i, B_i) \bigwedge \forall x \in \text{Source}(A_i, B_i): \text{if } x_e \rightarrow x \text{ then } x = x_e\}$;
4. Set $Y_e = \{y_e : y_e \in \text{Target}(A_i, B_i) \bigwedge \forall\, y \in \text{Target}(A_i, B_i): \text{if } y \rightarrow y_e \text{ then } y = y_e\}$;
5. For each pair $(x_e, y_e) \in X_e$ x Y_e: call QCheck(x_e, y_e);
6. Return $recommended$;
Function QCheck(concept x, concept y)
i. If $(x, y) \in visited$ then return;
ii. Add (x, y) to $visited$;
iii. If $\exists\, (x_r, y_r) \in recommended$: $x \rightarrow x_r \wedge y_r \rightarrow y$ then return;
iv. If x is a sub-concept of y according to the domain knowledge then
  Remove all (x_r, y_r) from $recommended$ for which $x_r \rightarrow x$ and $y \rightarrow y_r$;
  add (x, y) to $recommended$;
 else
  Let Y_{sup} be the set of direct super-concepts of y;
  For each $y_s \in Y_{sup} \cap \text{Target}(A_i, B_i)$: call QCheck$(x, y_s)$;
  Let X_{sub} be the set of direct sub-concepts of x;
  For each $x_s \in X_{sub} \cap \text{Source}(A_i, B_i)$: call QCheck$(x_s, y)$;

Fig. 5. Algorithm for recommending repairing actions

most informative repairing actions. For each of these elements (x,y) we check whether there is support in the domain knowledge in step 5. Steps i and ii in the function QCheck do bookkeeping regarding the already processed repairing actions. Step iii assures that we do not add recommended is-a relations that are less informative than others already recommended. In step iv we check whether there is support in the domain knowledge for the repairing action. If so, then the repairing action is recommended and all less informative repairing actions are removed from the recommendation set. If not, then we check whether there is support in the domain knowledge for the repairing actions that are less informative than (x,y). Among these we start with the most informative repairing actions.

3.3   Executing Repairing Actions

When a user has chosen a repairing action for a particular missing is-a relation, it may influence the set of possible repairing actions for other missing is-a relations. Therefore, the repairing actions for the other missing is-a relations need to be recomputed based on the ontology extended with the chosen repairing action.

For instance, figure 6 shows the new situation when choosing the repairing action (2,9) (shown in thick line) for repairing missing is-a relation (5,4) for the example in

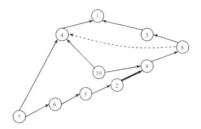

Fig. 6. Example 2 - update

figure 4. In this case the Source and Targets sets become the following for the basic algorithm: Source(8,4) = {8, 4, 1, 3} - {4, 1} = {8, 3}; Target(8,4) = {4, 8, 9, 10, 2, 5, 6, 7} - {8, 9, 10, 2, 5, 6, 7} = {4}; and the following for the extended algorithm: Source-ext(8,4) = {8, 4, 1, 3, x2, y2} - {4, 1, x2, y2} = {8, 3}; Target-ext(8,4) = {4, 8, 9, 10, 2, 5, 6, 7, x2, y2} - {8, 9, 10, 2, 5, 6, 7, x2, y2} = {4}. When we compare the computed repairing actions after the choice of (2,9) for repairing (5,4) with the repairing actions computed before the choice (see section 3.1), we note that the repairing actions that introduce equivalence relations (e.g. (8,6)) are removed after the choice of (2,9) (preference $\ll_{SH}$ in definition 4). However, before (2,9) is chosen these repairing actions do not necessarily introduce equivalence relations. For instance, we could have repaired (8,4) first using one of these actions, and afterwards repaired (5,4).

For small ontologies, computing the repairing actions does not take much time and the approach is feasible in a real setting. For large ontologies the computation time may not be small enough to guarantee immediate updates in an implemented tool for repairing. Therefore, in the algorithm[5] in figure 7 we have introduced a way to keep track of the influences between different missing is-a relations. The missing is-a relations for which the Source and Target sets can change are the missing is-a relations for which at least one of the concepts is a sub-concept or super-concept of at least one of the concepts in the chosen repairing action for the repaired missing is-a relation. We only update the Source and Target sets for these missing is-a relations. In addition, we also remove the other missing is-a relations that have been repaired by the current repairing action.

3.4   Ranking Missing is-a Relations

In general, there may be many missing is-a relations that need to be repaired. Although it is possible to repair the missing is-a relations in any order, it may be easier for the user to start with the ones where there are the fewest choices. We have therefore implemented an algorithm that ranks the missing is-a relations according to the size of the Source(A_i, B_i) x Target(A_i, B_i). The missing is-a relations with the fewest number of elements in its set are presented highest in the list of missing is-a relations.

---

[5] The algorithm in figure 7 deals with the case when we use the basic algorithm for finding repairing actions. We also have a version for when we use the extended algorithm.

Input
Ontology, the repaired missing is-a relation(A_r, B_r), the repair action (X_r, Y_r) taken for (A_r, B_r), the set of non-repaired missing relations $\mathcal{M}_r$.
Output
Updated Source and Target sets.
Algorithm
1. Add (X_r, Y_r) to the KB;
2. For each missing is-a relation (A_i, B_i) $\in \mathcal{M}_r$:
2.1 If $A_i \rightarrow X_r$ then recompute super-concepts(A_i);
2.2 If $B_i \rightarrow X_r$ then recompute super-concepts(B_i);
2.3 If $A_i \rightarrow X_r$ or $B_i \rightarrow X_r$ then Source(A_i, B_i) := super-concepts(A_i) - super-concepts(B_i);
2.4 If $Y_r \rightarrow A_i$ then recompute sub-concepts(A_i);
2.5 If $Y_r \rightarrow B_i$ then recompute sub-concepts(B_i);
2.6 If $Y_r \rightarrow A_i$ or $Y_r \rightarrow B_i$ then Target(A_i, B_i) := sub-concepts(B_i) - sub-concepts(A_i);

Fig. 7. Algorithm for updating repairing actions

4 Implemented System

We have implemented a prototype system that allows a user to repair the structure of an ontology using the algorithms described in section 3. We show its use using a piece of MA regarding the concept **Joint**. As input our system takes an ontology in OWL format as well as a list of missing is-a relations[6]. We use a framework and reasoner provided by Jena (version 2.5.7) [3]. The domain knowledge that we use is WordNet [15] and the Unified Medical Language System [14].

The ontology and missing is-a relations can be imported using the *Load/Derive Missing IS-A Relations* button. The user can see the list of missing is-a relations under the *Missing IS-A Relations* menu (see figure 8). In this case there are 7 missing is-a relations[7]. Clicking on the *Compute Repairing Actions* button, results in the computation of the Source and Target sets and the missing is-a relations in the list are ranked as described in section 3.4. The user can select which one to repair first. The first missing is-a relation in the list has the fewest possible repairing actions, and may therefore be a good starting point. When the user chooses a missing is-a relation, the Source and Target sets for the repairing actions are shown in the panels on the left and the right, respectively. The concepts in the missing is-a relation are highlighted in red.

Figure 9 illustrates the Source and Target sets for the missing is-a relation between **wrist joint** and **joint** as they were generated by our extended algorithm from figure 3. We see that, as the Target set displays x's and y's, there are a number of influences from other missing is-a relations. For instance, through x4 and y4, we see that repairing **(knee joint, joint)** may influence the repairing actions of the current missing is-a relation. The user can also ask for recommended repairing actions by clicking the

---

[6] We actually also allow to add two ontologies together with mappings. The system will then derive missing is-a relations for an ontology based on the other ontology and the mappings in a similar way as the approach described in [8].

[7] The missing is-a relations were actually derived using NCI-A and mappings between MA and NCI-A.

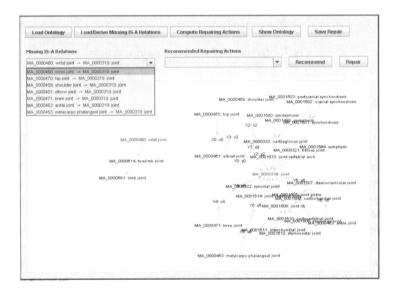

Fig. 8. Missing is-a relations

Fig. 9. Possible repairing actions for the selected missing is-a relation

Recommend button. In our case, the system recommends to add an is-a relation between limb joint and joint. In general, the system presents a list of recommendations. By selecting an element in the list, the concepts in the repairing action are highlighted in the panels. The user can repair a missing is-a relation by selecting a concept in the Source panel and a concept in the Target panel and clicking on the *Repair* button. The repairing action is then added to the ontology, and the relevant Source and Target sets and recommendations for other missing is-a relations are updated. At all times during the process the user can inspect the ontology by clicking the *Show Ontology* button. Newly added is-a relations will be highlighted (see figure 10). After adding the is-a relation between limb joint and joint, not only (wrist joint,joint) is repaired, but all other missing is-a relations as well, as they can be derived in the extended ontology. The list of missing is-a relations is therefore updated to be empty. After completing the repair of all missing is-a relations, the repaired ontology can be exported into an OWL file by clicking the *Save Repair* button.

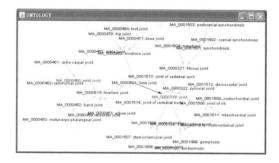

Fig. 10. The repaired ontology

5 Experiment

In our experiment we repair the two ontologies from the 2008 Anatomy track in OAEI. As described before, MA contains 2744 concepts and NCI-A contains 3304 concepts. Using the 988 mappings between the two ontologies, it can be derived that 178 is-a relations in MA and 146 in NCI-A are missing. After removing redundancy, we still have 121 missing is-a relations for MA and 83 for NCI-A. In the remainder we use these smaller sets of missing is-a relations.

Generating repairing actions. For MA our basic algorithm generates for 15 missing is-a relations only 1 repairing action (which is then the missing is-a relation itself). This means that these could be immediately repaired. For NCI-A this number is 8. Of the remaining missing is-a relations there are 65 missing is-a relations for MA that have only 1 element in the Source and 2 missing is-relations that have 1 element in the Target set. For NCI-A these numbers are 20 and 3, respectively. These are likely to be good starting points for repairing. Figure 11 shows for different ranges how many Source and Targets sets had a size in that range. We see that for most of the missing is-a relations these sets are small and thus can be easily visualized in the panels of our system.

Figure 12 shows the influences between different missing is-a relations that can be computed using our extended algorithm. In figure 12 the last column (ST) shows the number of missing is-a relations where x's and y's of other missing is-a relations occur in both Source and Target sets. For the other columns the x's and y's only occur in Source or Target, but not in both. For instance, for MA there are 23 missing is-a relations whose Source or Target set contain x and y from one other missing is-a

	total	1	2-10	11-20	21-30	31-40	41-50	51-100	101-200	201-300	301-400	>400
MA - Source	121	76	45	0	0	0	0	0	0	0	0	0
MA - Target	121	17	50	5	9	4	6	5	18	3	0	4
NCI-A - Source	83	28	55	0	0	0	0	0	0	0	0	0
NCI-A - Target	83	11	52	6	2	0	0	5	4	1	2	0

Fig. 11. Sizes of Source and Target sets

	total	1	2	3	4	5	6	7	8	9	10	11-15	16-35	ST
MA	92	23	5	3	0	25	9	9	0	4	0	13	0	1
NCI-A	67	15	21	3	1	2	0	0	0	0	0	6	6	13

Fig. 12. Influence between repairing actions of different missing is-a relations - in Source or Target

relation. We see that for a majority of the missing is-a relations (92/121 for MA and 67/83 for NCI-A) there are influences. An interesting observation is that in several cases missing is-a relations that have the same number of influences from other missing is-a relations, actually influence each other. For instance, in NCI-A we find missing is-a relations between each of Bronchus_Basement_Membrane, Bronchus_Cartilage, Bronchus_Lamina_Propria, Bronchus_Submucosa, and the concept Bronchus_Connective_Tissue. Repairing one of these missing is-a relations influences the repairing actions of all the others. We found several such clusters, among others for instance, in MA concerning body cavity/lining, lymphoid tissue, and brain nucleus with 7, 4 and 6 missing is-a relations, respectively.

Recommending repairing actions. In the experiment with the full ontologies we generated recommendations using WordNet only. The running time for generating recommendations for all missing is-a relations was circa 40 minutes for MA and circa 1 hour for NCI-A. In our tool, however, we do not generate recommendations for all missing is-a relations at once, but only on demand for a particular missing is-a relation.

For NCI-A the system recommended[8] repairing actions for only 5 missing is-a relations and each of those received one recommended repairing action. For MA 22 missing is-a relations received 1 recommended repairing action, 12 received 2 and 2 received 3. The recommendation can come from small sets of repairing actions or from large sets. For instance, for MA the system recommends for the missing is-a relation (mandible, bone) the three following repairing actions (oral region cartilage/bone, bone), (viscerocranium bone, bone), and (mandible, lower jaw). The repairing actions are recommended from a Source set of 177 concepts (and 15 influencing missing is-a relations) and a Target set of 3 concepts.

Executing repairing actions. To obtain information on the time it could take to repair real-case ontologies, as well as on the influences of the updates, we have run repairing sessions for MA and NCI-A with the basic algorithm. This test run was done by the authors. As we are not domain experts, we have used [2] to decide on possible choices and used the recommendation algorithm, although we cannot guarantee the correctness of our repairs. Clearly, we aim to redo this experiment with domain experts. However, this run already gives us some interesting information. After the ontologies were loaded and the first repairing actions were computed, the test run for NCI-A took about 40 minutes and for MA circa 90 minutes. In most cases the recommendations seemed useful. In the NCI-A session one missing is-a relation was removed as a result of repairing other is-a relations; in the MA session 18 were removed in three steps. Repairing influenced the number of repairing actions for other missing is-a relations.

---

[8] We do not count the missing is-a relation itself as a recommendation.

For the last 13 missing is-a relations for NCI-A (of 83 to start with) and 28 for MA (of 121 to start with) the Target set was too large to have a good visualization in the tool.

6 Related Work

We are not aware of other work that addresses the problem of repairing missing structure in ontologies. The closest is our work in [8] where we used structural repair in the context of ontology alignment. One of the methods included repairing the source ontologies by adding the missing is-a relations derived from a partial reference alignment, i.e. a set of given mappings, and the structure of the ontologies. Essentially, we used a least informative repair. However, in [8] there was no intention of trying to find better ways to repair the ontologies.

Other work that looks at the problem of repairing modeling defects is [9], where ontology repair is used when a formula can be derived from an ontology, but, in the words of the authors, it is not correct according to the world. In this case a mapping is computed such that the mapped formula is correct according to the world and can be derived from the mapped ontology, or such that the mapped formula cannot be derived from the mapped ontology. The setting where this is used is a framework where agents use ontologies and when certain tasks cannot be performed, communication between the agents takes place to identify mismatches between the ontologies and revise them.

There is more work that addresses repairing semantic defects in ontologies. In [13] minimal sets of axioms are identified which need to be removed to turn an ontology coherent. In [7,6,5] strategies are described for repairing unsatisfiable concepts, explanation of errors, ranking erroneous axioms, and generating repair plans. In [4] and [11] the setting is extended to repairing mapped ontologies. In this case semantic defects may be introduced by integrating ontologies. In [4] semantic defects are repaired by removing axioms in the source ontologies, while in [11] repairing removes mappings. The solutions are often based on the computation of minimal unsatisfiability-preserving sets or minimal conflict sets.

7 Conclusion

In this paper we introduced algorithms and a tool for repairing missing is-a relations in an ontology. We defined the notion of structural repairs and developed algorithms for generating, recommending and executing repairing actions. We also discussed an experiment for repairing the two ontologies of the Anatomy track of OAEI.

There are a number of directions that are interesting for future work. In our experiment we have repaired MA and NCI-A separately. However, as we have mappings between them, we want to investigate whether repairing them together could influence the quality of the generation or recommendation of repairing actions. Further, it may also be interesting to investigate possible influences between semantic defects and modeling effects. Regarding the user interface we intend two work on at least the following issues. For large ontologies with many missing is-a relations, the first generation of repairing actions may take time and thus we want to investigate ways to partition the set of missing is-a relations into parts that can be processed independently. Further, we want to investigate new ways to visualize the Source and Target sets.

References

1. AMA. Adult mouse anatomical dictionary,
   http://www.informatics.jax.org/searches/AMA_form.shtml
2. Feneis, F., Dauber, W.: Pocket Atlas of Human Anatomy, 4th edn. Thieme Verlag (2000)
3. Jena, http://jena.sourceforge.net/
4. Ji, Q., Haase, P., Qi, G., Hitzler, P., Stadtmuller, S.: RaDON - repair and diagnosis in ontology networks. In: Demo at the 6th European Semantic Web Conference, pp. 863–867 (2009)
5. Kalyanpur, A.: Debugging and Repair of OWL Ontologies. PhD thesis, University of Maryland, College Park (2006)
6. Kalyanpur, A., Parsia, B., Sirin, E., Cuenca-Grau, B.: Repairing unsatisfiable concepts in OWL ontologies. In: Sure, Y., Domingue, J. (eds.) ESWC 2006. LNCS, vol. 4011, pp. 170–184. Springer, Heidelberg (2006)
7. Kalyanpur, A., Parsia, B., Sirin, E., Hendler, J.: Debugging unsatisfiable classes in OWL ontologies. Journal of Web Semantics 3(4), 268–293 (2006)
8. Lambrix, P., Liu, Q.: Using partial reference alignments to align ontologies. In: Aroyo, L., et al. (eds.) ESWC 2009. LNCS, vol. 5554, pp. 188–202. Springer, Heidelberg (2009)
9. McNeill, F., Bundy, A.: Dynamic, automatic, first-order ontology repair by diagnosis of failed plan execution. International Journal on Semantic Web & Information Systems 3(3), 1–35 (2007)
10. Meilicke, C., Stuckenschmidt, H.: Anatomy track at the 2008 Ontology Alignment Evaluation Initiative. Anatomy at: http://oaei.ontologymatching.org/2008/
11. Meilicke, C., Stuckenschmidt, H., Tamilin, A.: Repairing ontology mappings. In: Proceedings of the Twenty-Second National Conference on Artificial Intelligence - AAAI, pp. 1408–1413 (2007)
12. NCI-A. National cancer institute - anatomy,
    http://www.cancer.gov/cancerinfo/terminologyresources/
13. Schlobach, S.: Debugging and semantic clarification by pinpointing. In: Gómez-Pérez, A., Euzenat, J. (eds.) ESWC 2005. LNCS, vol. 3532, pp. 226–240. Springer, Heidelberg (2005)
14. UMLS. Unified medical language system,
    http://www.nlm.nih.gov/research/umls/about_umls.html
15. WordNet, http://wordnet.princeton.edu/

Entity Resolution in Texts Using Statistical Learning and Ontologies

Tadej Štajner and Dunja Mladenić

Jožef Stefan Institute, Jamova 39, 1000 Ljubljana, Slovenia
{tadej.stajner, dunja.mladenic}@ijs.si

Abstract. Ambiguities, which are inherently present in natural languages represent a challenge of determining the actual identities of entities mentioned in a document (e.g., *Paris* can refer to a city in France but it can also refer to a small city in Texas, USA or to a 1984 film directed by Wim Wenders having title *Paris, Texas*). Disambiguation is a problem that can be successfully solved by entity resolution methods.

This paper studies various methods for estimating relatedness between entities, used in collective entity resolution. We define a unified entity resolution approach, capable of using implicit as well as explicit relatedness for collectively identifying in-text entities. As a relatedness measure, we propose a method, which expresses relatedness using the heterogeneous relations of a domain ontology. We also experiment with other relatedness measures, such as using statistical learning of co-occurrences of two entities or using content similarity between them. Evaluation on real data shows that the new methods for relatedness estimation give good results.

Keywords: Entity resolution, text mining, semantic annotation, ontology mapping.

1 Introduction

Integration and sharing of data across different data sources is the basis for an intelligent and efficient access to multiple heterogeneous resources. Since a lot of knowledge is present in plain text rather than a more explicit format, an interesting subset of this challenge is integrating texts with structured and semi-structured resources, such as ontologies. This is especially interesting in the context of Open Linked Data, where the main motivation is to have cross-dataset mappings across as many datasets as possible. However, textual datasets have to be treated differently in some ways. This involves dealing with natural language ambiguities in names of entities. We formulate this as an entity resolution problem, where we are trying to choose the correct corresponding entities from the ontology for the entities mentioned in text.

Our goal is to explore possible improvements of entity resolution quality by using ontologies in different ways along with statistical knowledge. To achieve this, we experiment with using different kinds of available data that could help in improving in-text entity resolution quality. Since entities, which are related, tend to appear

A. Gómez-Pérez, Y. Yu, and Y. Ding (Eds.): ASWC 2009, LNCS 5926, pp. 91–104, 2009.

together in documents more often, we explore the possibilities of expressing relatedness in different ways, such as similarities of entities' descriptions, the entity graph topology and entity co-occurrence information.

For example, in the case when we have a document where there are two unknown entities referred to by the names "Elvis" and "Memphis". The first is a common personal name and the second one the name of several locations. We would like to use this relatedness information between those two entities to help in resolving "Elvis" as a well-known singer and "Memphis" as a city in Tennessee, where the identified singer lived.

A long-term goal of this work is to improve the quality of in-text entity resolution using existing ontologies and mappings between them. In other words, we would like to be able to bootstrap existing knowledge with the intention of obtaining new knowledge.

2 Related Work

Machine learning methods are successfully being used in text mining and analysis of documents [1]. Problems, analogous to entity resolution appear in many different areas. The theoretical foundations of entity resolution are defined in the theory of record linkage [2]. Related challenges can also be found in database integration [3,4], object identification [5], duplicate detection [6] and word sense disambiguation [7,8].

When observing our problem statement from a natural language processing perspective, we can describe our approach as disambiguation using background knowledge, which is a pattern, often found in literature [10,11,12]. For the purposes of this paper we use the ontology as background knowledge represented as a graph of entities, identified with URIs, described with attributes and interconnected with different relationships. Such models can be easily constructed from RDF data [13], which is general enough to describe other domains, such as entity-relational and class models [14]. We also require that we are aware of possible phrases that represent possible labels[1] of entities. As we will show in subsequent sections, we can also benefit from having descriptions[2] of entities, which can be used beneficially for entity resolution via vector space model similarity [11,15,16,17].

There also exist methods which use relational information for disambiguation, [18], which estimates relevance with a PageRank score over candidate meanings. A collective approach using Markov logic is shown in [19]. Since different relation types have different meaning, [20] suggests an adaptive method of determining relational significance.

When solving the entity resolution problem, the usual approach involves performing graph clustering over the entity graph using a certain similarity criterion [9]. In context of relational data, it is a combination of attribute similarity and relational similarity. However, such approaches are more often found in structured data, whereas our approach attempts to use these techniques on linking unstructured text with semi-structured data. Also, when using ontologies as a sense inventory,

---

[1] http://www.w3.org/2000/01/rdf-schema#label
[2] http://www.w3.org/2000/01/rdf-schema#comment

relationships between entities are heterogeneous. The proposed novel method for determining relatedness in collective entity resolution is based on using relational entity resolution. A distinction in entity resolution approaches can be made in regard to the entity resolution independence assumptions:

- Pair-wise resolution - decisions are being done independently for each mention of an entity in the document
- Collective resolution - decisions do not assume independence of resolution decisions, enabling us to use relatedness data in the subsequent decisions.

Since collective entity resolution can take relatedness between entities into account, we experiment with the following definitions of relatedness:

- **Content similarity as a relatedness measure** can be used in situations where only available data is in form of attributes and textual descriptions and no explicit relationships between entities, as shown in [22].
- **Entity co-occurrences as a relatedness measure** are useful in situations where we can obtain a corpus of documents, annotated with resolved in-text entities, which can be used as a training set for a supervised approach to entity resolution. In general, co-occurrences are a common source of training data for information retrieval problems, analogous to entity resolution. Use cases that apply this technique can be found in [23], who uses it for protein identification and [24], who successfully resolves geographical locations. Utilization of entity co-occurrences for identifying synonyms in a unsupervised approach, which is analogous to entity resolution, can be seen in [25]. Co-occurrences have also been used to construct a generative model [8] for entity resolution.
- **Explicit relationships as a relatedness measure:** relationships between entities are the most explicit form of relatedness. However, not all relationships have the same significance. This paper proposes one such possible approach to heterogeneous relational entity resolution which bases relational significance on the frequency of the relation appearing in the ontology with regard to entity types. This measure was suggested in [21] as one of the suggested methods of determining a minimal informative subgraph of a graph. Since this problem as well as multi-relational entity resolution both use the notion of relational significance, this paper will explore the possibilities of using this measure as a means of quantifying relatedness between entities.

3 Entity Resolution from Text

3.1 Treating Disambiguation as Entity Resolution

For representing the text as a collection of entities, the necessary first step is to identify potential entities in the text. However, since the entity resolution algorithm can benefit from better information on the in-text entity, we added a named entity extraction step. For this purpose we used the Stanford Named Entity Recognizer [26]. Before using our background knowledge base, we can still perform a part of co-reference resolution with the identified entities, such as canonicalization, partial name consolidation and acronym consolidation. Simple de-duplication of extracted entities also helps to reduce the search space when performing collective entity resolution. Once we have a basic understanding of which distinct entities we are trying to resolve, we can search our ontology for possible candidates that could match the named entities. We then perform a series of decisions, where entities from the document are matched with the most relevant ontology entity based on some relevance criteria. This is then repeated as long as there are unmatched entities in the documents or none of the remaining candidates fulfill the minimum criteria for matching.

3.2 Pair-Wise Entity Matching

When matching an entity from the document to a candidate entity, we employ some heuristics to evaluate the confidence of their match. One such heuristic is description similarity. Note that since this scenario has no *a priori* matches of document entities with ontology entities, we have no use for relational information.

When a single entity has multiple documents, as shown in example in Fig.2, our task is to evaluate each candidate and finally match the document entity with the top candidate. Description similarity is defined as the cosine similarity of TF-IDF vectors of descriptions that represent the given entities. Since one of the entities is a document entity, its descriptions is the document text itself. We then resolve each entity in the document to its most similar candidate among the candidates from the ontology.

3.3 Collective Resolution with Relatedness

While leaving behind assumptions of independence, we can then benefit from using information on relatedness between entities. Collective candidate selection is performed with the following sequence of steps, adapted from the relational clustering algorithm [9] and adapted from the general dataset reconciliation domain to a text-ontology alignment scenario.

Required: document entities, candidate entities;
Initialize priority queue q, list *selected_matches*;

For each potential pair between document entity f and candidate entity e:
    Insert *(pairwise_relevance(f,e), f, e)* into q;
While q is not empty:
    Pop *(relevance$_{f,e}$, f, e)* from q;
        Add *(relevance$_{f,e}$, f, e)* to *selected_matches*;
    For each *entry* in q containing f:
        Remove *entry* from q;
For each *entry* in q,:
    Update *collective_relevance(e$_{entry}$, f$_{entry}$, selected_matches)*;
Return *selected_matches*

Fig. 1. Collective in-text entity resolution algorithm

$$relevance_{collective}(f, e, S)$$
$$= relevance_{pairwise}(f, e) + \lambda \cdot \frac{\sum_{e_s \in S} related(e, e_s)}{|\{e_s \in S; \; related(e, e_s) \neq 0\}|}$$

Fig. 2. Collective relevance estimate as a combination of pair-wise and relational similarity

Fig. 1 describes the adapted entity resolution algorithm. The high-level operation is the same for all of the described approaches.

The three approaches differ only in the calculation details of the *relatedness* estimate, which is used in collective relevance calculation, as seen in Fig. 2. The following chapters will describe the respective relatedness estimation approaches.

3.3.1 Using Semantic Relations from the Ontology

In Fig. 3, the blue nodes (Elvis and Memphis on the left) represent the document entities, whereas all the other nodes (colored pink) represent entities from the ontology. In this case, the relatedness between the entities is expressed explicitly in the form of RDF statements in the background knowledge - as shown in Figure 3. Consider the case where the subject »Elvis Presley in relation »Hometown« (as his »origin«) to the subject »Memphis, Tennessee«. For use in our resolution model, we interpret relations as links with a specified weight. If the relations in the ontology are only of a single type, they can all be treated equivalently. However, when dealing with heterogeneous ontologies, as is often the case, one has to estimate the importance of each link. For instance, if the ontology contained the RDF statement <*Elvis Presley, type, Person*>, this would not be too useful, since it would likely encompass every entity called "Elvis" since they are mostly of the type »Person«. On the other hand, the relation <*x, Hometown, Memphis_Tennessee*> is a strong indicator, because it covers a much smaller set of entities. This property is defined as selectivity, and its value can be used as a weighting of links in the graph. Determining the selectivity of the links is a problem, similar to finding the most informative subgraph in a given semantic graph, described in [21]. The authors wanted to find the smallest subgraph, which would be sufficiently informative. For the purposes of determining subsets of

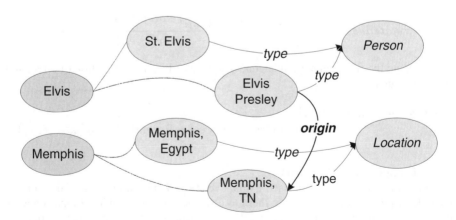

Fig. 3. Using different semantic relations as a relatedness measure

the links they have developed a few metric to estimate the selectivity. One of the proposed metrics, which is also suitable for our domain, is Instance Participation Selectivity, which stipulates that the selectivity of the assertion $<s, p, o>$ is inversely proportional the number of statements RDF which correspond to the $<type\ (s),\ p,\ type$ $(o)\ >$ where the predicate "$type(x)$" is defined as the relation of $rdf{:}type$ of the entity. Let $\pi(type(s),p,type(o))$ be the set of all statements in the domain ontology, where type of subject is $type(s)$, the predicate is p and type of object is $type(o)$.

$$IPS(s,p,o) = \frac{1}{|\pi(type(s),p,type(o))|}$$

To balance the estimate values for our use case, this paper modifies the equation slightly to:

$$IPS_{log}(s,p,o) = \frac{1}{\log(1 + |\pi(type(s),p,type(o))|)}$$

The consequence is that the link type $<Person,\ Origin,\ Area>$ is less selective than $<Person,\ Origin,\ City>$, which is also what we want to model. This approach therefore enables us to quantify the relatedness of a pair of entities based on ontology data. The direct relatedness score is then calculated as:

$$relatedness_{direct}(e_i, e_j) = \frac{\sum_{<e_i,p,e_j>\in KB} IPS(e_i,p,e_j)}{|<e_i,p,e_j>\in KB|}$$

However, when considering actual relatedness, we also take into account not only direct relations, but also indirect ones – the relations to entities that are in the common neighborhood. We define this as:

$$Nbr(e) = \{f; relatedness_{direct}(e,f) \neq 0\}$$

$$Nbr(e_i, e_j) = Nbr(e_i) \cup Nbr(e_j)$$

We define indirect relatedness as an average of paths between both entities:

$$relatedness_{indirect}(e_i, e_j) = \frac{\sum_{f\in Nbr(e_i,e_j)} relatedness_{direct}(e_i,e_j)}{|Nbr(e_i,e_j)|}$$

We compute the final semantic relatedness score as a linear combination of direct and indirect relatedness:

$$relatedness_{ontology}(e_i, e_j)$$
$$= \lambda_1 relatedness_{direct}(e_i,e_j) + \lambda_2 relatedness_{indirect}(e_i,e_j)$$

3.3.2 Using Content Similarity

In some situations, we do not have explicit relations between entities. If the entities have descriptive attributes, we use them to estimate relatedness with comparing their content similarity, as illustrated in Fig. 4. One advantage of such approach is that we do not require any more data than with pair-wise resolution, which adds to the

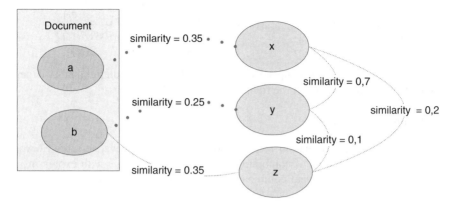

Fig. 4. Using content similarity as a relatedness measure (the green dotted lines represent selected entities)

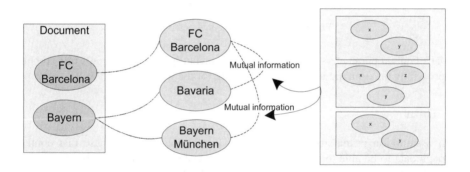

Fig. 5. Using mutual information form entity co-occurrences as relatedness

flexibility of this method. This approach was first explored in [22] and is formulated as:

$$relatedness_{content}(e_i, e_j) = sim_{content}(e_i, e_j)$$

3.3.3 Using Co-occurrences

We can also represent relatedness between entities as co-occurrences, as shown in Fig. 3.Data on the co-occurrences of two events are successfully used in information retrieval problems such as cross-language information retrieval [29] and determining the importance of words [30,31], which is a problem related to entity resolution. Intuition for the use of the co-occurrences is, the more often that the two events occur together more frequently than by chance, the more likely is that they are related. This principle was also demonstrated in [11] with a collective generative model. In our domain, we can model relatedness with point-wise mutual information [32] of two entities occurring in the same document.

$$relatedness_{co-occurences}(e_i, e_j) = SI(e_i, e_j) = log \frac{p(e_i, e_j)}{p(e_i)p(e_j)}$$
$$8(y)$$

Since this procedure requires supervised statistical learning, its output quality depends on the quality and coverage of the training corpus.

3.4 Combining Methods

Since each relevance estimation method produces its own relevance score, it would make sense to have means of combining them. This can be done with expressing the relatedness function as a linear combination of all relatedness estimation functions.

$$
\begin{aligned}
relatedness(e, f) \\
= \lambda_1 relatedness_{content}(e, f) \\
+ \lambda_2 relatedness_{co-occurences}(e, f) \\
+ \lambda_3 relatedness_{ontology}(e, f)
\end{aligned}
$$

Fig. 6. Combining relatedness estimations

The lambda parameters in Fig. 6. are experimentally obtained using a hill-climbing approach by maximizing the average $F_{0.2}$ score for the test set.

4 Data

Our assumption is that the ontology consists of knowledge database that contains enough data to be able to perform the following tasks. First, it should be able to refer to each entity with multiple aliases to facilitate candidate retrieval., Second, it should be able to provide enough additional entity features, which we can use to compare those entities to each other and to article anchors that we attempt to link to. Following these requirements, we chose to use a part of DBpedia, as described in [34] for the facts that it provides both description and attribute data from Wikipedia, as well as references to other ontologies that describe other aspects of the same real-world objects. For the purpose of having rich heterogeneous relational data, we also used the Yago ontology, defined [35], which maps Wikipedia concepts to corresponding WordNet classes. Since a direct mapping from Yago to DBpedia exists, merging the two together is trivial. However, both ontologies are much broader than what our approach requires – we currently only use information on aliases, textual descriptions, rdf:type attributes and Yago categories of entities.

5 Evaluation

5.1 Methodology

For determining the quality of the methods we have used precision and recall, measured at a certain level of confidence in the suggested entities for a given article.

We then compared the suggested entities for those articles with manually identified entities of those articles.

Precision and recall are balanced with a relevance score threshold, selecting only those entities whose relevance score is above this threshold. This serves as a useful balancing tool, since in many examples the entity cannot be correctly resolved because they do not even exist in the domain ontology. In those cases, even the best candidate has a relatively low score.

We report the final results the value of F_α, which is the weighted harmonic average of precision and recall. Namely, in some applications we want to rate precision higher than recall, as false positives are much less desired than false negatives. Therefore, we provide results for two α values, one with equally weighted precision and recall ($\alpha=1$) and one that weights precision higher than recall ($\alpha=0.2$).

We perform evaluation using the New York Times article corpus [33], using 39953 articles from January 2007 to April 2007 as training data for construction of TF/IDF weighted vectors. The articles were then processed with an implementation of the described algorithm. For evaluating the performance of different approaches we manually selected and evaluated 945 entity resolution decisions from 79 articles as either correct or incorrect. Those articles were then used as a test set on which we based our quality estimation. Since the methods of pair-wise content comparison, collective content comparison and collective relational comparison are unsupervised, they do not require any pre-labeled articles as training data. On the other hand, using co-occurrences as a relatedness measure requires training data for statistical learning. For this purpose, we take the remainder of the articles that we did not use as a test set and process them with the collective relational comparison method. Since we wish to maximize the training data quality with our best effort, we use only entities whose relevance estimate is above a certain threshold. We used the same threshold which gives us 95% precision and 45% recall on our test data. The collective relational comparison is used because it gives the highest quality output for this purpose. We experimentally determined the parameters for the methods to maximize the $F_{0.2}$. These values depend on a specific ontology and text corpus, so they are not necessarily universally applicable.

5.2   Results

Results show that additional information does indeed show improvement in $F_{0.2}$, as can be seen in Table 1. However, on higher recall (on values over 0.55), collective methods show a tendency for having performance barely similar to the baseline method of pair-wise resolution. This is evident in relatively low $F_{1.0}$ scores. The reason for this behavior is that because collective resolution depends on earlier

Table 1. F-scores of respective methods

Method	Relatedness	max $F_{1.0}$	max $F_{0.2}$
Pair-wise		0.749	0.772
Collective	Content similarity	0.750	0.789
Collective	Co-occurrences	0.721	0.747
Collective	Relations	0.728	0.789
Collective	Combined	0.741	**0.799**

Table 2. Precision and recall at max $F_{0.2}$

Method	Relatedness	Precision at max $F_{0.2}$	Recall at max $F_{0.2}$
Pair-wise		0.784	0.717
Collective	Content similarity	0.836	0.616
Collective	Co-occurrences	0.818	0.522
Collective	Relations	0.868	0.541
Collective	Combined	0.882	0.543

decisions when deciding on an entity candidate, it is sensitive to the case of misjudging an early decision within a document. However, this high precision at low recall comes at the expense of precision at high recall. In that case, it is merely comparable to that of the baseline method of pair-wise entity resolution. This is also the cause of the small differences we see in the $F_{1.0}$ score.

Further observations in Table 2. confirm that while precision successfully increases for the max $F_{0.2}$ scenario, there is something to be desired regarding recall at that point.

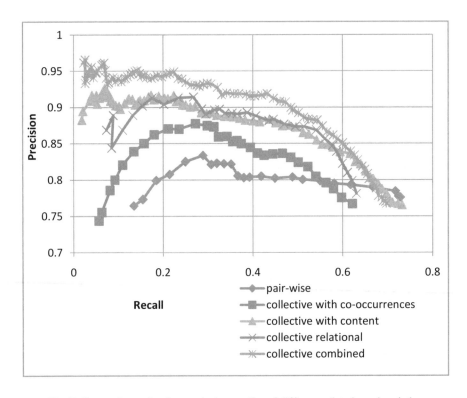

Fig. 7. Comparison of entity resolution quality of different relatedness heuristics

Table 3. Recall at two levels of high precision: 80% and 90% precision

Method	Relatedness	Recall at 80% prec.	Recall at 90% prec.
Pair-wise		0.51	/
Collective	Content similarity	0.66	0.28
Collective	Co-occurrences	0.55	/
Collective	Relations	0.61	0.27
Collective	Combined	0.65	0.48

Fig. 7 shows that additional precision can in fact be obtained by performing relational entity resolution and this even applies to scenarios where we do not have homogenous relations between entities.

As is confirmed in Table 3., all collective methods have an advantage over the baseline when looking at recall at 90% precision. Here we can also demonstrate improvement with combining all of the aforementioned methods, which yields the best overall result. We can also state that we have reached our goal of operating with higher precision, which would not be possible at all with simple pair-wise resolution.

When observing all the collective methods that we discuss in this paper, the best performing ones were collective resolution with content similarity and collective resolution with relational weighing. However, we are not able to confidently prove whether any of them is significantly better than the other. On the other hand, both outperformed the statistical learning method of counting entity co-occurrences. The cause of the lower performance of collective resolution with co-occurrences as relatedness is most likely the choice of the training set. Since it was not feasible to manually construct a training set of sufficient size, we decide to automatically construct a training corpus with the best performing method without using prior co-occurrences. For this purpose, we used collective resolution with combined multi-relational and content similarity. We selected only those entities, whose estimate was greater than the threshold that yielded 48% recall at 90% precision on the test set. The resulting performance is between the baseline and the performance of the training set for the greater part of the curve. This method of collective resolution with co-occurrences also exhibits a significant drop in precision at higher recall values. However, we can still conclude that even this method performs favorably to the baseline at higher thresholds. The best performance is obtained with collective combined method that is outperforming the other tested methods in the part with high precision and high recall. In the best performing range of recall between 0.3 and 0.5 this combined method is the only one that achieves precision over 0.9.

One of the causes for this sort of behavior is that some documents tend to discuss unrelated entities. Furthermore, in longer texts, the entities, mentioned at the beginning of the document are not necessarily related to the ones on the other parts of the document, which suggests that we should experiment with taking the document paragraph structure into account.

6 Conclusion

This paper proposes a framework for collective resolution of in-text entities on the basis of different notions of relatedness. As examples for this, we used three different relatedness estimation methods, each appropriate for a particular type of background knowledge. Among these methods we present and evaluate a novel method for determining relatedness based on commonness of ontological relations between two entity types and compare it to a supervised co-occurrence based approach and an approach using content similarity as relatedness. We confirm the previous related research that using collective resolution improves resolution precision and demonstrate this on various relatedness measures. Further improvement could be obtained by the use of machine learning on other segments of the problem, such as a means of determining the importance of relations rather than calculating their selectivity. A possible application in also in determining the significance of individual relevance estimates in the last step of calculation the total assessment.

The proposed solution capable of entity resolution from text is an important part of the knowledge extraction. The next level of this scenario would in addition to in-text entities, also identify the relations that occur between them. These newly identified relations between entities can be a basis for constructing new RDF statements, further building our ontology, thus closing the loop where we can use existing knowledge to obtain even more knowledge. This process brings new challenges, particularly in the field of selection of the appropriate statements on the basis of suitability for including them in the ontology, as discussed in [36]. Using this technology can also be useful for other purposes. Semantically expressed entities enable integration and interoperability with external data sources [37]. Also, visualization of the contents of the text in the format as described in [38] is also a use case for entity resolution.

On the other hand, our paper barely touches the possibilities that could be employed by using globally identified data approaches, opening way for better data integration, visualization and using annotated documents to enable semantic search. We expect that the proposed semantic article enrichment method to yield even more improvement on tasks that depend on the added semantic information, such as document summarization, triple extraction and recommendation systems. What all of those use cases have in common is dependence on a high quality output of the entity resolution phase.

References

1. Mladenić, D.: Text Mining: Machine Learning on Documents. In: Encyclopedia of Data Warehousing and Mining, pp. 1109–1112 (2006)
2. Fellegi, I., Sunter, A.: A theory for record linkage. Journal of the American Statistical Association, 1183–1210 (1969)
3. Haas, L., Miller, R., Niswonger, B., Roth, M., Schwarz, P., Wimmers, E.: Transforming heterogeneous data with database middleware: Beyond integration. IEEE Data Engineering Bulletin 22(1), 31–36 (1999)
4. Winkler, W.: The state of record linkage and current research problems. Statistical Research Division, US Bureau of the Census, Wachington, DC (1999)

5. Tejada, S., Knoblock, C., Minton, S.: Learning object identification rules for information integration. Information Systems 26(8), 607–633 (2001)
6. Elmagarmid, A., Ipeirotis, P., Verykios, V.: Duplicate Record Detection: A Survey (2006)
7. Yarowsky, D.: Unsupervised word sense disambiguation rivaling supervised methods. In: Proceedings of the 33rd annual meeting on Association for Computational Linguistics, pp. 189–196. Association for Computational Linguistics, Morristown (1995)
8. Kalashnikov, D., Mehrotra, S.: A probabilistic model for entity disambiguation using relationships. In: SIAM International Conference on Data Mining (SDM), Newport Beach, California, pp. 21–23 (2005)
9. Bhattacharya, I., Getoor, L.: Collective entity resolution in relational data (2007)
10. Schütze, H.: Automatic word sense discrimination. Computational Linguistics 24(1), 97–123 (1998)
11. Bunescu, R., Pasca, M.: Using encyclopedic knowledge for named entity disambiguation. In: Proceedings of the 11th Conference of the European Chapter of the Association for Computational Linguistics, pp. 3–7 (2006)
12. Cucerzan, S.: Large-scale named entity disambiguation based on Wikipedia data. In: Proceedings of the 2007 Joint Conference on Empirical Methods in Natural Language Processing and Computational Natural Language Learning (EMNLP-CoNLL), pp. 708–716 (2007)
13. Klyne, G., Carroll, J., McBride, B.: Resource description framework (RDF): Concepts and abstract syntax. W3C recommendation 10 (2004)
14. Bizer, C., Seaborne, A.: D2RQ-treating non-RDF databases as virtual RDF graphs. In: McIlraith, S.A., Plexousakis, D., van Harmelen, F. (eds.) ISWC 2004. LNCS, vol. 3298. Springer, Heidelberg (2004)
15. McCallum, A.: Information extraction: Distilling structured data from unstructured text. Queue 3(9), 48–57 (2005)
16. Lloyd, L., Bhagwan, V., Gruhl, D., Tomkins, A.: Disambiguation of references to individuals. IBM Research Report (2005)
17. Salton, G., Wong, A., Yang, C.: A vector space model for automatic indexing. Communications of the ACM 18(11), 613–620 (1975)
18. Mihalcea, R.: Unsupervised large-vocabulary word sense disambiguation with graph-based algorithms for sequence data labeling. In: Proceedings of the conference on Human Language Technology and EMNLP, pp. 411–418. Association for Computational Linguistics, Morristown (2005)
19. Singla, P., Domingos, P.: Entity resolution with markov logic. In: Proceedings of the Sixth IEEE International Conference on Data Mining, pp. 572–582 (2006)
20. Chen, Z., Kalashnikov, D., Mehrotra, S.: Adaptive graphical approach to entity resolution. In: Proceedings of the 7th ACM/IEEE-CS joint conference on Digital libraries, pp. 204–213. ACM, New York (2007)
21. Ramakrishnan, C., Milnor, W.H., Perry, M., Sheth, A.P.: Discovering informative connection subgraphs in multi-relational graphs. SIGKDD Explor. Newsl. 7(2), 56–63 (2005)
22. Štajner, T.: From unstructured to linked data: entity extraction and disambiguation by collective similarity maximization, Identity and reference in web-base knowledge representation workshop (2009)
23. Li, X., Morie, P., Roth, D.: Semantic integration in text: From ambiguous names to identifiable entities. AI Magazine. Special Issue on Semantic Integration 26(1), 45–58 (2005)

24. Bunescu, R., Mooney, R., Ramani, A., Marcotte, E.: Integrating co-occurrence statistics with information extraction for robust retrieval of protein interactions from Medline. In: Proceedings of the BioNLP Workshop on Linking NLP Processing and Biology at HLTNAACL, vol. 6, pp. 49–56 (2006)
25. Overell, S., Magalhaes, J., Ruger, S.: Place disambiguation with co-occurrence models. In: CLEF 2006 Workshop, Working notes (2006)
26. Yates, A., Etzioni, O.: Unsupervised resolution of objects and relations on the Web. In: Proceedings of NAACL HLT, pp. 121–130 (2007)
27. Finkel, J., Grenager, T., Manning, C.: Incorporating Non-local Information into Information Extraction Systems by Gibbs Sampling. Ann Arbor 100 (2005)
28. Cohen, W., Ravikumar, P., Fienberg, S.: A comparison of string distance metrics for name-matching tasks. In: Proceedings of the IJCAI-2003 Workshop on Information Integration on the Web, IIWeb 2003 (2003)
29. Jang, M., Myaeng, S., Park, S.: Using mutual information to resolve query translation ambiguities and query term weighting. In: Proceedings of the 37th annual meeting of the Association for Computational Linguistics on Computational Linguistics, pp. 223–229. Association for Computational Linguistics, Morristown (1999)
30. Church, K., Hanks, P.: Word association norms, mutual information, and lexicography. Computational Linguistics 16(1), 22–29 (1990)
31. Li, H., Abe, N.: Word clustering and disambiguation based on cooccurrence data. In: Proceedings of the 36th Annual Meeting of the Association for Computational Linguistics and 17th International Conference on Computational Linguistics, vol. 2, pp. 749–755. Association for Computational Linguistics, Morristown (1998)
32. Manning, C.D., Schütze, H.: Foundations of statistical natural language processing. MIT Press, Cambridge (1999)
33. Sandhaus, E: The New York Times Annotated Corpus, 2008.40
34. Auer, S., Bizer, C., Kobilarov, G., Lehmann, J., Cyganiak, R., Ives, Z.: Dbpedia: A nucleus for a web of open data. In: Aberer, K., Choi, K.-S., Noy, N., Allemang, D., Lee, K.-I., Nixon, L.J.B., Golbeck, J., Mika, P., Maynard, D., Mizoguchi, R., Schreiber, G., Cudré-Mauroux, P. (eds.) ASWC 2007 and ISWC 2007. LNCS, vol. 4825, pp. 722–735. Springer, Heidelberg (2007)
35. Suchanek, F., Kasneci, G., Weikum, G.: Yago: a core of semantic knowledge. In: Proceedings of the 16th international conference on World Wide Web, pp. 697–706. ACM, New York (2007)
36. Suchanek, F.M., Sozio, M., Weikum, G.: Sofie: a self-organizing framework for information extraction. In: WWW 2009: Proceedings of the 18th international conference on World wide web, pp. 631–640. ACM, New York (2009)
37. Decker, S., Melnik, S., Van Harmelen, F., Fensel, D., Klein, M., Broekstra, J., Erdmann, M., Horrocks, I.: The semantic web: The roles of XML and RDF. IEEE Internet Computing 4(5), 63–73 (2000)
38. Fortuna, B., Grobelnik, M., Mladenić, D.: Visualization of text document corpus. Special Issue: Hot Topics in European Agent Research 29, 497–502 (2005)

An Effective Similarity Propagation Method for Matching Ontologies without Sufficient or Regular Linguistic Information

Peng Wang[1,2] and Baowen Xu[2,3]

[1] College of Software Engineering, Southeast University, China
[2] State Key Laboratory for Novel Software Technology, Nanjing University, China
[3] Department of Computer Science and Technology, Nanjing University, China
pwang@seu.edu.cn, bwxu@nju.edu.cn

Abstract. Most existing ontology matching methods are based on the linguistic information. However, some ontologies have not sufficient or regular linguistic information such as natural words and comments, so the linguistic-based methods can not work. Structure-based methods are more practical for this situation. Similarity propagation is a feasible idea to realize the structure-based matching. But traditional propagation does not take into consideration the ontology features and will be faced with effectiveness and performance problems. This paper analyzes the classical similarity propagation algorithm *Similarity Flood* and proposes a new structure-based ontology matching method. This method has two features: (1) It has more strict but reasonable propagation conditions which make matching process become more efficient and alignments become better. (2) A series of propagation strategies are used to improve the matching quality. Our method has been implemented in ontology matching system Lily. Experimental results demonstrate that this method performs well on the OAEI benchmark dataset.

1 Introduction

Currently more and more ontologies are used distributedly and built by different communities. Many ontologies describe similar domains but use different terminologies. Such ontologies are referred to as heterogeneous ontologies. It is the major obstacle to realize semantic interoperation. Ontology matching, which captures relations between ontologies, aims to provide a common layer from which heterogeneous ontologies could exchange information in semantically sound manners.

Some ontology matching methods have been proposed in recent years. In these methods, calculating linguistic similarity is the most popular way to discover alignments. However, not all ontologies provide sufficient and regular linguistic information. For example, the adult mouse anatomy ontology[1] uses codes like MA_0000436 to name the concepts. Some ontologies have few comments

---

[1] http://webrum.uni-mannheim.de/math/lski/anatomy09/mouse_anatomy_2008.owl

A. Gómez-Pérez, Y. Yu, and Y. Ding (Eds.): ASWC 2009, LNCS 5926, pp. 105–119, 2009.
© Springer-Verlag Berlin Heidelberg 2009

and labels to help the readers to understand their elements, and the 248-266 ontologies in the OAEI benchmark dataset are such extreme cases. For this situation, linguistic-based methods would miss a lot of alignments. Therefore, a practical matching system should consider the structure similarity to compensate for the disadvantages of linguistic-based methods.

The structure-based ontology matching is different from the geometrical graph matching because the latter can not reflect the semantic matching. So the traditional graph matching algorithms [1] are not suitable here. For ontology matching, the structure-based methods usually employ the similarity propagation idea "*similar objects are related to similar objects*". Several similarity propagation matching algorithms have been used for database or XML schema matching [2,3]. However, in our practice, we find these traditional similarity propagation matching algorithms can not be used for ontology matching directly. For example, we implement Blondel's graph matching algorithm [4] for ontology matching and can not obtain good results, but a modified method [5] performs well on the OAEI benchmark.

This paper analyzes the classical similarity propagation matching algorithm *Similarity Flood* [2], then proposes an effective similarity propagation method according to the ontology features. The new method can avoid some disadvantages of the previous one has, and can solve the ontology structure matching problem efficiently. The original contributions of this paper include: (1) We propose an effective similarity propagation method for matching ontologies, especially for the ontologies without sufficient and regular linguistic information; (2) The new method has more strict but reasonable propagation condition which makes matching process become more efficient and alignments become better; (3) A series of similarity propagation strategies are used in the method to improve the matching quality; (4) We implement the new method and the experimental results show the method is effective for ontology matching.

The remainder of this paper is organized as follows: Section 2 discusses the structure similarity problem in ontology matching. Section 3 presents the new similarity propagation method. Section 4 describes the propagation strategies. Some experimental results and discussions are presented in Section 5. Section 6 is a brief overview of related work and section 7 is the conclusion.

2   Structure Similarity Problem in Ontology Matching

Usually, an ontology contains concepts, relations, instances and axioms. The ontology matching can be defined as follows:

Definition 1 (*Ontology Matching*). *The matching between two ontologies O_1 and O_2 is a set of quadruples: $\mathcal{M} = \{m_k | m_k = <se_i, te_j, r, s>\}$, where m_k denotes an alignment, se_i and te_j represent the expressions which are composed of elements from O_1 and O_2 respectively; r is the semantic relation between se_i and te_j, and r could be equivalence($=$), generic/specific($\sqsupseteq/\sqsubseteq$), disjoint($\perp$) and overlap ($\sqcap$), etc.; s is the confidence about an alignment and typically in the $[0, 1]$ range.*

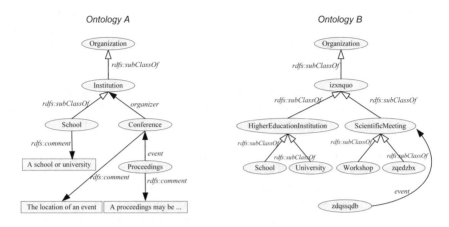

Fig. 1. An ontology matching example

This paper focuses on the matching about *concept − concept* and *relation − relation* with equivalence semantic relation.

Usually if we have enough and regular linguistic information about the elements, the alignments can be discovered easily. But the real world ontologies can not always provide sufficient and regular linguistic information. For example, Fig.1 is a matching snapshot in OAEI benchmark. ontology A has necessary comments for the concepts. The labels and comments in Ontology A are the regular natural words. But in ontology B some concepts have meaningless label and there is no any comments to explain these concepts. In our view, two reasons may cause this phenomenon. Firstly, some ontology engineers do not provide sufficient annotation for each element. Secondly, the ontology engineers would use some particular labels to name the elements. For instance, concept "Address" may be nameds as "Add" , "Adr" or "Dizhi" (in Chinese spelling). Therefore, it is necessary to find a way to discover the alignments for the ontologies without sufficient and regular linguistic information.

When the ontologies lack of linguistic information, matching methods usually utilize ontology structure information to find alignments. Although an ontology can be described as a graph, ontology matching is not equal to a graph matching problem. In graph matching, two elements are matched means they are similar in geometrical view other than they have semantic relation. Moreover, graph matching is a N–P problem [1], so it can not process ontology matching efficiently. For example, when we try to use the graph matching API provided by *SOQA − SimPack* [6] to match ontology graphs, we find it needs more than several days or even several weeks for a normal matching task. Most importantly, only the graph topology can not represent the semantic information in ontology, so the geometrical graph similarity can not imply the corresponding elements are semantically similar. Therefore, it is not suitable to treat ontology matching as a simple graph matching problem.

Currently most structure-based ontology matching methods are inspired by the simple idea: "similar objects are related to similar objects". This idea also derives some heuristic rules, such as "concepts may be similar when their super/sub concepts are similar" and "concepts may be similar when they have similar instances". These rules have been used by some matching system [7, 8]. But when the ontologies have little regular linguistic information, the heuristic rules usually can not work. It is because that the similarity between elements' neighbors can not be determined by linguistic information, the similarity between elements can not be determined too.

A feasible way is similarity propagation, namely, current similarity can propagate to neighbors in the graph to get more similarity results. After each propagation, the similarity results are normalized. The propagation process is terminated until the similarity results is convergent. Based on such similarity propagation idea, researchers have proposed some similarity propagation models [2, 3, 4, 5, 9]. Among these models, similarity flood is the most influential one. This paper will modify the similarity flood to solve the matching problem for the ontologies without sufficient or regular linguistic information.

The similarity flood includes three steps: (1) constructing pairwise connectivity graph; (2) constructing induced propagation graph; (3) computing fixpoint values for matching. Similarity flood is a versatile matching algorithm and can be implemented easily, but it is not sensitive for the initial similarity. Similarity flood algorithm has been used for schema matching in database and XML data. However, similarity flood is not a perfect algorithm. Melnik and his colleagues summarize six disadvantages [2], such as the neighbors have similarity is the necessary precondition of this algorithm. After we try to use similarity flood to match ontologies directly, we also find the algorithm can not work smoothly for ontology matching. First, similarity flood does not consider the similarity between edges, so the edge matching between ontologies can not be determined. Secondly, the maximum pairwise connectivity graph is $N_A * N_B$ (N_A and N_B are the numbers of edges in two ontologies), and it will greatly increase the time complexity for fixpoint computing and space complexity for storing the pairwise connectivity graphs. In real world matching tasks, the ontology graph may be thousands scale, so the corresponding pairwise connectivity graphs would become very large. For the above reasons, the similarity flood algorithm can not be used directly for ontology matching.

3   Similarity Propagation Method with Strong Constraint Condition

Ontology graph consists of the triples like $<s_i, p_i, o_i>$. The propagation condition is the core for a similarity propagation method. In ontology graph matching, a reasonable similarity propagation should consider both vertexes (s_i and o_i) and edges (p_i) in the triples. As far as similarity flood, the propagation condition presumes that all edge pairs (p_x, p_y) have 1.0 similarity value, and the similarity of one vertex pair (s_x, s_y) will be propagated to another vertex pair (o_x, o_y). This propagation condition obviously has the disadvantages: (1) It would produce a large number of alignment candidates and generate a large scale pairwise connectivity graph; (2) The propagation condition would produce many incorrect alignment candidates.

To provide a new similarity propagation method for dealing with ontology matching, this paper proposes a new propagation condition for ontology triples as definition 2, namely, the strong constraint condition for similarity propagation.

Definition 2 (*Strong Constraint Condition for Similarity Propagation in Triples*). *Given two triples $t_i = <s_i, p_i, o_i>$ and $t_j = <s_j, p_j, o_j>$, and let S_s, S_p and S_o denote the corresponding similarities of (s_i, s_j), (p_i, p_j) and (o_i, o_j) for the two triples. The similarity can be propagated iff t_i and t_j satisfy the following three conditions:*

(1) *In S_s, S_p and S_o, at least two similarities must be large than threshold θ;*
(2) *If t_i includes ontology language primitives, the corresponding positions of t_j must be the same primitives;*
(3) *t_i or t_j has at most one ontology language primitive.*

Condition (1) ensures the final similarity result is creditable after propagating. We set $\theta = 0.005$ in the implementation. The ontology language primitives refer to RDF vocabularies and OWL vocabularies. Condition (2) ensures two triples use same ontology language primitive to describe the facts. For example, $<Conference\_Paper, rdfs:subClassOf, Paper>$ and $<Paper, rdfs:subClassOf, Document>$ use the RDF primitive $rdfs:subClassOf$ as predicate, so the similarity can be propagated between them. Condition (3) ensures there is no definition and declaration triples during propagating, because such triples may cause incorrect matching results. For example, two triples $<PhDStu, rdf:type, rdfs:Class>$ and $<Paper, rdf:type, rdfs:Class>$ will cause wrong alignment: $PhDStu = Paper$.

After once propagation, the similarity of one element pair will be increased by the amount of other two pairs. Taking the similarity S_s as an example after i^{th} propagation, its new similarity is:

$$S_s^i = S_s^{i-1} + w_{po} \times S_p^{i-1} \times S_o^{i-1} \tag{1}$$

Analogously, the S_p^i and S_o^i are:

$$S_p^i = S_p^{i-1} + w_{so} \times S_s^{i-1} \times S_o^{i-1} \tag{2}$$

$$S_o^i = S_o^{i-1} + w_{sp} \times S_s^{i-1} \times S_p^{i-1} \tag{3}$$

The w_{po}, w_{so} and w_{sp} are propagation factors, and we will discuss them later.

All similarities will be normalized after each similarity propagation.

Based on the strong constraint condition, the new similarity propagation method still can be divided into three steps as follows:

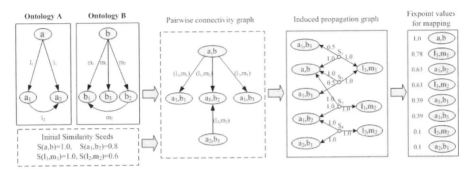

Fig. 2. Similarity propagation method with strong constraint condition

(1) *Constructing pairwise connectivity graph*

Traditional similarity flood is not sensitive to the initial similarity seeds, so all initial similarity values can be set to 1.0. However, in ontology matching, similarity propagation can not use the same setting, because it would not only cause very large pairwise connectivity graph but also generate many wrong alignments. In our view, the quality of the initial similarity seeds is very important for matching ontologies. We believe correct alignments would generate more correct alignments during similarity propagation, but wrong alignments would be noise in similarity propagation. Therefore, this paper try to use some high quality alignments as the initial similarity seed.

The seeds can be calculated by other linguistic-based matching methods or provided manually. This paper uses a method called *semantic description document* to produce the initial seeds. The detail about this linguistic-based matching method can be referred to our other work [10].

According to the initial similarity seed and the strong constraint condition, the pairwise connectivity graph can be constructed as Fig. 2 shows. Obviously, the pairwise connectivity graph is influenced by similarity seeds. Different seeds would cause different pairwise connectivity graphs.

(2) *Constructing propagation graph*

According to formula (1)-(3), the similarity from two element pairs always be propagated to the third pair. The propagation factor measures how many similarity can be propagated. There are three kinds of propagation factor: w_{sp}, w_{so} and w_{po}. Take w_{sp} as an example, it denotes how many similarity come from S_s and S_p can be propagated to S_o. Let f_{sp} denote the number of the triple pairs having $(s_i, s_j) \xrightarrow{(p_i, p_j)} (o_x, o_y)$ style in pairwise connectivity graph, then $w_{sp} = 1/f_{sp}$. w_{so} and w_{po} can be defined and calculated similarly.

Propagation graph can be represented by a bipartite graph as Fig. 2 shows. The weights of edges denotes the propagation factors. In the implementation, for the reason that the propagation factors can be directly obtained according to the pairwise connectivity graph, we can just record the propagation factors but need not to store the propagation graph.

(3) *Computing fixpoint*

The similarity propagation between ontology graphs can be computed iteratively until the final similarity matrix converges. Under the strong constraint condition, the fixpoint can be computed by formula (4), where normalization is omitted for clarity.

Actually, formula (4) is the synthesized style for formula (1)-(3). For each element pair (x, y), it would be subject pair, predicate pair or object pair, so its new similarity in the $(i + 1)^{th}$ propagation would come from four parts: (1) the similarity in i^{th} propagation; (2) the propagating similarity when (x, y) is object pair; (3) the propagating similarity when (x, y) is subject pair; (4) the propagating similarity when (x, y) is predicate pair.

$$
\begin{aligned}
s^{i+1}(x, y) = s^i(x, y) &+ \sum_{\substack{<a_u,p_u,x>\in A \\ <b_u,q_u,y>\in B}} s^i(a_u, b_u) \cdot s^i(p_u, q_u) \cdot w_{sp} \\
&+ \sum_{\substack{<x,p_v,a_v>\in A \\ <y,q_v,b_v>\in B}} s^i(a_v, b_v) \cdot s^i(p_v, q_v) \cdot w_{po} \\
&+ \sum_{\substack{<a_t,x,c_t>\in A \\ <b_t,y,d_t>\in B}} s^i(a_t, b_t) \cdot s^i(c_t, d_t) \cdot w_{so}
\end{aligned} \tag{4}
$$

4   Propagation Strategies

To improve the efficiency of similarity propagation and the quality of propagation results, some reasonable strategies are adopted in the propagation. In general, this paper uses six strategies to make the matching results better and to improve the matching efficiency.

I. Propagation Scale Strategy

We should select the right parts in ontologies for similarity propagation. This paper studies four propagation scale strategies as follows:

(1) *Full graph propagation*

In similarity propagation, full ontology graph is the most direct propagation scale. It can assure that there is no ontology information to be missed. However, this propagation scale strategy also has obvious disadvantages: (a) For the large scale ontology graph, it is possible to cause large pairwise connectivity graph. (b) More triples do not mean better propagation results. Some triples are not important for describing the semantics. So too many triples may increase the uncertainty in propagation and bring negative affection for matching results.

(2) *Independent semantic subgraph propagation*

In an ontology, a semantic subgraph of an element is used to describe the element's meaning precisely. The definition and extracting algorithm of the semantic subgraph can be found in our other work [10]. If we constrain the propagation scale in the semantic subgraphs, the propagation can avoid the triples without important semantics. Therefore, the similarity propagation result would be determined by the semantic subgraphs. Given two elements a and b and the corresponding semantic subgraphs G_a^s and G_b^s, the similarity $S(a,b)$ is obtained by the similarity propagating between G_a^s and G_b^s. If two ontologies have n and m elements respectively, this strategy needs $n \times m$ times propagation.

(3) Combined semantic subgraph propagation
The semantic subgraphs are combined in this strategy. We implement two combining ways: (a) combine all semantic subgraphs; (b) combine all semantic subgraphs of concepts to a graph G_C^c, and then combine semantic subgraphs of relations to another graph G_R^c. When we match concepts, we just consider G_C^c. Similarly, the matching between relations just uses G_R^c.

(4) Hybrid semantic subgraph propagation
This strategy is a mix of strategy (2) and strategy (3). In the propagation, one side is a semantic subgraph of an element e, and another side is the combined graph G_C^c or G_R^c. After a propagation, we can get the similarity about e to all the elements in another ontology. Obviously, this strategy needs n times propagation.

II. Incremental Updating for Pairwise Connectivity Graph
The strong constraint condition greatly reduces the scale of pairwise connectivity graph. After once similarity propagation, the similarity matrix would change and new similarity values would appear. Therefore, we need to construct a new pairwise connectivity graph for the next propagation. It is a time consuming process.

To reduce the constructing cost, we adopt an incremental updating way. After a propagation, the new pairwise connectivity graph need not to be reconstructed, but it can be extended based on the previous one. Namely, we just update the parts in the pairwise connectivity graph whose similarities have been changed.

III. Trust the Credible Seeds
In the initial similarity seeds, we regards the alignments having high similarity value as right alignments. Therefore, we keep these alignments during the propagation. The first advantage of this strategy is assuring some correct alignment can not be changed. Another advantage is avoiding some unnecessary similarity propagation computing. If $S(a_i, b_j)$ is a credible seed, then all similarity propagation like $S(a_i, b_x)$ and $S(a_y, b_j)$ can be skipped. In short, credible seeds not only can reduce the propagation cost, but also decrease the negative affection in propagation.

IV. Cross Validation for Propagation Result
This strategy only works for hybrid semantic subgraph propagation scale strategy. Given an element a_i, we can get a set of similarities $\{S(a_i, b_x)\}(x = 1, ..., n)$. Given another element b_j in the opponent ontology, we also can get another similarity set $\{S(a_y, b_j)\}(y = 1, ..., m)$. Therefore, we will have two similarity matrices. The similarity value at $S(a_i, b_j)$ may be different. This paper calculates the average of two similarity matrices as the final propagation result. The two similarity matrices have the function of validating crossly, so it can improve the quality of propagation result.

V. Penalty for Propagation

For an ideal similarity matrix, correct alignments should have higher confidence values and incorrect alignments should have lower confidence values. The real world similarity matrix is far away from the perfect one. So it is necessary to penalize the propagation result. The penalty should make little influence for alignments having high confidence value and make the potential misalignments have lower confidence value.

We provide two penalty factors p_a and p_b as follows:

$$p_a = \frac{s(a_i, b_j)}{max(s_{max}(a_i, b_x), s_{max}(a_y, b_j))} \tag{5}$$

$$p_b = \frac{1}{1 + e^{-\alpha t}}, t = (\frac{N+1}{n_i + 1}/log(N+1)), \alpha \geq 1 \tag{6}$$

N is the sum of columns and rows of similarity matrix; n_i is the number of the alignments whose confidence values are large than 0 in i^{th} column and j^{th} row. After being penalized, the new similarity value is:

$$S'(a_i, b_j) = S(a_i, b_j) \cdot p_a \cdot p_b \tag{7}$$

p_a penalizes the alignments having low similarity values, and p_b penalizes the alignments whose column and row have too many alignments with $S(x, y) > 0$. We set $\alpha = 3$ in the implementation.

VI. Termination Condition

Our propagation should satisfy two termination conditions: (1) The cosine between two sequential similarity matrices is not bigger than the given threshold. Propagation should assure the final similarity matrix is convergent. Melnik and his colleagues have proved that fixpoint computing can be convergent if the pairwise connectivity graph is a strongly connected graph [2]. (2) There is no updating for the pairwise connectivity graph. Besides the two conditions, to avoid the matrix needs too many times propagation to converge, we also set the maximum propagation times as 8 in the implementation.

5   Experimental Evaluation

We have implemented the new similarity propagation method in ontology mapping system Lily. Lily is implemented by Java and C++. More information about Lily can be found at http://ontomappinglab.googlepages.com/lily.htm.

The dataset is OAEI benchmark[2]. It includes more than 50 matching tasks having non-sequential number from 101 to 304. According to the dataset feature, we divide the dataset into 5 groups: (1) 101-104: this group contains same, irrelevant, language generalized and restricted ontologies. (2) 201-210: the ontology structure is preserved, but the labels and identifiers are replaced

---

[2] http://oaei.ontologymatching.org

by random names, misspellings, synonyms and foreign names. The comments have been suppressed in some cases. (3) 221-247: This group can be divided into two subgroups: 221-231 and 232-247. The first subgroup contains 11 kinds of modifications, such as the hierarchy is flattened or expanded, and individuals, restrictions and data types are suppressed. In the second subgroup, the modifications are the combinations of the ones used in 221-231. (4) 248-266: This is the most difficult test set. All labels and identifiers are replaced by random names, and the comments are also suppressed. (5) 301-304: This group contains 4 real matching tasks.

This paper uses the classical criterion: precision, recall and F-measure to evaluate the matching results. Let Q is the real matching result and T is the reference result, then the precision, recall and F-measure are:

$$P = \frac{|Q \cap T|}{|Q|}, R = \frac{|Q \cap T|}{|T|}, F\text{-}measure = \frac{2PR}{P+R} \qquad (8)$$

5.1 Evaluating the Propagation Scale Strategies

This experiment aims to compare different propagation scale strategies. The dataset is 248 task. The experimental result (F-measure) is showed in Table 1. *Size* is the semantic subgraph size. *Seed* denotes the initial similarity seeds obtained by the linguistic matching method. C1, C2, C3 and C4 represent the four kinds of propagation scale strategies. Notice that C3A denotes all semantic subgraphs are combined; C3B denotes concept semantic subgraphs and relation semantic subgraphs are combined independently. C4A and C4B are both hybrid semantic subgraph propagation scale strategy, but in C4A the ontology has been enriched. The last row of Table 1 provides the average values for the semantic subgraphs from size 5 to 35.

Comparing with the seeds, Table 1 shows the similarity propagation can improve the quality of matching results. For all propagation scale strategies, their matching qualities can rank as: $C4B > C4A > C3B > C3A > C1 > C2$. Through

Table 1. Comparison of different propagation scale strategies

Size	Seed	C1	C2	C3A	C3B	C4A	C4B
0	0.020	0.020	0.020	0.020	0.020	0.020	0.020
1	0.371	0.603	0.604	0.422	0.246	0.537	0.496
2	0.431	0.653	0.604	0.547	0.352	0.570	0.552
3	0.418	0.531	0.476	0.541	0.400	0.715	0.736
5	0.493	0.608	0.529	0.658	0.607	0.761	0.802
10	0.536	0.587	0.592	0.675	0.693	0.828	0.828
15	0.586	0.643	0.586	0.662	0.658	0.849	0.837
20	0.557	0.610	0.630	0.671	0.675	0.822	0.785
25	0.561	0.629	0.598	0.662	0.731	0.789	0.832
30	0.561	0.648	0.690	0.658	0.706	0.753	0.879
35	0.561	0.620	0.651	0.621	0.653	0.716	0.826
Avg.	0.531	0.621	0.611	0.658	0.675	0.788	0.827

analyzing the experimental data, we have the conclusions: (1) Full graph strategy does not produce good results as we expect. (2) Independent semantic subgraph strategy causes the worst results. The reason is that the misalignments would have high similarity values when the similarity matrix is normalized after the propagation. So it is difficult to determine the correct alignments. (3) The results of C3A and C3B are very close; (4) C4A and C4B produce the best results. Surprisingly, the C4B without ontology enrich preprocess performs well than C4A with original ontology. The fact implies that some ontology preprocess may cause negative affection for similarity propagation.

5.2 Initial Similarity Seeds

We need validate whether and how our similarity propagation method is sensitive to the initial similarity seeds. In this experiment, we modified the seeds manually to keep the seeds always have the feature: $Precision = Recall = F-measure$. The seed quality F-measure decreases from 1.0 to 0 with step 0.1. The dataset is 248 task too. For an experiment at $F-measure=x$, we execute the matching three times. In each time, we modifies the seed randomly. We treat the average of the three results as the final $F-measure$ value.

The experimental result is showed in Fig. 3, where line B denotes the seed quality; C1, C3A and C4B are F-measure lines for the corresponding propagation scale strategies. We can draw the conclusions: (1) The initial seed influences the matching result greatly. With the change of seed quality, the matching result quality changes monotonously. (2) After propagating, the result is usually better than the initial seed. (3) C4B scale strategy is influenced by the seed slightly, so C4B is the preferred propagation scale strategy in the implementation.

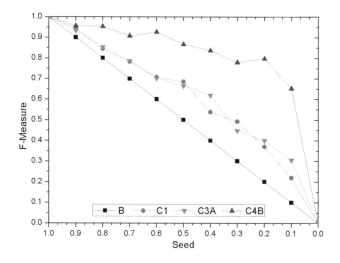

Fig. 3. Influence of initial seed to matching results

5.3   Overall Matching Results

Fig. 4 compares the matching results with similarity propagation and the results without similarity propagation. P1, R1 and F1 denote the precision, recall and F-measure for the method without similarity propagation. P2, R2 and F2 are the quality criterions about the method in this paper.

According to Fig. 4: (1) The similarity propagation method proposed in this paper improves the matching result quality, especially for the 248-266 dataset. (2) When the ontologies have not sufficient or regular linguistic information, we also find that our similarity propagation method can increase the recall greatly. Therefore, the similarity propagation method can discover more matching results using limited linguistic information.

Ontology mapping system Lily has implemented the similarity propagation methods in this paper, and Lily is one of the best systems in the OAEI benchmark evaluation in recent years. Table 2 shows results of some matching systems in OAEI-2008 benchmark evaluation [11, 12]. The evaluation divides the dataset

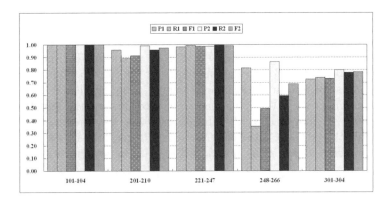

Fig. 4. Matching without similarity propagation VS with similarity propagation

Table 2. Overall matching results of some systems (OAEI 2008)

Systems	ASMOV		DSSim		Anchor-Flood		RiMOM		AROMA	
	Prec.	Rec.	Prec.	Rec.	Prec.	Rec.	Prec.	Rec.	Prec.	Rec.
1xx	1.00	1.00	1.00	1.00	1.00	1.00	1.00	1.00	1.00	1.00
2xx	0.95	0.85	0.97	0.64	0.98	0.59	0.96	0.82	0.96	0.70
3xx	0.81	0.77	0.90	0.71	0.95	0.31	0.80	0.81	0.82	0.71
H-mean	0.95	0.86	0.97	0.67	0.98	0.62	0.96	0.84	0.95	0.70
Systems	CIDER		GeRoMe		SPIDER		SAMBO		Lily	
	Prec.	Rec.	Prec.	Rec.	Prec.	Rec.	Prec.	Rec.	Prec.	Rec.
1xx	0.99	0.99	0.96	0.79	0.99	0.99	1.00	0.98	1.00	1.00
2xx	0.97	0.57	0.56	0.52	0.97	0.57	0.98	0.54	0.97	0.86
3xx	0.90	0.75	0.61	0.40	0.15	0.81	0.95	0.80	0.87	0.81
H-mean	0.97	0.62	0.60	0.58	0.81	0.63	0.99	0.58	0.97	0.88

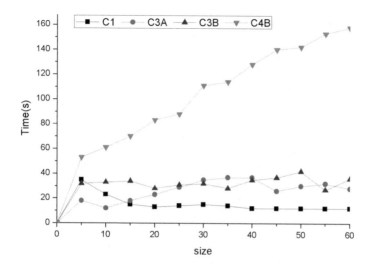

Fig. 5. Performance under different propagation scale strategies

into three groups: 1XX, 2XX and 3XX. The evaluation results show our system performs well on the benchmark dataset.

5.4 Performance

In Lily, we find the similarity propagation process occupies about 30%-50% matching time. It is necessary to analyze the major performance factors in propagation. In practical matching tasks, small semantic subgraphs would cause small pairwise connectivity graph, and the iteration process for calculating fixpoint would also terminate quickly. On the contrary, big semantic subgraphs would increase the burden in the propagation. So we believe the semantic subgraph size is a key factor for the performance. Fig. 5 demonstrates the running time with various subgraph size under different propagation scale strategies.

According to Fig. 5, (1) For full graph or combined subgraph scale strategies, the running time of propagation has no direct relevance to the semantic subgraph size. So C1, C3A and C3B are almost steady. When the semantic subgraph size is large than 5, the size of pairwise connectivity graph will keep stable. So the running time of C1, C3A and C3B would have little correlation with semantic subgraph size. (2) But for C4 propagation scale strategy, the running time increases with the semantic subgraph size quickly. It means we should set suitable semantic subgraph size for hybrid semantic subgraph scale strategy.

6   Related Work

Many ontology matching approaches are proposed in recent years. Some researchers have gave several excellent reviews for this topic [13, 14, 15]. This paper mainly focuses on the matching problem for the ontologies without sufficient or regular linguistic information. Structure-based matching method is a feasible way to deal with this special matching situation. For the reason that the ontology graph topology can not represent the semantics reasonably, the way using classical graph matching algorithm can not obtain good alignments, and it also have the serious performance problem due to high time complex of graph matching. Therefore, the method based on similarity propagation idea is the feasible way to solve the problem.

Blondel and his colleagues proposed a iteration equation for measuring the similarity between directed graphs [4]. This method is based on the *Hub − Authority* idea. Based on the similar idea, in [9] the authors proposed a more universal measurement for the vertex similarity in network, and they also pointed out that Blondel's method is a special case of their method. These graph matching algorithms can only calculate the vertex similarity in graph, but they can not deal with the edge similarity. To overcome the problem, the bipartite graph is used to represent the ontology graph [5]. Tous and Delgado represent the ontology graph as the vector space method [16]. Similarity flood [2] is the most popular algorithm inspired by the similarity propagation, but it can not be directly used for ontology matching. In ontology mapping system RiMOM, three propagation strategies are used for structure matching: (1) propagation between concepts; (2) propagation between relations; (3) propagation between concepts and relations [17]. In another matching system PROMPT [8], a similarity propagation algorithm called AnchorPROMPT is used to find alignment.

The structure-based matching method based on our similarity propagation method is an effective solution for the ontologies without sufficient or regular linguistic information. Especially, our propagation method is reasonable for the ontology. We also propose some strategies to improve the propagation results and accelerate the matching process such as the propagation scale is constrained in the semantic subgraphs.

7 Conclusion

This paper proposes an effective similarity propagation method for matching the ontologies without sufficient or regular linguistic information. The new method is based on the strong constrained condition, and it is reasonable for ontology model. Some useful propagation strategies are also adopted to improve the matching results. Experiments shows that this structure-based matching method performs well on OAEI benchmark dataset.

References

1. Conte, D., Foggia, P., Sansone, C., Vento, M.: Thirty years of graph matching in pattern recognition. International Journal of Pattern Recognition and Articial Intelligence 18(3), 265–298 (2004)

2. Melnik, S., Garcia-Molina, H., Rahm, E.: Similarity flooding: A versatile graph matching algorithm and its application to schema matching. In: Proceeding of the 18th International Conference on Data Engineering (ICDE), San Jose, CA (2002)
3. Jeh, G., Widom, J.: Simrank: A measure of structural-context similarity. In: Proceedings of the Eighth ACM SIGKDD International Conference on Knowledge Discovery and Data Mining, Edmonton, Alberta, Canada (2002)
4. Blondel, V.D., Gajardo, A., Heymans, M., Senellart, P., Van Dooren, P.: A measure of similarity between graph vertices: Applications to synonym extraction and web searching. SIAM Review 46(4), 647–666 (2004)
5. Hu, W., Jian, N., Qu, Y., Wang, Y.: Gmo: A graph matching for ontologies. In: Integrating Ontologies Workshop, Banff, Alberta, Canada (2005)
6. Ziegler, P., Kiefer, C., Sturm, C., Dittrich, K.R., Bernstein, A.: Generic similarity detection in ontologies with the soqa-simpack toolkit. In: Proceedings of the ACM SIGMOD International Conference on Management of Data (SIGMOD 2006), Chicago, Illinois, USA (2006)
7. Ehrig, M., Staab, S.: QOM – quick ontology mapping. In: McIlraith, S.A., Plexousakis, D., van Harmelen, F. (eds.) ISWC 2004. LNCS, vol. 3298, pp. 683–697. Springer, Heidelberg (2004)
8. Noy, N.F., Musen, M.A.: The prompt suite: Interactive tools for ontology merging and mapping. International Journal of Human-Computer Studies 59(6), 983–1024 (2003)
9. Leicht, E.A., Holme, P., Newman, M.E.J.: Vertex similarity in networks. Physical Review E 73 (2006)
10. Wang, P.: Research on the Key Issues in Ontology Mapping. PhD thesis, Southeast University, China (2009)
11. Caracciolo, C., Euzenat, J., Hollink, L., Ichise, R., et al.: Results of the ontology alignment evaluation initiative 2008. In: The Third International Workshop on Ontology Matching (OM 2008), Karlsruhe, Germany (2008)
12. Wang, P., Xu, B.: Lily: Ontology alignment results for oaei 2008. In: The Third International Workshop on Ontology Matching, OM 2008 (2008)
13. Kalfoglou, Y., Schorlemmer, M.: Ontology mapping: the state of the art. The Knowledge Engineering Review 18(1), 1–31 (2003)
14. Shvaiko, P., Euzenat, J.: A survey of schema-based matching approaches. In: Spaccapietra, S. (ed.) Journal on Data Semantics IV. LNCS, vol. 3730, pp. 146–171. Springer, Heidelberg (2005)
15. Rahm, E., Bernstein, P.A.: A survey of approaches to automatic schema matching. The VLDB Journal 10, 334–350 (2001)
16. Tous, R., Delgado, J.: A vector space model for semantic similarity calculation and OWL ontology alignment. In: Bressan, S., Küng, J., Wagner, R. (eds.) DEXA 2006. LNCS, vol. 4080, pp. 307–316. Springer, Heidelberg (2006)
17. Li, J., Tang, J., Li, Y., Luo, Q.: Rimom: A dynamic multistrategy ontology alignment framework. IEEE Transactions on Knowledge and Data Engineering 21(8), 1218–1232 (2009)

Deciding Query Entailment for Fuzzy $\mathcal{SHIN}$ Ontologies

Jingwei Cheng, Z.M. Ma, Fu Zhang, and Xing Wang

Northeastern University, Shenyang, 110004, China
cjingwei@gmail.com, mazongmin@ise.neu.edu.cn

Abstract. Significant research efforts in the Semantic Web community are recently directed toward the representation and reasoning with fuzzy ontologies. As the theoretical counterpart of fuzzy ontology languages, fuzzy Description Logics (DLs) have attracted a wide range of concerns. With the emergence of a great number of large-scale domain ontologies, the basic reasoning services cannot meet the need of dealing with complex queries (mainly conjunctive queries), which are indispensable in data-intensive applications. Conjunctive queries (CQs), originated from relational databases, play an important role as an expressive reasoning service for ontologies. Since, however, the negation of a role atom in a CQ is not expressible as a part of a knowledge base, existing tableau algorithms cannot be used directly to deal with the issue. In this paper, we thus present a tableau-based algorithm for deciding query entailment of fuzzy conjunctive queries w.r.t. fuzzy $\mathcal{SHIN}$ ontologies. Moreover, the data complexity problem was still open for answering CQs in expressive fuzzy DLs. We tackle this issue by proving a tight CONP upper bound for the problem in f-$\mathcal{SHIN}$, as long as only simple roles occur in the query. Regarding combined complexity, we prove that the algorithm for query entailment is CO3NEXPTIME in the size of the knowledge base and the query.

1 Introduction

In order to achieve reusability and a high level of interoperability of knowledge, ontologies are commonly used to express domain knowledge in the context of the Semantic Web. A key component of the Semantic Web is thus the representation and reasoning of ontologies. Description logics (DLs, for short) [1] are the logical foundation of the Semantic Web, which support knowledge representation and reasoning by means of the concepts and roles. As the logic underpinnings of Web Ontology Languages (OWLs)[1], DLs have attracted much more attentions due to their inherently built reasoning services.

In the real world, there exists a great deal of uncertainty and imprecision which is likely the rule than an exception. Based on Zadeh's fuzzy set theory[2], there have been substantial amounts of work carried out in the context of fuzzy DLs [3][4], and fuzzy ontology knowledge bases [5] are thus established.

---

[1] http://www.w3.org/submission/owl11-overview/

A. Gómez-Pérez, Y. Yu, and Y. Ding (Eds.): ASWC 2009, LNCS 5926, pp. 120–134, 2009.

Conjunctive queries originated from research in relational databases, and, more recently, have also been identified as a desirable form of querying DL knowledge bases. Conjunctive queries provide an expressive query language with capabilities that go beyond standard instance retrieval. There are close ties among conjunctive query answering, conjunctive query entailment and conjunctive query containment in the sense that they can be transformed into one another. The first conjunctive query algorithm [6] over DLs was actually specified for the purpose of deciding conjunctive query containment for $\mathcal{DLR}_{reg}$. Recently, query entailment and answering have been extensively studied for tractable DLs, i.e., DLs that have reasoning problems of at most polynomial complexity. For example, the constructors provided by DL-Lite family [7] are elaborately chosen such that the standard reasoning tasks are PTIME-COMPLETE and query entailment is in LOGSPACE with respect to data complexity. Moreover, in DL-Lite family, as TBox reasoning can usually be done independently of the ABox, ABox storage can be transformed into database storage, thus knowledge base users can achieve efficient queries by means of well-established DBMS query engines. Another tractable DL comes from $\mathcal{EL}$ with PTIME-COMPLETE reasoning complexity. It was shown that union of conjunctive queries (UCQs) entailment in $\mathcal{EL}$ and in its extensions with role hierarchies is NP-complete regarding the combined complexity [8]. The data complexity of UCQ entailment in $\mathcal{EL}$ is PTIME-COMPLETE [9]. Allowing, additionally, role composition in the logic as in $\mathcal{EL}^{++}$, leads to undecidability [10]. Query answering algorithms for expressive DLs are being tracked with equal intensity. CARIN system [11], the first framework for combining a description logic knowledge base with rules, provided a decision procedure for conjunctive query entailment in the description logic $\mathcal{ALCNR}$, where $\mathcal{R}$ stands for role conjunction. The conjunctive query entailment algorithms for more expressive DLs, e.g., $\mathcal{SHIQ}$ and $\mathcal{SHOQ}$, are presented in [12][13].

When querying over fuzzy DL KBs, as in the crisp case, same difficulties emerged in that existing fuzzy DL reasoners, such as fuzzyDL[2] and FiRE[3], are not capable of dealing with CQs either. Some work has been done in a relative narrow range, mainly focused on lightweight fuzzy ontology languages, e.g. [14] and [15] on f-DL-Lite, [16] on f-$\mathcal{ALC}$ and [17] on f-$\mathcal{ALCN}$. In [18], A fuzzy extension of CARIN is provided, along with an fuzzy version of existential entailment algorithm for answering conjunctive queries.

In this paper, we extend the results obtained in [18] for fuzzy $\mathcal{ALCNR}$ and in [19] for crisp $\mathcal{SHIN}$ to fuzzy $\mathcal{SHIN}$. This paper makes the following major contributions:

- It presents a tableau-based algorithm for deciding query entailment over f-$\mathcal{SHIN}$ KBs.
- It provides complexity upper bounds w.r.t both combined complexity and data complexity.

The remainder of this paper is organized as follows. Section 2 briefly reviews the necessary background knowledge of fuzzy sets, fuzzy logics, the syntax and

---

[2] http://gaia.isti.cnr.it/~ straccia/software/fuzzyDL/fuzzyDL.html
[3] http://www.image.ece.ntua.gr/~nsimou

semantics of $f\text{-}\mathcal{SHIN}$, and the formal definition of a fuzzy query language. Section 3 presents the conjunctive query entailment algorithm for $f\text{-}\mathcal{SHIN}$ along with complexity analysis. Section 4 concludes this paper.

For the lack of the space, we omit most of the proofs, which can be found in a accompanying technical report [20].

2   Preliminaries

2.1   Fuzzy Set and Fuzzy Logic

In classical set theory, an element in a set either belongs or0 does not belong to the set. By contrast, in fuzzy set theory, an element belongs to a set with certain degree, which is described with the aid of a membership function valued in the real unit interval $[0,1]$.

A fuzzy set A with regard to a universe U is characterized by a membership function $\mu_A : U \to [0,1]$ (or simply $A(x) \in [0,1]$), which assigns a membership degree to each element u in U, denoted by $\mu_A(u)$. $\mu_A(u)$ gives us an assessment of the degree that u belongs to A. Typically, if $\mu_A(u) = 1$ then u definitely belongs to A, whereas $\mu_A(u) \geq 0.8$ means that u is "likely" to be an element of A. In addition, when using Gödel T-norm, T-conorm and Lukasiewicz negation for interpreting conjunctions, disjunctions and complements respectively, we have: for all $u \in U$ and for all fuzzy sets A_1, A_2 with respect to U, $\mu_{A_1 \cap A_2}(u) = \min\{\mu_{A_1}(u), \mu_{A_2}(u)\}$, $\mu_{A_1 \cup A_2}(u) = \max\{\mu_{A_1}(u), \mu_{A_2}(u)\}$, and $\mu_{\bar{A}}(u) = 1 - \mu_A(u)$, where $\bar{A}$ is the complement of A in U.

As for fuzzy logics, the degree of membership $\mu_A(u)$ of an element $u \in U$ w.r.t. the fuzzy set A over U is regarded as the truth-value of the statement "u is in A". Accordingly, in fuzzy DL, (i) a concept C, rather than being interpreted as a classical set, will be interpreted as a fuzzy set and, thus, concepts become imprecise; and, consequently, (ii) the statement "o is in C", i.e. $o : C$, will have a truth-value in $[0,1]$ given by the degree of membership of being the individual o a member of the fuzzy set C.

2.2   Fuzzy $\mathcal{SHIN}$

We introduce the basic terms and notations used throughout this paper. In particular, we introduce the syntax and semantics for fuzzy DL $f\text{-}\mathcal{SHIN}$ [4].

Let N_C, N_R, and N_I be countable infinite and pairwise disjoint sets of concept, role and individual names, respectively. We assume that the set of role names N_R can be divided into two disjoint subsets, N_{tR} and N_{nR}, where the former stands for the subset of *transitive role names*, and the latter stands for the subset of *non-transitive role names*. This assumption can be written as $N_R = N_{tR} \cup N_{nR}$ and $N_{tR} \cap N_{nR} = \emptyset$.

$f\text{-}\mathcal{SHIN}$ roles (or roles for short) are defined as $R ::= R_N | R^-$, where $R_N \in N_R$, R^- is called the *inverse role* of R.

A role inclusion axiom is of the form $R \sqsubseteq S$, with R, S roles. A role hierarchy (also called a RBox) $\mathcal{R}$ is a finite set of role inclusion axioms.

For the sake of brevity and clarity, we use following notations:

1. To avoid using verbose role expressions of the form R^{--} or even R^{---}, we use an abbreviation $\mathsf{Inv}(R)$ to denote inverse role of R, i.e. $\mathsf{Inv}(R) = R^-$ if $R \in N_R$, and $\mathsf{Inv}(R) = S$ if $R = S^-$ with $S \in N_R$.
2. For a RBox $\mathcal{R}$, we define $\sqsubseteq^*_\mathcal{R}$ as the reflexive transitive closure of $\sqsubseteq$ over $\mathcal{R} \cup \{\mathsf{Inv}(R) \sqsubseteq \mathsf{Inv}(S) | R \sqsubseteq S \in \mathcal{R}\}$. We use $R \equiv^*_\mathcal{R} S$ as an abbreviation for $R \sqsubseteq^*_\mathcal{R} S$ and $S \sqsubseteq^*_\mathcal{R} R$.
3. For a RBox $\mathcal{R}$ and a role S, we define the set $\mathsf{Trans}_\mathcal{R}$ of transitive roles as $\{S|$ there is a role R with $R \equiv^*_\mathcal{R} S$ and $R \in N_{tR}$ or $\mathsf{Inv}(R) \in N_{tR}\}$.
4. A role S is called simple w.r.t. a RBox $\mathcal{R}$ if, for each role R such that $R \sqsubseteq^*_\mathcal{R} S$, $R \notin \mathsf{Trans}_\mathcal{R}$.

The subscript $\mathcal{R}$ of $\sqsubseteq^*_\mathcal{R}$ and $\mathsf{Trans}_\mathcal{R}$ is dropped if clear from the context.

f-$\mathcal{SHIN}$ concepts (or concepts for short) are formed out of concept names according to the following abstract syntax, where $A \in N_C, R \in N_R, p \in \mathbb{N}$, and S a simple role:

$$C, D ::= \top | \bot | A | \neg A | C \sqcap D | C \sqcup D | \forall R.C | \exists R.C | \leq pS | \geq pS$$

A TBox is a finite set of *concept definition axioms* of the form $A \equiv D$ and *general concept inclusion* axioms (GCIs) of the form $C \sqsubseteq D$.

An ABox consists of fuzzy assertions of the form $C(o) \bowtie n$, $R(o, o') \rhd n$, or $o \not\approx o'$, where $o, o' \in N_I$, $\bowtie$ stands for any type of inequality, i.e., $\bowtie \in \{\geq, >, \leq, <\}$. We use $\rhd$ to denote $\geq$ or $>$, and $\lhd$ to denote $\leq$ or $<$. We call ABox assertions defined by $\rhd$ *positive assertions*, while those defined by $\lhd$ *negative assertions*. Note that, we consider only positive fuzzy role assertions, since negative role assertions would imply the existence of role negation, which would lead to undecidability [18].

An f-$\mathcal{SHIN}$ knowledge base (KB) $\mathcal{K}$ is a triple $(\mathcal{T}, \mathcal{R}, \mathcal{A})$ with $\mathcal{T}$ a TBox, $\mathcal{R}$ a RBox and $\mathcal{A}$ an ABox.

Let C be a fuzzy concept, $\mathcal{R}$ a RBox, we denote by $sub(C, \mathcal{R})$ the set of subconcepts of C. We define $sub(\mathcal{A}, \mathcal{R}) = \bigcup_{C(o) \bowtie n \in \mathcal{A}} sub(C, \mathcal{R})$. $\mathcal{R}$ is dropped if clear from the context.

The semantics of f-$\mathcal{SHIN}$ are provided by a fuzzy interpretation, which is a pair $\mathcal{I} = (\Delta^\mathcal{I}, \cdot^\mathcal{I})$. Here $\Delta^\mathcal{I}$ is a non-empty set of objects, called the domain of interpretation, and $\cdot^\mathcal{I}$ is an interpretation function which maps different individual names into different elements in $\Delta^\mathcal{I}$, concept name A into membership function $A^\mathcal{I} : \Delta^\mathcal{I} \to [0,1]$, role R into membership function $R^\mathcal{I} : \Delta^\mathcal{I} \times \Delta^\mathcal{I} \to [0,1]$. The semantics of f-$\mathcal{SHIN}$ concepts and roles are depicted as follows.

- $\top^\mathcal{I}(o) = 1$
- $\bot^\mathcal{I}(o) = 0$
- $(C \sqcap D)^\mathcal{I}(o) = \min\{C^\mathcal{I}(o), D^\mathcal{I}(o)\}$
- $(C \sqcup D)^\mathcal{I}(o) = \max\{C^\mathcal{I}(o), D^\mathcal{I}(o)\}$
- $(\neg C)^\mathcal{I}(o) = 1 - C^\mathcal{I}(o)$
- $(\forall R.C)^\mathcal{I}(o) = \inf_{o' \in \Delta^\mathcal{I}}\{\max\{1 - R^\mathcal{I}(o, o'), C^\mathcal{I}(o')\}\}$
- $(\exists R.C)^\mathcal{I}(o) = \sup_{o' \in \Delta^\mathcal{I}}\{\min\{R^\mathcal{I}(o, o'), C^\mathcal{I}(o')\}\}$

- $(\geq pS)^{\mathcal{I}}(o) = \sup_{o_1,\dots,o_p \in \Delta^{\mathcal{I}}} \{\min_{i=1}^{p} \{R^{\mathcal{I}}(o, o_i)\}\}$
- $(\leq pS)^{\mathcal{I}}(o) = \inf_{o_1,\dots,o_{p+1} \in \Delta^{\mathcal{I}}} \max_{i=1}^{p+1} \{1 - R^{\mathcal{I}}(o, o_i)\}$
- $\mathsf{Inv}(R)^{\mathcal{I}}(o, o') = R^{\mathcal{I}}(o', o)$

Given an interpretation $\mathcal{I}$ and an inclusion axiom $C \sqsubseteq D$, $\mathcal{I}$ is *a model of* $C \sqsubseteq D$, if $C^{\mathcal{I}}(o) \leq D^{\mathcal{I}}(o)$ for any $o \in \Delta^{\mathcal{I}}$, written as $\mathcal{I} \models C \sqsubseteq D$. Similarly, for ABox assertions, $\mathcal{I} \models B(o) \bowtie n$ (resp. $\mathcal{I} \models R(o, o') \bowtie n$), iff $B^{\mathcal{I}}(o^{\mathcal{I}}) \bowtie n$ (resp. $R^{\mathcal{I}}(o^{\mathcal{I}}, o'^{\mathcal{I}}) \bowtie n$), and $\mathcal{I} \models o \not\approx o'$ iff $o^{\mathcal{I}} \neq o'^{\mathcal{I}}$. As for RBox, $\mathcal{I} \models R \sqsubseteq S$ iff $\forall \langle o, o' \rangle \in \Delta^{\mathcal{I}} \times \Delta^{\mathcal{I}}$, $R^{\mathcal{I}} \langle o, o' \rangle \leq S^{\mathcal{I}} \langle o, o' \rangle$, and $\mathcal{I} \models \mathsf{Trans}(R)$, iff $\forall o, o', o'' \in \Delta^{\mathcal{I}}$, $R^{\mathcal{I}}(o, o'') \geq \sup_{o' \in \Delta^{\mathcal{I}}} \{\min(R^{\mathcal{I}}(o, o'), R^{\mathcal{I}}(o', o''))\}$. Note that the semantics of transitive roles result from the definition of *sup-min transitive* relation in fuzzy set theory.

If an interpretation $\mathcal{I}$ is a model of all the axioms and assertions in a KB $\mathcal{K}$, we call it *a model of* $\mathcal{K}$. A KB is *satisfiable* iff it has at least one model. A KB $\mathcal{K}$ *entails* (logically implies) a fuzzy assertion φ, written as $\mathcal{K} \models \varphi$, iff all the models of $\mathcal{K}$ are also models of φ.

Given a KB $\mathcal{K}$, we can w.l.o.g assume that the following conditions hold.

1. All concepts are in their negative normal forms (NNFs), i.e. negation occurs only in front of concept names. Through de Morgan law, the duality between existential restriction ($\exists R.C$) and universal restriction ($\forall R.C$), and the duality between atmost restriction ($\leq pS$) and atleast restriction ($\geq pS$), each concept can be transformed into its equivalent NNF by pushing negation inwards.

2. All fuzzy concept assertions are in their positive inequality normal forms (PINFs). A negative concept assertion can be transformed into its equivalent PINF by applying fuzzy complement operation on it. For example, $C(o) < n$ is converted to $\neg C(o) > 1 - n$.

3. All fuzzy assertions are in their normalized forms (NFs). As is shown in [21], by introducing a positive, infinite small value ϵ, a fuzzy assertion of the form $C(o) > n$ can be normalized to $C(o) \geq n + \epsilon$. The model equivalence of KB $\mathcal{K}$ and $\mathcal{K}$'s normalized form is also proved in [21].

4. There are only fuzzy GCIs in the TBox. A fuzzy concept definition axiom $A \equiv D$ can be eliminated by replacing every occurrence of A with D. The elimination is also known as knowledge base expansion. Note that the size of the expansion can be exponential in the size of the TBox. But if we follow the principle of "Expansion is done on demand" [22], the expansion will have no impact on the algorithm complexity of deciding fuzzy query entailment.

2.3   Fuzzy Conjunctive Queries

In general, existing fuzzy DL reasoners can provide most of the basic fuzzy inference services [3], such as checking of fuzzy concept satisfiability, fuzzy concept subsumption, and ABox consistency. In addition, some fuzzy DL reasoners support different kinds of simple queries over a KB $\mathcal{K}$ for obtaining assertional knowledge. These queries include

- *retrieval* given a fuzzy KB $\mathcal{K}$, a fuzzy concept C, and $n \in (0,1]$, to retrieve all instances o occurring in the ABox, such that $\mathcal{K}| = C(o) \geq n$ holds,
- *realisation* given a fuzzy KB $\mathcal{K}$, a individual name o, and $n \in (0,1]$, to determine the most specific concept C, such that $\mathcal{K}| = C(o) \geq n$ holds. In other words, for any fuzzy concept D, $\mathcal{K}| = D(o) \geq n$ implies $\mathcal{K}| = C \sqsubseteq D$.
- *instantiation* given a fuzzy KB $\mathcal{K}$, a fuzzy concept assertion $C(o) \geq n$, to decide whether or not $\mathcal{K}| = C(o) \geq n$.

In fact, fuzzy DL reasoners deal with these queries by transferring them into basic inference tasks. For example, the instantiation problem $\mathcal{K}| = C(o) \geq n$ can be reduced to the (un)satisfiability problem of the KB $\mathcal{K} \cup \{\neg C(o) < n\}$, while the latter one is a basic inference problem. There is, however, no support for queries that ask for n-tuples of related individuals or for the use of variables to formulate a query, just as conjunctive queries do.

Conjunctive queries stemmed from the domain of relational databases, and have attracted much attention recently in Semantic Web. With the emergence of a good number of large-scale domain ontologies, it is of particular importance to provide users with expressive querying service.

Recently there have been quite a lot of work on answering Conjunctive queries over fuzzy DLs. In [14], Straccia defined a conjunctive query language over fuzzy KBs. Conjunctive queries over a fuzzy KB $\mathcal{K}$ are expressions of the form $q(\boldsymbol{x}) \leftarrow \exists \boldsymbol{y}.conj(\boldsymbol{x}, \boldsymbol{y})$, where vector $\boldsymbol{x}$ is constituted by *distinguished variables* (also known as *answer variables*), which can be bound with individual names in a given knowledge base to answer the conjunctive queries, $\boldsymbol{y}$ is constituted by *non-distinguished* variables, which are treated as existential quantified, i.e., we just require the existence of a suitable element in the model, but this element does not have to correspond to an individual explicitly named in the ABox. $conj(\boldsymbol{x}, \boldsymbol{y})$ is a conjunction of atoms of the form $B(z)$ or $R(z_1, z_2)$, where B and R are basic concept and role in $\mathcal{K}$ respectively. The concept and role atoms are syntactically equal to concept and role assertions except that z, z_1 and z_2 may be variables in $\boldsymbol{x}$ or $\boldsymbol{y}$, besides constants in $\mathcal{K}$.

The query language in [14] has the same syntax as that of crisp DLs, and thus cannot express queries such as "find me hotels that are very close to the conference venue (with membership degree at lest 0.9) and offer inexpensive (with membership degree at lest 0.7) rooms". For this reason, a query language, which allows a threshold for every query atom, was proposed in [23]. Clearly, threshold queries give users more flexibility in that users can specify different thresholds for different atoms.

In this study, we mainly deal with conjunctive query entailment problem. As is generally believed in database community, however, there is a tight connection among the problems of conjunctive query containment, conjunctive query answering, and query entailment [24].

2.4 Fuzzy Query Language

Following [18], we provide the formal definition of the syntax and semantics of the fuzzy querying language used in this paper.

Let N_V be a countable infinite set of variables and is disjoint from N_C, N_R, and N_I. A *term* t is either an individual name from N_I or a variable name from N_V. A *fuzzy query atom* is an expression of the form $\langle C(t) \geq n \rangle$ or $\langle R(t, t') \geq m \rangle$ with C a concept, R a simple role, and t, t' terms. As with fuzzy assertions, we refer to these two different types of atoms as *fuzzy concept atoms* and *fuzzy role atoms*, respectively. We also w.l.o.g. assume all the query atoms are in their NNFs, PINFs, and NFs.

Definition 1. *(Fuzzy Boolean Conjunctive Queries) A fuzzy boolean conjunctive query q is a non-empty set of fuzzy query atoms of the form $q = \{\langle at_1 \geq n_1 \rangle, \ldots, \langle at_k \geq n_k \rangle \}$. Then for every fuzzy query atom, we can say $\langle at_i \geq n_i \rangle \in q$.*

We use $\mathsf{Var}(q)$ to denote the set of variables occurring in q, $\mathsf{Ind}(q)$ to denote the set of individual names occurring in q, and $\mathsf{Term}(q)$ for the set of terms in q, i.e. $\mathsf{Term}(q) = \mathsf{Var}(q) \cup \mathsf{Ind}(q)$.

The semantics of a fuzzy query is given in the same way as for the related fuzzy DL by means of fuzzy interpretation consisting of an interpretation domain and a fuzzy interpretation function.

Definition 2. *(Models of Fuzzy Queries) Let $\mathcal{I} = (\Delta^{\mathcal{I}}, \cdot^{\mathcal{I}})$ be a fuzzy interpretation of an f-$\mathcal{SHIN}$ KB, q a fuzzy boolean conjunctive query, and t, t' terms in q. We say $\mathcal{I}$ is a model of q, if there exists a mapping $\pi : \mathsf{Term}(q) \to \Delta^{\mathcal{I}}$ such that $\pi(a) = a^{\mathcal{I}}$ for each $a \in \mathsf{Ind}(q)$, $C^{\mathcal{I}}(\pi(t)) \geq n$ for each fuzzy concept atom $C(t) \geq n \in q$, $R^{\mathcal{I}}(\pi(t), \pi(t')) \geq n$ for each fuzzy role atom $R(t, t') \geq n \in q$.*

If $\mathcal{I} \models^{\pi} at$ for every atom $at \in q$, we write $\mathcal{I} \models^{\pi} q$. If there is a π, such that $\mathcal{I} \models^{\pi} q$, we say $\mathcal{I}$ satisfies q, written as $\mathcal{I} \models q$. We call such a π a *match* of q in $\mathcal{I}$. If $\mathcal{I} \models q$ for each model $\mathcal{I}$ of a KB $\mathcal{K}$, then we say $\mathcal{K}$ entails q, written as $\mathcal{K} \models q$. The *query entailment problem* is defined as follows: given a knowledge base $\mathcal{K}$ and a query q, decide whether $\mathcal{K} \models q$.

3   Query Entailment Algorithm

As for basic inference services and simple queries, our algorithm for deciding fuzzy query entailment is also based on tableau algorithms. The query entailment problem is, however, not reducible to the knowledge base satisfiability problem, since the negation of a fuzzy conjunctive query is not expressible with existing constructors provided by an f-$\mathcal{SHIN}$ knowledge base. For this reason, tableau algorithms for reasoning over knowledge bases is not sufficient. A knowledge base $\mathcal{K}$ may have infinitely many possibly infinite models, whereas tableau algorithms construct only a subset of finite models of the knowledge base. As is defined in Section 2.4, the query entailment holds only if the query is true in all models of the knowledge base, we thus have to show that inspecting only a subset of the models, namely the *canonical* ones, suffices to decide query entailment.

As with the tableau algorithm for f-$\mathcal{SHIN}$ [4], our algorithm works on a data structure called *completion forest*. A completion forests is a finite relational structure capturing sets of models of a KB $\mathcal{K}$. Roughly speaking, models of $\mathcal{K}$

are represented by an initial completion forest $\mathcal{F}_\mathcal{K}$. Then, by applying *expansion rules* repeatedly, new completion forests are generated. Since every model of $\mathcal{K}$ is preserved in some completion forest that results from the expansion, $\mathcal{K} \models q$ can be decided by considering a set $\mathbb{F}_\mathcal{K}$ of sufficiently expanded forests. From each such $\mathcal{F}$ a single *canonical model* is constructed. Semantically, the finite set of these canonical models is sufficient for answering all queries q of bounded size. Furthermore, we prove that entailment in the canonical model obtained from $\mathcal{F}$ can be checked effectively via a syntactic mapping of the terms in q to the nodes in $\mathcal{F}$.

3.1   Completion Forests

Definition 3. *(Completion Forest) A completion tree T for a $f\text{-}\mathcal{SHIN}$ KB is a tree all whose nodes are generated by expansion rules, except for the root node which might correspond to a individual name in N_I. A completion forest $\mathcal{F}$ for a $f\text{-}\mathcal{SHIN}$ KB consists of a set of completion trees whose nodes correspond to individual names in the ABox, an equivalent relation $\approx$ among nodes, and an inequivalent relation $\not\approx$ among nodes.*

Each node x in a completion forest (which is either a root node or a node in a completion tree) is labelled with a set $\mathcal{L}(x) = \{\langle C, \geq, n \rangle\}$, where $C \in sub(\mathcal{A})$, $n \in (0,1]$. Each edge $\langle x, y \rangle$ (which is either one between root nodes or one inside a completion tree) is labelled with a set $\mathcal{L}(\langle x, y \rangle) = \{\langle R, \geq, n \rangle\}$.

   If $\langle x, y \rangle$ is an edge in a completion forest with $\langle R', \geq, n \rangle \in \mathcal{L}(\langle x, y \rangle)$ and $R' \sqsubseteq^* R \in \mathcal{R}$, then y is called an $R_{\geq,n}$-*successor* of x and x is called an $R_{\geq,n}$-*predecessor* of y. Ignoring the inequality and membership degree, we can also call y an R-successor of x and x an R-predecessor of y. *Ancestor* and *descendant* are the transitive closure of predecessor and successor, respectively. The union of the successor and predecessor relation is the *neighbor* relation. The *distance* between two nodes x, y in a completion forest is the shortest path between them.

   Starting with an $f\text{-}\mathcal{SHIN}$ KB $\mathcal{K} = \langle \mathcal{T}, \mathcal{R}, \mathcal{A} \rangle$, the completion forest $\mathcal{F}_\mathcal{K}$ is initialized such that it contains a root node o, with $\mathcal{L}(o) = \{\langle C, \geq, n \rangle \mid C(o) \geq n \in \mathcal{A}\}$, for each individual name o occurring in $\mathcal{A}$, and an edge $\langle o, o' \rangle$ with $\mathcal{L}(\langle o, o' \rangle) = \{\langle R, \geq, n \rangle \mid \langle R(o, o') \geq n \rangle \in \mathcal{A}\}$, for each pair $\langle o, o' \rangle$ of individual names for which the set $\{R \mid R(o, o') \geq n \in \mathcal{A}\}$ is non-empty. We initialize the relation $\not\approx$ as $\{\langle o, o' \rangle \mid o \not\approx o' \in \mathcal{A}\}$, and the relation $\approx$ to be empty.

   Now we can formally define a new blocking condition, called *k-blocking*, for fuzzy query entailment depending on a depth parameter $k \geq 0$.

Definition 4. *(k-tree equivalence) The k-tree of a node v in T, denoted as T_v^k, is the subtree of T rooted at v with all the descendants of v within distance k. We use $\mathsf{Nodes}(T_v^k)$ to denote the set of nodes in T_v^k. Two nodes v and w in T are said to be k-tree equivalent in T, if T_v^k and T_w^k are isomorphic, i.e., there exists a bijection $\psi : \mathsf{Nodes}(T_v^k) \to \mathsf{Nodes}(T_w^k)$ such that (i) $\psi(v) = w$, (ii) for every node $o \in \mathsf{Nodes}(T_v^k)$, $\mathcal{L}(o) = \mathcal{L}(\psi(o))$, (iii) for every edge connecting two nodes o and o' in T_v^k, $\mathcal{L}(\langle o, o' \rangle) = \mathcal{L}(\langle \psi(o), \psi(o') \rangle)$.*

Definition 5. *(k-witness) A node w is a k-witness of a node v, if v and w are k-tree equivalent in T, w is an ancestor of v in T and v is not in T_w^k. Furthermore, T_w^k tree-blocks T_v^k and each node o in T_w^k tree-blocks node $\psi^{-1}(o)$ in T_v^k.*

Definition 6. *(k-blocking) A node o is k-blocked in a completion forest $\mathcal{F}$ iff it is not a root node and it is either directly or indirectly k-blocked. Node o is directly k-blocked iff none of its ancestors is k-blocked, and o is a leaf of a tree-blocked k-tree. Node o is indirectly k-blocked iff one of its ancestors is k-blocked or it is a successor of a node o' and $\mathcal{L}(\langle o', o \rangle) = \emptyset$.*

An initial completion forest is expanded according to a set of *expansion rules* that reflect the constructors allowed in *f-SHIN*. The expansion rules, which syntactically decompose the concepts in node labels, either infer new constraints for a given node, or extend the tree according to these constraints (see Table 1). Termination is guaranteed by k-blocking. We denote by $\mathbb{F}_\mathcal{K}$ the set of all completion forests obtained this way.

There should be some explanation for the $\sqsubseteq$-rule, which is applicable for fuzzy GCIs in the TBox. In the context of crisp DLs, the solution to GCIs is, for each $C \sqsubseteq D$, to construct an universal concept $\neg C \sqcup D$ and let the label of every node within the completion forests contain this universal concept, thus ensuring the model satisfies the TBox. In fuzzy cases, however, this solution is not feasible in that $\neg C \sqcup D$ cannot capture the semantics of $C \sqsubseteq D$. Li et al. [25] and Stoilos et al. [21] proposed two similar methods for dealing with fuzzy GCIs in parallel. In the $\sqsubseteq$-rule, $N^\mathcal{A}$ and N^q denotes the sets of membership degrees in the ABox $\mathcal{A}$ and in CQ q, respectively, i.e., $N^\mathcal{A} = X^\mathcal{A} \cup \{1 - n | n \in X^\mathcal{A}\}$, where $X^\mathcal{A} = \{0, 0.5, 1\} \cup \{n | C(o) \geq n \in \mathcal{A}, \text{ or } R(o, o') \geq n \in \mathcal{A}\}$, $N^q = \{n | C(t) \geq n \in q, \text{ or } R(t, t') \geq n \in q\}$. In addition, for each fuzzy concept C occurring in q, we augment the TBox with a fuzzy GCI $C \sqsubseteq C$. This clearly has no logical impact on the knowledge base, but it ensures that, for each node o in the completion forest, a decision is made as to whether $C(o) \geq n$ or $\neg C(o) \geq 1 - n + \epsilon$ ($C(o) < n$) holds, according to the $\sqsubseteq$-rule in Table 1.

For a node o, $\mathcal{L}(o)$ is said to contain a *clash*, if it contain one of the following: (i) a pair of triples $\langle C, \geq, n \rangle$ and $\langle \neg C, \geq, m \rangle$ with $n + m > 1$, (ii) one of the triples: $\langle \bot, \geq, n \rangle$ with $n > 0$, $\langle C, \geq, n \rangle$ with $n > 1$, (iii) some triple $\langle \leq pR, \geq, n \rangle$, and o has $p+1$ $R_{\geq, n'}$-neighbors $o_1, \ldots, o_{p+1}$, with $o_i \neq o_j$ for all $1 \leq i < j \leq p+1$ and $n' = 1 - n + \epsilon$.

Definition 7. *(clash-free completion forest) A completion forest $\mathcal{F}$ is called clash free if none of its nodes and edges contains a clash.*

Definition 8. *(k-complete completion forest) A completion forest is called k-complete if (under k-blocking) no rule can be applied to it. We denote by ccf_k $(\mathbb{F}_\mathcal{K})$ the set of k-complete and class-free completion forests in $\mathbb{F}_\mathcal{K}$.*

Example 1. Let $\mathcal{K} = (\mathcal{T}, \mathcal{R}, \mathcal{A})$ be an *f-SHIN* KB with $\mathcal{T} = \{C \sqsubseteq \exists R.C\}$, $\mathcal{A} = \{C(a) \geq n\}$ and $\mathcal{R} = \emptyset$. Figure 1 shows a 1-complete and clash-free completion forest $\mathcal{F}$ for $\mathcal{K}$. The node x_4 in $T_{x_3}^1$-tree is directly blocked by x_2 in $T_{x_1}^1$-tree, indicated by the dashed line.

Table 1. Expansion rules

Rule	Description
$\sqcap_{\geq}$	if 1. $\langle C \sqcap D, \geq, n \rangle \in \mathcal{L}(x)$, x is not indirectly k-blocked, and 2. $\{\langle C, \geq, n \rangle, \langle D, \geq, n \rangle\} \nsubseteq \mathcal{L}(x)$ then $\mathcal{L}(x) \to \mathcal{L}(x) \cup \{\langle C, \geq, n \rangle, \langle D, \geq, n \rangle\}$
$\sqcup_{\geq}$	if 1. $\langle C \sqcup D, \geq, n \rangle \in \mathcal{L}(x)$, x is not indirectly k-blocked, and 2. $\{\langle C, \geq, n \rangle, \langle D, \geq, n \rangle\} \cap \mathcal{L}(x) = \emptyset$ then $\mathcal{L}(x) \to \mathcal{L}(x) \cup \{C'\}$, where $C' \in \{\langle C, \geq, n \rangle, \langle D, \geq, n \rangle\}$
$\exists_{\geq}$	if 1. $\langle \exists R.C, \geq, n \rangle \in \mathcal{L}(x)$, x is not k-blocked. 2. x has no $R_{\geq,n}$-neighbor y s.t. $\langle C, \geq, n \rangle \in \mathcal{L}(y)$, then create a new node y with $\mathcal{L}(x,y) = \{\langle R, \geq, n \rangle\}$ and $\mathcal{L}(y) = \{\langle C, \geq, n \rangle\}$
$\forall_{\geq}$	if 1. $\langle \forall R.C, \geq, n \rangle \in \mathcal{L}(x)$, x is not indirectly k-blocked. 2. x has an $R_{\geq,n'}$-neighbor y with $\langle C, \geq, n \rangle \notin \mathcal{L}(y)$, where $n' = 1 - n + \epsilon$, then $\mathcal{L}(y) \to \mathcal{L}(y \cup \{\langle C, \geq, n \rangle\}$
$\forall_{+}$	if 1. $\langle \forall R.C, \geq, n \rangle \in \mathcal{L}(x)$ with $\mathsf{Trans}(R)$, x is not indirectly k-blocked, and 2. x has an $R_{\geq,n'}$-neighbor y with $\langle \forall R.C, \geq, n \rangle \notin \mathcal{L}(y)$, where $n' = 1 - n + \epsilon$, then $\mathcal{L}(y) \to \mathcal{L}(y \cup \{\langle \forall R.C, \geq, n \rangle\}$
$\forall'_{+}$	if 1. $\langle \forall S.C, \geq, n \rangle \in \mathcal{L}(x)$, x is not indirectly k-blocked, and 2. there is some R, with $\mathsf{Trans}(R)$ and $R \sqsubseteq^* S$, 3. x has an $R_{\geq,n'}$-neighbor y with $\langle \forall R.C, \geq, n \rangle \notin \mathcal{L}(y)$, where $n' = 1 - n + \epsilon$, then $\mathcal{L}(y) \to \mathcal{L}(y \cup \{\langle \forall R.C, \geq, n \rangle\}$
$\geq_{\geq}$	if 1. $\langle \geq pR, \geq, n \rangle \in \mathcal{L}(x)$, x is not k-blocked, $\sharp\{x_i \in N_I | \langle R, \geq, n \rangle \in \mathcal{L}(x, x_i)\} < p$, then introduce new nodes, s.t. $\sharp\{x_i \in N_I \mid \langle R, \geq, n \rangle \in \mathcal{L}(x, x_i)\} \geq p$
$\leq_{\geq}$	if 1. $\langle \leq pR, \geq, n \rangle \in \mathcal{L}(x)$, x is not indirectly k-blocked, 2. $\sharp\{x_i \in N_I | \langle R, \geq, 1 - n + \epsilon \rangle \in \mathcal{L}(x, x_i)\} > p$ and 3. there exist x_l and x_k, with no $x_l \not\approx x_k$, 4. x_l is neither a root node nor an ancestor of x_k. then (i) $\mathcal{L}(x_k) \to \mathcal{L}(x_k) \cup \mathcal{L}(x_l)$ (ii) $\mathcal{L}(x, x_k) \to \mathcal{L}(x, x_k) \cup \mathcal{L}(x, x_l)$ (iii) $\mathcal{L}(x, x_l) \to \emptyset$, $\mathcal{L}(x_l) \to \emptyset$ (iv) set $x_i \not\approx x_k$ for all x_i with $x_i \not\approx x_l$
$\leq_{r\geq}$	if 1. $\langle \leq pR, \geq, n \rangle \in \mathcal{L}(x)$, 2. $\sharp\{x_i \in N_I | \langle R, \geq^-, 1 - n \rangle \in \mathcal{L}(x, x_i)\} > p$ and 3. there exist x_l and x_k, both root nodes, with no $x_l \not\approx x_k$, then 1. $\mathcal{L}(x_k) \to \mathcal{L}(x_k) \cup \mathcal{L}(x_l)$ 2. For all edges $\langle x_l, x' \rangle$, i. if the edge $\langle x_k, x' \rangle$ does not exist, create it with $\mathcal{L}(\langle x_k, x' \rangle) = \emptyset$, ii. $\mathcal{L}(\langle x_k, x' \rangle) \to \mathcal{L}(\langle x_k, x' \rangle) \cup \mathcal{L}(\langle x_l, x' \rangle)$. 3. For all edges $\langle x', x_l \rangle$, i. if the edge $\langle x', x_k \rangle$ does not exist, create it with $\mathcal{L}(\langle x', x_k \rangle) = \emptyset$, ii. $\mathcal{L}(\langle x', x_k \rangle) \to \mathcal{L}(\langle x', x_k \rangle) \cup \mathcal{L}(\langle x', x_l \rangle)$. 4. Set $\mathcal{L}(x_l) = \emptyset$ and remove all edges to/from x_l. 5. Set $x'' \not\approx x_k$ for all x'' with $x'' \not\approx x_l$ and set $x_l \approx x_k$
$\sqsubseteq$	if 1. $C \sqsubseteq D \in \mathcal{T}$ and 2. $\{\langle \neg C, \geq, 1 - n + \epsilon \rangle, \langle D, \geq, n \rangle\} \cap \mathcal{L}(x) = \emptyset$ for $n \in N^{\mathcal{A}} \cup N^q$, then $\mathcal{L}(x) \to \mathcal{L}(x) \cup \{C'\}$ for some $C' \in \{\langle \neg C, \geq, 1 - n + \epsilon \rangle, \langle D, \geq, n \rangle\}$

$$\begin{array}{l} \langle R,\geq,n\rangle \\ \\ \langle R,\geq,n\rangle \\ \\ \langle R,\geq,n\rangle \\ \\ \langle R,\geq,n\rangle \end{array} \quad \begin{array}{l} a \ \langle C,\geq,n\rangle, \langle \exists R.C,\geq,n\rangle \\ x_1 \ \langle C,\geq,n\rangle, \langle \exists R.C,\geq,n\rangle \\ x_2 \ \langle C,\geq,n\rangle, \langle \exists R.C,\geq,n\rangle \\ x_3 \ \langle C,\geq,n\rangle, \langle \exists R.C,\geq,n\rangle \\ x_4 \ \langle C,\geq,n\rangle, \langle \exists R.C,\geq,n\rangle \end{array}$$

Fig. 1. A 1-complete and clash-free completion forest $\mathcal{F}$ for $\mathcal{K}$

3.2 Models of a Completion Forest

We now show that every model of a KB $\mathcal{K}$ is preserved in some complete and clash-free completion tree $\mathcal{F}$. We first define models of $\mathcal{F}$, then prove that, for each model $\mathcal{I}$ of $\mathcal{K}$, there exists some $\mathcal{F}$, such that a extended model $\mathcal{I}'$ of $\mathcal{I}$ is a model of $\mathcal{F}$.

If we view all the nodes (either root nodes or generated nodes) in a completion forest $\mathcal{F}$ as individual names, we can define models of $\mathcal{F}$ in terms of models of $\mathcal{K}$ over an extended vocabulary.

Definition 9. *(Models of completion forests) An interpretation $\mathcal{I}$ is a model of a completion forest $\mathcal{F}$ for $\mathcal{K}$, denoted $\mathcal{I} \models \mathcal{F}$, if $\mathcal{I} \models \mathcal{K}$ and for all nodes v,w in $\mathcal{F}$ it holds that (i) $C^{\mathcal{I}}(v^{\mathcal{I}}) \geq n$ if $\langle C,\geq,n\rangle \in \mathcal{L}(v)$, (ii) $R^{\mathcal{I}}(v^{\mathcal{I}},w^{\mathcal{I}}) \geq n$ if there exists an edge $\langle v,w\rangle$ in $\mathcal{F}$ and $\langle R,\geq,n\rangle \in \mathcal{L}(\langle v,w\rangle)$, (iii) $v^{\mathcal{I}} \neq w^{\mathcal{I}}$ if $v \not\approx w \in \mathcal{F}$.*

Apparently, the initial completion forest $\mathcal{F}_{\mathcal{K}}$ and $\mathcal{K}$ share the same models in that there are only root nodes in $\mathcal{F}_{\mathcal{K}}$, which correspond individual names in $\mathcal{K}$. Then, each time an expansion rule is applied, every model of $\mathcal{K}$ is preserved in some expanded completion forest [4]. It thus holds that for each model $\mathcal{I}$ of $\mathcal{K}$, there exists some $\mathcal{F} \in \mathsf{ccf}_k(\mathcal{F}_{\mathcal{K}})$ and a model of $\mathcal{F}$ which extends $\mathcal{I}$ (for any $k > 0$). Since the set of k-complete and clash-free completion forests for $\mathcal{K}$ semantically captures $\mathcal{K}$ (modulo new generated nodes), query entailment $\mathcal{K} \models q$ can be transferred to logical consequence of q from completion forests as follows. For any completion forest $\mathcal{F}$ and CQ q, let $\mathcal{F} \models q$ denote that $\mathcal{I} \models q$ for every model $\mathcal{I}$ of $\mathcal{F}$.

Proposition 1. *Let $k > 0$ be arbitrary. Then $\mathcal{K} \models q$ iff $\mathcal{F} \models q$ for each $\mathcal{F} \in \mathsf{ccf}_k(\mathbb{F}_{\mathcal{K}})$.*

3.3 Checking Query Entailment within Completion Forest

Now we will show that, if k is large enough, we can decide $\mathcal{F} \models q$ for each $\mathcal{F} \in \mathsf{ccf}_k(\mathbb{F}_{\mathcal{K}})$ by syntactically mapping the query q into $\mathcal{F}$.

Definition 10. *(Query mapping) A fuzzy query q can be mapped into $\mathcal{F}$, denoted $q \hookrightarrow \mathcal{F}$, if there is a mapping $\mu : \mathsf{Terms}(q) \to \mathsf{Nodes}(\mathcal{F})$, such that*

– $\mu(a) = a$ for every individual name a,
– for each fuzzy concept atom $C(x) \geq n$ in q, $\langle C(x), \geq, n \rangle \in \mathcal{L}(\mu(x))$,
– for each fuzzy role atom $\langle R(x, y) \geq n \rangle$ in q, $\mu(y)$ is a $R_{\geq n}$-neighbor of $\mu(x)$.

We use n_q to denote the number of fuzzy role atoms in a fuzzy query q.

Theorem 1. *Let $k \geq n_q$, where n_q denote the number of fuzzy role atoms in a fuzzy query q. Then $\mathcal{K} \models q$ iff for each $\mathcal{F} \in \mathsf{ccf}_k(\mathbb{F}_\mathcal{K})$, it holds that $q \hookrightarrow \mathcal{F}$.*

The if direction is easy. If q can be mapped to $\mathcal{F}$ via μ, then q is satisfied in each model $\mathcal{I}$ of $\mathcal{F}$ by assigning to each variable x in q the value of its image $\mu^\mathcal{I}(x)$. By Proposition 1, $\mathcal{K} \models q$.

To prove that the converse also holds, we have to show that if k is large enough, a mapping of q into $\mathcal{F} \in \mathsf{ccf}_k(\mathbb{F}_\mathcal{K})$ can be constructed from a distinguished canonical model of $\mathcal{F}$. The canonical model $\mathcal{I}_\mathcal{F}$ of $\mathcal{F}$ is constructed by unravelling the forest $\mathcal{F}$ in the standard way , where the blocked nodes act like 'loops' [4]. Its domain comprises the set of all paths from some root in $\mathcal{F}$ to some node of $\mathcal{F}$ (thus, it can be infinite). Note that in order for $\mathcal{I}_\mathcal{F}$ to be a model, $\mathcal{F}$ must be in $\mathsf{ccf}_k(\mathbb{F}_\mathcal{K})$ for some $n \geq 1$. For the complexity of the formal definition of $\mathcal{I}_\mathcal{F}$, we provide an example.

Example 2. By unravelling $\mathcal{F}$ in Figure 1, we obtain a model $\mathcal{I}_\mathcal{F}$ that has as domain the infinite set of paths from a to each x_i. Note that a path actually comprises a sequence of pairs of nodes, in order to witness the loops introduced by the blocked variables. When a node is not blocked, like x_1, the pair $\frac{x_1}{x_1}$ is added to the path. Since $T^1_{x_1}$ tree-blocks $T^1_{x_3}$, every time a path reaches x_4, which is a leaf of a blocked tree, we add $\frac{x_2}{x_4}$ to the path and loop back to the successors of x_2. In this way, we obtain the following infinite set of paths:

$$p_0 = \begin{bmatrix} a \\ a \end{bmatrix}, p_1 = \begin{bmatrix} a & x_1 \\ a & x_1 \end{bmatrix}, p_2 = \begin{bmatrix} a & x_1 & x_2 \\ a & x_1 & x_2 \end{bmatrix},$$

$$p_3 = \begin{bmatrix} a & x_1 & x_2 & x_3 \\ a & x_1 & x_2 & x_3 \end{bmatrix}, p_4 = \begin{bmatrix} a & x_1 & x_2 & x_3 & x_2 \\ a & x_1 & x_2 & x_3 & x_4 \end{bmatrix},$$

$$p_5 = \begin{bmatrix} a & x_1 & x_2 & x_3 & x_2 & x_3 \\ a & x_1 & x_2 & x_3 & x_4 & , & x_3 \end{bmatrix}, \dots$$

This set of paths constitute the domain $\Delta^{\mathcal{I}_\mathcal{F}}$. For each concept name A, we have $A^{\mathcal{I}_\mathcal{F}}(p_i) \geq n$, if $\langle A, \geq, n \rangle$ occurs in the label of the last node in p_i. For each role R, $R(p_i, p_j) \geq n$ if the last node in p_j is an R successor of p_i. If role $R \in \mathsf{Tran}$, the extension of R is expanded according to the sup-min transitive semantics. In the following, let n_q denote the number of fuzzy role atoms in q, and let $k \geq n_q$. Since $\mathcal{I}_\mathcal{F} \models q$, there exists a mapping $\sigma : \mathsf{Nodes} \to \Delta^{\mathcal{I}_\mathcal{F}}$ s.t. for each fuzzy concept atom $\langle C(x) \geq n \rangle$ in q, $C^{\mathcal{I}_\mathcal{F}}(\sigma(x)) \geq n$, and for each fuzzy role atom $R(x, y) \geq n$ in q, $R^{\mathcal{I}_\mathcal{F}}(\sigma(x), \sigma(y)) \geq n$. For any k-complete and clash free completion forest $\mathcal{F}$, a mapping μ of q into $\mathcal{F}$ can be obtained from σ. We use $G_{\mathcal{I}_\mathcal{F}}$ to denote the graph that has as nodes the domain of $\mathcal{I}_\mathcal{F}$, and as arcs the

R-successor edges of $\mathcal{I}_{\mathcal{F}}$ for each role occurring in q. For any two nodes q_i and q_j in $\mathcal{I}_{\mathcal{F}}$, let $d(q_i, q_j)$ denote the distance between q_i and q_j in $G_{\mathcal{I}_{\mathcal{F}}}$. Let the image of q under σ as a graph G_q, then the length of a path in G_q connecting the images $\sigma(x)$ and $\sigma(y)$ of any two variables x and y in q will be at most n_q. If $\mathcal{F}$ is k-complete with $k \geq n_q$, then for every path in G_q there will be an isomorphic one in $\mathcal{F}$. Therefore, a k-complete completion forest is large enough to find a mapping whose image is isomorphic to G_q.

We can, from the only if direction of Theorem 1, establish our key result, which reduce query entailment $\mathcal{K} \models q$ to finding a mapping of q into every $\mathcal{F}$ in $\mathsf{ccf}_k(\mathbb{F}_{\mathcal{K}})$.

Example 3. Given $\mathcal{F}$ and $q = \langle R(\langle x, y \rangle) \geq n \rangle \wedge C(x) \geq n$, we can easily recognize a mapping $q \hookrightarrow \mathcal{F}$.

3.4 Complexity Analysis

For the standard reasoning tasks, e.g., knowledge base consistency, the combined complexity is measured in the size of the input knowledge base. For query entailment, the size of the query is additionally taken into account. The size of a knowledge base $\mathcal{K}$ or a query q is simply the number of symbols needed to write it over the alphabet of constructors, concept, role, individual, and variable names that occur in $\mathcal{K}$ or q, where numbers are encoded in binary.

Theorem 2. *Given an f-$\mathcal{SHIN}$ KB $\mathcal{K}$ and a fuzzy conjunctive query q all of whose roles are simple, deciding whether $\mathcal{K} \models q$ is in* CO-3NEXPTIME.

Proof. (*sketch*) The proof is quite similar with the one presented in[26]. We use $||\mathcal{K}, q||$ to denote the total size of the string encoding the knowledge base $\mathcal{K}$ and the query q in a query entailment $\mathcal{K} \models q$. The branches in each completion tree within a completion forest $\mathcal{F} \in \mathbb{F}_{\mathcal{K}}$ is polynomially bounded in $||\mathcal{K}, q||$, and the maximal height of a non-isomorphic k-tree is double exponential in $||\mathcal{K}, q||$, if k is polynomial in $||\mathcal{K}, q||$. $\mathcal{F}$ thus has at most triple exponentially many nodes. Since each expansion rule can be applied only polynomially often to a node, the expansion of the initial completion forest $\mathcal{F}_{\mathcal{K}}$ into some $\mathcal{F} \in \mathbb{F}_{\mathcal{K}}$ terminates in nondeterministic triple exponential time in $||\mathcal{K}, q||$ for $k = n_q$. Checking whether $q \hookrightarrow \mathcal{F}$ is thus in triple exponential time in $||\mathcal{K}, q||$.

As for data complexity, we consider the ABox as the only input for the algorithm, i.e., the size of the TBox, the role hierarchy, and the query is fixed. Therefore, each completion forest $\mathcal{F} \in \mathbb{F}_{\mathcal{K}}$ has linearly many nodes in $|\mathcal{A}|$ and any expansion of $\mathcal{F}_{\mathcal{K}}$ terminates in polynomial time. Deciding whether $q \hookrightarrow \mathcal{F}$ is thus polynomial in the size of $\mathcal{F}$.

Theorem 3. *Given an f-$\mathcal{SHIN}$ KB $\mathcal{K}$ and a fuzzy conjunctive query q all of whose roles are simple, deciding whether $\mathcal{K} \models q$ is in* CO-NP *w.r.t. data complexity.*

4   Conclusion

Fuzzy Description Logics-based knowledge bases are envisioned to be useful in the Semantic Web. Existing fuzzy DL reasoners either are not capable of answering complex queries (mainly conjunctive queries), or only apply to DLs with less expressivity. We thus present an algorithm for answering expressive fuzzy conjunctive queries, which allow the occurrence of both lower bound and the upper bound of threshold in a query atom, over the relative expressive DL, namely fuzzy $\mathcal{SHIN}$. The algorithm we suggest here can easily be adapted to existing (and future) DL implementations. Future direction concern applying the proposed technique to even more expressive logics, for example fuzzy DLs additionally extended with nominals and datatype groups [27], or to more expressive fuzzy query language as suggested in [15].

Acknowledgments. This work was supported by the National Natural Science Foundation of China (60873010).

References

1. Baader, F., Calvanese, D., McGuinness, D.L., Nardi, D., Patel-Schneider, P.F. (eds.): The description logic handbook: theory, implementation, and applications. Cambridge University Press, New York (2003)
2. Zadeh, L.A.: Fuzzy sets. Information and Control 8(3), 338–353 (1965)
3. Straccia, U.: Reasoning within fuzzy description logics. J. Artif. Intell. Res. (JAIR) 14, 137–166 (2001)
4. Stoilos, G., Stamou, G., Pan, J., Tzouvaras, V., Horrocks, I.: Reasoning with very expressive fuzzy description logics. Journal of Artificial Intelligence Research 30(8), 273–320 (2007)
5. Stoilos, G., Simou, N., Stamou, G.B., Kollias, S.D.: Uncertainty and the semantic web. IEEE Intelligent Systems 21(5), 84–87 (2006)
6. Calvanese, D., De Giacomo, G., Lenzerini, M.: On the decidability of query containment under constraints. In: Proc. of the 17th ACM SIGACT SIGMOD SIGART Sym. on Principles of Database Systems (PODS 1998), pp. 149–158 (1998)
7. Calvanese, D., De Giacomo, G., Lembo, D., Lenzerini, M., Rosati, R.: Tractable reasoning and efficient query answering in description logics: The dl-lite family. J. of Automated Reasoning 39(3), 385–429 (2007)
8. Rosati, R.: On conjunctive query answering in EL. In: Proceedings of the 2007 International Workshop on Description Logic (DL 2007), CEUR Electronic Workshop Proceedings (2007)
9. Rosati, R.: The limits of querying ontologies. In: Schwentick, T., Suciu, D. (eds.) ICDT 2007. LNCS, vol. 4353, pp. 164–178. Springer, Heidelberg (2006)
10. Krotzsch, M., Rudolph, S., Hitzler, P.: Conjunctive queries for a tractable fragment of OWL 1.1. In: Aberer, K., Choi, K.-S., Noy, N., Allemang, D., Lee, K.-I., Nixon, L.J.B., Golbeck, J., Mika, P., Maynard, D., Mizoguchi, R., Schreiber, G., Cudré-Mauroux, P. (eds.) ASWC 2007 and ISWC 2007. LNCS, vol. 4825, pp. 310–323. Springer, Heidelberg (2007)
11. Levy, A.Y., Rousset, M.C.: Combining horn rules and description logics in carin. Artif. Intell. 104(1-2), 165–209 (1998)

12. Glimm, B., Horrocks, I., Lutz, C., Sattler, U.: Conjunctive query answering for the description logic shiq. In: IJCAI, pp. 399–404 (2007)
13. Glimm, B., Horrocks, I., Sattler, U.: Conjunctive query entailment for shoq. In: Proceedings of the 2007 International Workshop on Description Logic, DL 2007 (2007)
14. Straccia, U.: Answering vague queries in fuzzy dl-lite. In: Proceedings of the 11th International Conference on Information Processing and Management of Uncertainty in Knowledge-Based Systems (IPMU 2006), pp. 2238–2245 (2006)
15. Pan, J.Z., Stamou, G., Stoilos, G., Thomas, E.: Expressive Querying over Fuzzy DL-Lite Ontologies. In: Proc. of 2007 International Workshop on Description Logics, DL 2007 (2007)
16. Cheng, J.W., Ma, Z.M., Zhang, F., Wang, X.: Conjunctive query answering over an f-alc knowledge base. In: Web Intelligence/IAT Workshops, pp. 279–282 (2008)
17. Cheng, J.W., Ma, Z.M., Zhang, F., Wang, X.: Deciding query entailment in fuzzy description logic knowledge bases. In: Bhowmick, S., Küng, J., Wagner, R. (eds.) DEXA 2009. LNCS, vol. 5690, pp. 830–837. Springer, Heidelberg (2009)
18. Mailis, T.P., Stoilos, G., Stamou, G.B.: Expressive reasoning with horn rules and fuzzy description logics. In: Marchiori, M., Pan, J.Z., Marie, C.d.S. (eds.) RR 2007. LNCS, vol. 4524, pp. 43–57. Springer, Heidelberg (2007)
19. Ortiz, M., Calvanese, D., Eiter, T.: Data complexity of query answering in expressive description logics via tableaux. J. Autom. Reasoning 41(1), 61–98 (2008)
20. Cheng, J.W., Ma, Z.M., Zhang, F., Wang, X.: Deciding query entailment for fuzzy shin ontologies. Technical report, Northeastern University (2009), ftp://202.118.18.134/
21. Stoilos, G., Straccia, U., Stamou, G.B., Pan, J.Z.: General concept inclusions in fuzzy description logics. In: ECAI, pp. 457–461 (2006)
22. Baader, F., Nutt, W.: Basic description logics. In: Description Logic Handbook, pp. 43–95 (2003)
23. Lukasiewicz, T., Straccia, U.: Top-k retrieval in description logic programs under vagueness for the semantic web. In: Prade, H., Subrahmanian, V.S. (eds.) SUM 2007. LNCS (LNAI), vol. 4772, pp. 16–30. Springer, Heidelberg (2007)
24. Chandra, A.K., Merlin, P.M.: Optimal implementation of conjunctive queries in relational data bases. In: STOC 1977: Proceedings of the ninth annual ACM symposium on Theory of computing, pp. 77–90. ACM, New York (1977)
25. Li, Y., Xu, B., Lu, J., Kang, D.: Discrete tableau algorithms for *shi*. In: Description Logics (2006)
26. Ortiz, M., Calvanese, D., Eiter, T.: Data complexity of answering unions of conjunctive queries in *shiq*. In: Description Logics (2006)
27. Wang, H., Ma, Z.M.: A decidable fuzzy description logic f-alc(g). In: Bhowmick, S.S., Küng, J., Wagner, R. (eds.) DEXA 2008. LNCS, vol. 5181, pp. 116–123. Springer, Heidelberg (2008)

Merging and Ranking Answers in the Semantic Web: The Wisdom of Crowds

Vanessa Lopez, Andriy Nikolov, Miriam Fernandez, Marta Sabou, Victoria Uren, and Enrico Motta

Knowledge Media Institute, The Open University, Walton Hall, MK76AA, United Kingdom
{v.lopez, a.nikolov, m.fernandez, r.m.sabou,
e.motta}@open.ac.uk, v.uren@dcs.shef.ac.uk

Abstract. In this paper we propose algorithms for combining and ranking answers from distributed heterogeneous data sources in the context of a multi-ontology Question Answering task. Our proposal includes a merging algorithm that aggregates, combines and filters ontology-based search results and three different ranking algorithms that sort the final answers according to different criteria such as popularity, confidence and semantic interpretation of results. An experimental evaluation on a large scale corpus indicates improvements in the quality of the search results with respect to a scenario where the merging and ranking algorithms were not applied. These collective methods for merging and ranking allow to answer questions that are distributed across ontologies, while at the same time, they can filter irrelevant answers, fuse similar answers together, and elicit the most accurate answer(s) to a question.

Keywords: Merging, Ranking, Fusion, Question Answering, Semantic Web.

1 Introduction

Large-scale, open-domain question answering has been addressed with a variety of approaches in the last decades. Firstly, open domain Question Answering (QA) across unstructured Web data has been stimulated since 1999 by the TREC QA track evaluations. Secondly, the intuition that it would be easier to obtain answers from structured data (i.e., an ontology) where ambiguities in the queries can be resolved using reasoning techniques, has lead to much interest in Natural Language Interfaces (NLI) to knowledge bases, and in particular on NLI systems that directly query a given ontology [2, 4, 15]. However, although existing ontology-based NLI systems are generally domain independent or portable across domains, their scope is limited to one (or a set of) a-priori selected domain(s) at a time. A third recent trend in obtaining structured answers in a an open domain scenario has seen several industrial startups such as Powerset, START, Wolfram Alpha, True Knowledge[1], among others. A well-established approach for these systems is to semi-automatically build their own comprehensive factual knowledge bases. For example, similarly to OpenCyc and

---

[1] http://www.powerset.com/, http://start.csail.mit.edu/, http://www.wolframalpha.com/index.html, http://www.trueknowledge.com/ respectively

A. Gómez-Pérez, Y. Yu, and Y. Ding (Eds.): ASWC 2009, LNCS 5926, pp. 135–152, 2009.
© Springer-Verlag Berlin Heidelberg 2009

Freebase[2], the Wolfram Alpha knowledge inference engine builds a broad trusted knowledge base about the world by ingesting massive amounts of information (currently storing approximately 10TBs, still a tiny fraction of the Web). True Knowledge relies on users to add and curate its information, while PowerSet uses Freebase and annotates Wikipedia with its semantic resources.

Differently from this last trend, PowerAqua [9] attempts to perform open domain QA by taking advantage of the freely available structured information on the Semantic Web (SW)[3]. This is a key difference as, unlike the previous systems, PowerAqua does not impose an internal structure on its knowledge nor does it claim ownership of its knowledge base, but rather explores the increasing number of multiple, heterogeneous knowledge sources available on the Web. As such, PowerAqua supports users in searching and exploring information on the SW. Users introduce a factual query in natural language and PowerAqua is able to match it into one or many ontological facts, from which an answer can be inferred.

A major challenge faced by PowerAqua is that answers to a query may need to be derived from different ontological facts and even different semantic sources and domains. Often, multiple, redundant information needs to be combined (or *merged*) to obtain a reduced number of answers. Then, because different semantic sources have varying levels of quality and trust, when multiple answers are derived to a query it is important to be able to *rank* them in terms of their relevance to the query at hand.

In this paper we present merging and ranking methods for combining results (answers given as ontological facts) across ontologies. These methods have been integrated in PowerAqua and evaluated in the context of a multi-ontology QA task. The question that we try to answer here is: are aggregated answers from many heterogeneous independent semantic sources better than answers derived from single ontological facts? Or, similarly to the hypothesis of *Wisdom of Crowds*[4], are the many smarter than the few? These initial experiments confirm that the quality of derived answers can be improved by cross-ontology merging and ranking techniques.

The rest of the paper is structured as follows. Section 2 introduces a motivating scenario, Sections 3 and 4 describe the merging and ranking algorithms respectively. Section 5 describes the evaluation setup and the analysis of the results. We present related work in Section 6, and conclude in Section 7.

2 Motivating Scenario: Question Answering on the Semantic Web

Because PowerAqua derives answers from multiple online semantic resources, thus operating in a highly heterogeneous search space, it requires mechanisms for merging

---

[2] www.opencyc.org, http://www.freebase.com

[3] To be precise, PowerAqua accesses multiple ontologies through the Watson semantic gateway (http://watson.kmi.open.ac.uk/WatsonWUI/) or by exploring information stored in online servers.

[4] A book written by Jame Surowiecki in 2004, primarily on the fields of economic and psychology, stating that "a diverse collection of independently-deciding individuals is likely to make certain types of decisions and predictions better than individuals or even experts." It is also connected to social and collective intelligence on the Web (http://en.wikipedia.org/wiki/The_Wisdom_of_Crowds).

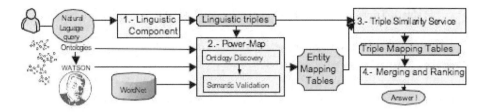

Fig. 1. PowerAqua components

and ranking answers to generate a commonly agreed set of answers across ontologies. Consider, for example, the query "Which languages are spoken in Islamic countries?". PowerAqua is designed following a cascade model (see Fig. 1). Steps 1, 2 and 3 are have been detailed in [9], so here we only summarize the key aspects of its behavior. At the first stage, the linguistic component using the GATE NL processing tool [5] transforms the NL query into an intermediate format called Query-Triples (QT). These QTs relate words together and mimic the structure of triples in the ontology but using the NL terms in the user query. For instance, our example query is initially translated into the QT <languages, spoken, Islamic countries>.

At the next step, the QTs are passed on to the PowerMap component, which identifies potentially suitable ontologies to answer a query, producing initial element level mappings between the QT terms and the entities in these sources. The output of PowerMap is a set of Entity Mapping Tables (EMTs), where each table associates each QT term with a set of entities found on the SW. To identify all semantic sources that are likely to describe QT terms, PowerMap maximizes recall by searching for approximate (lexical overlap) and exact (lexical equality) mappings. These are jointly referred to as equivalent mappings. PowerMap also uses both WordNet and the SW itself as sources of background knowledge to perform query expansion and to find lexically dissimilar (but semantically similar) matches – including synonyms, hypernyms and hyponyms. A semantic validation component attempts to generate WordNet synsets for all classes and individuals included in the EMTs. PowerMap uses the Watson[5] semantic search engine as a gateway to the SW. In addition, PowerMap can also query its own repositories and offers the capability to index and add new online ontologies[6].

In the third step, the Triple Similarity Service (TSS) matches the QTs to ontological expressions. The TSS takes as input the EMTs returned by PowerMap and the initial QTs, and returns a set of Triple Mapping Tables (TMTs), which define a set of complete mappings between a QT, and the appropriate Onto-Triples (OTs). The TSS chooses whenever possible the ontologies that better cover the user query and domain. In our example, as PowerMap does not find any covering ontology with mappings for both arguments in the QT: "languages" and "Islamic countries", the TSS algorithm reiterates again by splitting the compound term "Islamic countries", and consequently modifying the QT into: <languages, spoken, countries / Islamic> and creating a new QT for the compound <Islamic, ?, countries>. For the QTs

---

[5] http://watson.kmi.open.ac.uk/WatsonWUI
[6] PowerMap uses Lucene (lucene.apache.org) for the offline creation of the inverted indexes in order to provide efficient keyword searches to an ontology store in platforms such as Sesame.

Table 1. Triple Mapping Tables returned by PowerAqua for the example query

QT₁: <languages, spoken, countries / islamic>	
Dbpedia_infoboxes	OT_1 *<language, regionalLanguage, Country>* - 164 answers. E.g.:
	English (Pakistan), Arabic (Somalia), French (Algeria), Kurdish (Iran), Pashto (Pakistan), Welsh (United Kingdom), Albanian (Serbia), Catalan (Spain), Munji (Afghanistan).
Dbpedia_infoboxes	OT_2 *<language, states, country >* - 713 answers. E.g.:
	Canadian_French (Canada), Japanese (Japan), Russian (Poland), Bukawa (Papau New Guinea, Malay (Philippines), Filipino (Philippines), Hindi (India), Wakhi (Pakistan)
QT₂: <Islamic, ?, countries>	
Dbpedia_infoboxes	OT_3 *<country, governmenttype, Islamic_republic>* - 3 answers :
	Afghanistan (Afghanistan), Islamic_State_of_Afghanistan, Iran, Pakistan
Dbpedia_infoboxes	OT_4 *<Country, country, Islamic University>* - 1answer: Bangladesh
SWETO ontology	OT_5 *<Country, occurred_in, Terrorist_Attack>* *<Terrorist_Attack, responsible_for, Armed Islamic Group>* - 2 answers:
	Algeria (Aug31,1998,ArmedIslamicGroup_Bombing), France (Jul11,1995,ArmedIslamicGroup_Shooting)

obtained in this second iteration, the TSS extracts, by analyzing the ontology relations, a small set of covering ontologies containing the valid OTs that jointly cover the user query and produce an answer. The TMTs generated for each QT by the TSS are presented in Table 1.

Finally, because each resultant OT only leads to partial answers, they need to be combined into one complete answer. The goal of the fourth component is to merge and rank the various interpretations that different ontologies may produce. In our example, this is achieved by intersecting the answers from both QTs to obtain as a final set of answers those in "languages spoken in a country" whose "country" is shared with the answers obtained from "countries that are Islamic". Among other things, merging requires to identify similar entities across ontologies, e.g., "France" and "French republic".

In our example, from a total of 885 partial answers retrieved by PowerAqua the final set of answers obtained after merging contains 63 answers (e.g.: Aramaic (Iran), Abduyi_dialect (Iran), Kurdish (Iran), Pashto (Pakistan), Wakhi (Pakistan), among others. However, from those 63 answers 57 are correct. The 7 incorrect answers are derived from the partial answer "France" from the SWETO ontology, namely the languages, regional languages or extinct languages in France: French, Breton, Zarphatic, Balearic, Shuadit, Judeo-Spanish, Basque. Nevertheless, ranking measures can be applied to sort the answers and filter these results. As will be explained in Section 4, a ranking measure based on OTs confidence is capable of providing a lower confidence to answers derived from the SWETO OT than OTs formed from other ontologies by a direct relationship. This example illustrates how merging and ranking algorithms can enhance the quality of the search results. We now describe these algorithms.

3   Merging Algorithm

A side effect of the fact that PowerAqua explores multiple knowledge sources to obtain an answer is that, the TSS frequently associates the query to several OTs from different ontologies, each OT generating an answer. Depending on the complexity of the query, i.e., the number of QTs it has been translated to and the way each QT was matched to OTs, these individual answers may fall in one of the following categories: a) valid but duplicated answers, b) part of a composite answer and c) alternative answers derived from different ontological interpretations of the QTs. Different merging scenarios suit different categories: some cases require the intersection of the partial answers while other cases require their union. In this section we discuss the various merging scenarios (Section 3.1) and the fusion algorithm they rely on (Section 3.2).

3.1 Merging Scenarios

Scenario 1. A query translates into one QT only. These are the simplest queries, and therefore the easiest ones to merge, as each OT provides an answer on its own. The final set of answers is the union of all OTs across ontologies. E.g., "find me cities in Virginia".

Scenario 2. A query translates into two QTs that are linked together by the first QT term (the subject). Because each QT only leads to partial answers, they need to be merged to generate a complete response. This is achieved by intersecting the answers from both QTs. E.g., for the question "which Russian rivers flow into the Azov sea?" the final answers are composed by intersecting the results obtained with "*rivers* in Russia" and "*rivers* that flow in the Azov sea".

Scenario 3. A query translates into two QTs that are linked together by the object of the first one and the subject of the second one. In this scenario a complete answer can only be assigned by merging the partial answers. The answers for the first main QT are conditioned by the answers for the second QT. E.g., for the query "which rivers flow in European countries?" the final set of answers are the set of all countries in which rivers flow, <rivers, flow, *country*>, which are linked to the set of European countries <*countries*, ?, European>.

Scenario 4. Complex queries which are translated into multiple QTs. These queries are solved in a similar way as a combination of scenarios 1, 2 and 3. E.g., for the query "What are the main cities located in US states bordering Georgia?", where Georgia is an ambiguous term that can represent a state in the USA or a country in the Caucasus, the valid answers come from the intersection or condition between first "cities in US states" and second "cities bordering Georgia" (both as a state and as a country) or "US states bordering Georgia". Both alternative paths result in cities in the state of Georgia (USA), rather than those in the country of Georgia.

In sum, the merging procedure deals with these four scenarios by applying three types of operators over the set of retrieved answers: *union, intersection* and *condition*. The union operator combines answers related to the same QT but coming from different ontologies (scenario 1). The intersection operator is needed when a query specifies more than one constraint for a single first query term (scenario 2). The intersection operator merges the answers from two corresponding QTs and removes those, which only occur in one answer set. The condition operator is similar to the intersection one; however, the condition operator filters the answers not by the first query term but by the second one (scenario 3). These operators can be applied to any complex cases in which the query is translated into several combinations of QTs (scenario 4) so all answers produced by alternative paths are merged.

3.2 The Co-reference Fusion Algorithm

The merging procedure assigns the individuals returned as answers from different ontologies into subsets of answers that represent identical entities. The union operation processes a set of answers from a single QT and merges the similar answers representing identical entities. For example, the QT: <countries, locatedIn, Asia> returns, among its answers, "Thailand" from the TAP ontology and "Kingdom of Thailand" from the KIM ontology. These answers need to be grouped into a single subset as they refer to the same entity. As described above, depending on the query type, these subsets of answers can afterwards be combined by various operations.

The atomic procedure performed by all of these operations is matching. Two answers are compared and a decision is made about whether or not they are identical. To increase the speed, initial matching is performed only between the labels and local names of the returned entities. The entities are considered identical either if they are WordNet synonyms or if one of the used string similarity functions (Jaro, edit distance) returns a value above a certain threshold. A special case is the processing of ambiguity, which occurs when an entity has two potentially identical matching entities that belong to the same ontology. E.g., in "Give me all cities in the USA," a single entity, "arlington" from the *FAO* ontology, has two potential matches, "arlingtonVa" and "arlingtonTx" from the *UTexas geographic* ontology. Assuming that individuals belonging to the same ontology are distinct, the system tries to choose the best match out of the two using additional context data from the ontologies. The system receives, for each entity, all of their property values from their respective ontologies and compares these sets using the same similarity functions as above on their elements. Thus, in our example, context sets for both of the entities "arlington" and "arlingtonVa" mention "Virginia", while "arlingtonTx" mentions "Texas" instead. The similarity between the context sets of "arlington" and "arlingtonVa" is greater and, therefore, these entities are merged.

Pairwise comparison of entities would make the complexity of the procedure N^2 with respect to the input set size. In order to avoid this, candidate matches are selected using a search over the indexes and the comparison focuses only on the entities that appear among the search results. This makes the complexity linear with respect to the answer set size.

4 Ranking Algorithms

As we can see in Fig. 2, a filtered set of answers for each query is obtained after the merging step. While an unsorted list of answers can be manageable in some cases, the search system may become useless if the retrieval space is too big. In these cases a clear ranking criteria is needed to sort the final list of answers. The aim of the ranking measures presented here is to: a) assign a score to each individual answer and b) cluster the set of answers according to their score. Cluster analysis of ranking data attempts to identify typical groups of rank choices. In our case, according to the

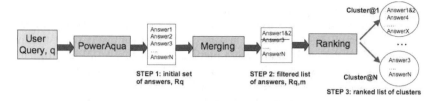

Fig. 2. Flow of the data retrieval, merging and ranking process

chosen ranking criteria, clusters identify results of different quality, popularity or meaning. The cluster ranked at position one (C@1) represents the best subset of results according to the chosen ranking method.

The ranking component defines three different ranking algorithms:

- *Ranking by semantic similarity*: this ranking criterion exploits a semantic standard (WordNet) to compute the distance between OTs (Section 4.1).
- *Ranking by confidence*: this ranking criterion is based on the confidence of the OTs from which the answer is extracted. The quality of the OT depends on the type of the mapping between the OT and the QT (e.g., direct or indirect - Section 4.2).
- *Ranking by popularity:* this ranking criterion is based on the popularity of the answer, defined as the number of ontologies from which this answer can be derived (Section 4.3).

4.1 Ranking by Semantic Similarity

Answers are ranked according to the popularity of the semantic interpretation of the OT they belong to. The hypothesis behind this is that if an answer is derived from an OT that has similar interpretations to other OTs from different ontologies, it is more likely to be correct than answers coming from unique semantically different interpretations. This criterion takes advantage of the knowledge inherent in the ontology and WordNet high quality descriptions, and combines some well-founded ideas from the Word Sense Disambiguation community to compute semantic similarity distance across ontological entities as detailed in [8].

Let's clarify this idea with the example "Give me cities in Virginia", a query matched by PowerAqua into eight ontologies (and eight OTs). The final set of answers obtained after merging should be the union of all the answers describing cities in Virginia. However, an instance labeled "Copenhagen" appears between the set of merged answers. In order to rank last this inaccurate answer, semantic similarity between OTs is computed by comparing the distance path and common ancestors between the WordNet synsets for each ontological concept representing the subject and object of the triple (predicates are not well covered in WordNet, and in the case of instances we look at its type). The WordNet synset (i.e. the true meaning) of an ontological term A, is determined by its parents in the hierarchy of the ontology (that is, those synsets of A that are similar to at least one synset of its ancestors in the ontology), and by its intended meaning in the user query (those synsets of A that are similar to at least one synset of the user term it matches to, if their labels differ). Having said that, while "city" has similar meanings in all its eight ontological matches (the synsets are semantically similar, even if they are not exactly the same), the ontological meaning of "Virginia" differs. Indeed, seven of the ontologies are referring to Virginia as an instance of an *state* or *province* (in USA), while the answer "Copenhagen" is derived from an eighth ontology about *film festivals* with the only semantically different OT, namely <city, hasActorWinner, VirginiaMadsen>, where "VirginiaMadsen" is classified as *person* in the ontology and not as a *state*, and therefore the intended meaning of the OT differs from the previous ones.

Ranking among answers is then calculated according to the popularity of the interpretation of the OT they belong to. Therefore, the first complete set of ranked answers comes from the union of the answers from the seven semantically similar OTs referring to cities in the *state* or *province* of Virginia. The answer, labeled "Copenhagen" because its derived from the only semantically different OT, it would be ranked lower than the previous answers.

To conclude with, the score for each answer is the number of ontologies that share the semantic interpretation of the OT they belong to, or -1 if an answer is coming from two OTs with different semantic interpretation. The C@1 groups all the answers ranked with score 2 or highest (at least two ontologies with the same semantic interpretation).

4.2 Ranking by Confidence

The quality of the matching between a QT and one (or, in some cases, two) OTs often has an influence on the quality of the derived answers. We identified a set of rules to predict which of these OTs are likely to be more reliable and potentially lead to a better set of answers. These rules are listed in the same order as they are applied, i.e., from the most to the least significant. The rules we use can be seen as nodes in a decision tree. Their order of preference is discriminative in order to avoid conflicts.

1) OTs that are based on only equivalent (i.e., exact and approximate match) or synonym type mappings to the corresponding QT terms are ranked highest. E.g., for QT <capitals, ?, USA> ("Find me capitals in the USA") the OT1 = <capital (exact), isCityOf, State> <State, isStateOf, USA (exact)> with only equivalent mappings is ranked higher than OT2 =<City (hypernym), attribute_country, USA (exact)> which contains an hypernym.

2) OTs that link the two arguments in the QT through an IS-A relation (instead of an ad-hoc relationship) are ranked lower than any other triples. The reason for this is that many online ontologies misuse IS-A relations to model other types of relations (e.g., partonomy). We do not apply this rule when the original question contains an IS-A relation, as this is an indication that such a relation is expected (e.g. "which animals ARE reptiles?"). For instance, the QT <person, plays, Nirvana> ("Who play in Nirvana?") is matched to OT1 = <person, hasMember, MusicianNirvana> and OT2 = <Nirvana Meratnia, IS-A, person>. Note that while rule 1 ranks these two triples equally, this rule ranks OT1 higher (even if, in this particular case, OT2 complies with correct modeling).

3) OTs that cover not only all of the terms in the QT, but also the linguistic relation (mapped as an ontological entity), are ranked first over triples that do not cover the relation. E.g., for the QT <states, bordering, Colorado> ("what are the states

bordering Colorado?") the OT <state, borders, Colorado (state)> is ranked higher than <state, runsThrough, Colorado (river)>.

4) The OTs containing more exact mappings are preferred. E.g., for QT <london, capital, country> ("is London the capital of any country?"), the OT <London (exact), hasCapitalCity, Country(exact)> is preferred over <capital_city (synonym), has_capital_city, country (exact)>. This rule is similar to rule 1, but because it is applied at a later stage, it is more restrictive.

5) For "who queries", OTs formed with "person" are preferred over "organization".

6) OTs based on direct mappings (1:1 mapping between a QT and an OT) are preferred to those relying on indirect mappings (1:2 mapppings). E.g, <person, works, open university> ("who works in the open university?") is translated to both OT1= <person, memberOf, openUniversity> and to OT2 = <person, mentions-person, kmi-planet-news (subclassOf publication)>, <kmi-planet-news, mentions-organization, the-open-university>. OT1 is ranked higher than OT2.

Once the score is assigned to each answer the clusters are created as follow: C@1 is all the answers ranked highest (score 1), C@2 is all the answers ranked in position 2, and so on.

4.3 Ranking by Popularity

Finally, answers are ranked according to their popularity, i.e., the number of individual ontologies from which they are derived. For instance, "where is Paris?" produces two answers: France (or French Republic) and United States (as Paris is a city in the state of Texas). In this case, France is the most popular answer across ontologies and therefore is ranked first. The number of ontologies for a given answer is provided by the merging algorithm described in section 3. An answer is C@1 if its popularity is higher than 1 (more than 1 ontology).

4.4 Ranking by Combination

Finally, we propose a last strategy to improve ranking, by the combined use of all the ranking methods presented before. We argue that, due to the different nature of these approaches, relevant answers not selected and irrelevant answers not filtered by one specific method, are suitable to be selected or filtered by the others. For the combination strategy we have used the weighted Borda method [1], in which votes are weighted taking into account the quality of the source. The combined weight for the answer i within the context of the query q, $W_{i,q}$, is therefore computed as: $W_{i,q} = 2*x$ (x=1, $i \in$ confidence C@1) $+ 1*x$ (x=1, $i \in$ confidence C@2) $+ 1*x$ (x=1, $i \in$ semantic similarity C@1) $+ 1*x$ (x=1, $i \in$ popularity C@1).

We have empirically tested that the most important ranking algorithm is confidence. With the proposed combination we attempt, on the one hand, to provide this measure for a significant number of answers (selecting C@1 and C@2) and, on the other hand, to provide a higher score to those answers with a higher confidence value. Once the scores are computed each answer is then clustered according to: a) its final score value and b) the selected degree of relevance for precision and recall measures in the final answer. To maximize precision, C@1 is generated with all the answers for which Wi,q > 1. To maximize recall, C@1 is generated with all the answers for which Wi,q >2.

5 Evaluation

In this section we describe the evaluation of PowerAqua's merging and ranking capabilities for queries that require to be answered by combining multiple facts from the same or different ontologies. The design of this evaluation is focused around two main questions:

a)   *How do we measure if the quality of the collective results obtained after merging and ranking are better than the individual answers?*
b)   *Which datasets are more suitable to be used for this evaluation?*

Because there are different steps in the merging and ranking process that can influence the final quality of the answers, we have divided the evaluation in three main stages:

- Evaluation of the efficiency and effectiveness of the fusion algorithm.
- Evaluation of the level of filtering performed by the merging algorithm over the initial set of answers retrieved by PowerAqua.
- Evaluation of the three proposed ranking algorithms applied to the final set of answers obtained after the merging process.

This initial evaluation is conducted using our own benchmark which comprises:

- *Ontologies and Knowledge Bases*: We collected around 4GBs of data stored in 130 Sesame repositories. Each repository contains one or more semantic sources. We have collected in total more than 700 documents. The dataset includes high-level ontologies, e.g., ATO, TAP, SUMO, DOLCE and very large ontologies, e.g., SWETO (around 800,000 entities and 1,600,000 relations) or the DBPedia Infoboxes (around 1GB of metadata). This set of ontologies is stored in several online Sesame repositories. Even though PowerAqua can access larger amounts of SW data through Watson, in this experiment we decided to use a substantial static dataset in order to make these experiments reproducible.
- *Queries*: We collected a total of 40 questions selected from the PowerAqua website[7] and from previous PowerAqua evaluations that focused on its mapping capabilities [10]. These are factual questions[8] that PowerAqua maps into several OTs, each of them producing partial answers. Merging and ranking is needed for these queries to generate a complete answer, or to rank between the different interpretations.
- *Judgments*: In order to evaluate the merging and ranking algorithms a set of judgments over the retrieved answers is needed. To perform this evaluation two ontology engineers provided a True/False manual evaluation of answers for each query.

---

[7] http://technologies.kmi.open.ac.uk/poweraqua/fusion-evaluation.html
[8] Factual queries formed with wh-terms (which, what, who, when, where) or commands (give, list, show, tell,..) vary in length and complexity:from simple queries, with adjunct structures or modifiers, to complex queries with relative sentences and conjuctions/disjunctions.

The construction of this benchmark was needed due to the lack of SW standard evaluation benchmarks comprising all the required information to judge the quality of the current semantic search methods [7].

5.1 Evaluating the Fusion Algorithm

The gold standard for the evaluation of the merging algorithm was created by manually annotating the answer sets produced by the 40 test queries. For each answer set, subsets of identical answers were identified. The generated gold standard was compared to the fusion produced by the merging algorithm and standard precision and recall measures were calculated. Each pair of answers correctly assigned to the same subset was considered a "true positive" result, each pair erroneously put into the same subset constituted a "false positive" result, and each pair of individuals, which were assigned to different subsets, while being in the same subset in the gold standard, represented a "false negative" result. The results are shown in Table 2.

Table 2. Test results of the co-reference resolution stage

Gold standard size	Precision	Recall	F1-measure
1006	0.946	0.931	0.939

When analyzing the results we found that most errors of the merging stage were caused by:

- Syntactically dissimilar labels for which no synonyms could be obtained from WordNet, e.g: #SWEET_17874 (Longview/Gladewater), or grammatical mistakes (like "she_sthe_one" instead of "she_the_one").
- Homonymous or syntactically similar labels for different entities.
- Incorrectly modeled ontologies, which contain duplicate instances under different URIs: e.g., in SWETO the city of Houston, Texas has 5 distinct URIs. Since such errors were not caused by the merging algorithm, they were not counted during the evaluation experiments.

5.2 Evaluating the Level of Filtering Performed by the Merging Algorithm

The major advantage of merging the multiple answers derived by PowerAqua is that irrelevant answers are filtered out (eliminated). The filtering obtained by the merging algorithm described in this paper allows, on the one hand to eliminate duplicated information by means of fusing redundant answers together and, on the other hand, to compose a complete answer using different subsets of partial responses. The filtering of duplicated and partial information helps to eliminate non relevant responses from the initial set of results. The following measure is used to compute the level of non relevant results filtered by the merging algorithm.

$$f_q = \frac{|R_q| - |R_{qm}|}{|R_q|}$$

Where f_q is the percentage of filtering for the query q, R_q is the set of initial results retrieved by PowerAqua for the query q and $R_{q,m}$ is the set of answers that remain after merging. Note that, for simplicity, we consider that all the eliminated answers are irrelevant. This is not necessarily true when the merging algorithm intersects partial answers. For those cases, the rate of false positives (or number of relevant results lost in the filtering process) has been computed (section 5.1) and discarded as irrelevant. Results are presented in Section 5.4.

5.3 Evaluating the Three Proposed Ranking Algorithms

Here we present the evaluation of the three ranking algorithms detailed in Section 4 in terms of Precision and Recall. As the golden standard for the evaluation we consider the completed list of answers for query q including all the potential relevant and irrelevant results as the unsorted list of answers obtained after the merge step, $R_{q,m}$ (see Fig. 2 for further details). For each ranking metric we consider as retrieved list of answers for the query q the first ranked cluster (C@1). Taking into account this, we define Precision and Recall as:

$$P_q = \frac{|\{Rel_q \cap C@1_q\}|}{|C@1_q|}, R_q = \frac{|\{Rel_q \cap C@1_q\}|}{|Rel_q|}$$

Where: P_q is precision for query q, R_q is recall for the query q, Rel_q is the set of relevant answers included in $R_{q,m}$ for the query q and $C@1_{,q}$ is the set of retrieved answers, or answers included in the first ranked cluster.

Once these measures have been defined we compare the results obtained by our three different ranking metrics against our baseline, $R_{q,m}$. For the ranking based on confidence the precision is computed not just for the first ranked cluster C@1 but also for the union of the first two clusters C@1 U C@2. As explained in Section 4.4, the most accurate ranking algorithm is confidence, therefore both confidence clusters are used in the combined ranking.

5.4 Results

The results of our experiments are reported in Tables 3 and 4 for the 40 selected queries. Table 3 contains the queries merged by union while Table 4 contains the results for the queries merged by intersection and condition. The different columns of the table represent:

1) The type of merging done for that query (U=union, I=intersection, C=condition)/ the number of ontologies involved in the merging process.
2) The percentage of irrelevant queries filtered by the merging algorithm.
3) The precision obtained for the set of answers returned after the merging process (the baseline ranking).
4) The error type as explained below.
5) Precision/Recall measures for the confidence ranking at the level of the first cluster C@1.
6) Precision/Recall measures for the confidence ranking at the level of the first two clusters C@1 U C@2.
7) Precision/Recall measures for the popularity ranking at the level of the first cluster C@1.

8) Precision/Recall measures for the semantic similarity ranking at the level of the first cluster C@1.
9) Precision/Recall measures for the combined approach at the level of the first cluster C@1 with the target of optimizing recall.
10) Precision/Recall measures for the combined approach at the level of the first cluster C@1 with the target of optimizing precision.

An empty sets {} represent that no answer was retrieved for that cluster while – indicates that the query generates only 1 unique answer after merging, and therefore there is nothing to rank.

As we can see in the tables, the merging component is able to filter an average of 93% of irrelevant answers for intersections/conditions and 32% for unions. For instance, in Q14: *find me university cities in Japan*, 20 final answers are selected out of 991 partial answers, by intersecting 417 from cities with a university from *dbpedia ontology* and 574 from cities in Japan from *fao, ato, KIM, tap, SWETO*. The average recall of the fusion algorithm, as shown in Section 5.1, is 0.93, i.e., a 0.07 loss in recall occurs in the case of intersections/conditions when partial answers representing the same individual are not recognized. The average precision of the fusion algorithm is 0.94, which indicates that most of the answers are correctly fused. The high precision and recall values obtained for the fusion algorithm, as well as the high percentage of filtering of irrelevant answers performed by this method, reflect PowerAqua's ability to derive valid semantic interpretations to a query across ontologies.

The causes of the merging algorithm leading to irrelevant results in the final answers are:

- Incorrect modeling of the ontological elements in the OTs that lead to the answer (M). For instance in Q30: *what mountains are in Alaska?*, the instance Germany is given as an answer because it is defined as rdf:type {country, mountain} in one of the ontologies.
- An inaccurate semantic interpretation given by PowerAqua (I). For instance Q36: *who belongs to the Open University?*. Among OTs representing people that work for the Open University, there is an OT : <organization, type, open universities>.
- Retrieval of irrelevant answers (R). E.g., the answer Houston to Q29: *Where is Houston?*

These sets of errors are often filtered out afterwards by the ranking algorithms. As we can see in the tables, for the union queries all ranking methods are able to provide better precision than the baseline, with an increase of 0.22 points of precision for the best ranking algorithm, in this case ranking by confidence at C@1. This increase in precision is usually translated in a recall loss as in the case of the popularity ranking algorithm where recall drops to 0.31. However, the rest of the ranking metrics are able to keep the recall measure between 0.77 and 0.94. Finally, the best combined approach is able to enhance with 0.12 points the precision of the baseline without causing a drop in recall.

5.5 Discussion on the Results

For the intersection and condition queries all the ranking methods are able to keep or increase the precision, except in the case of the popularity algorithm that decreases precision to 0.31 points. The same effect occurs with recall. All the ranking algorithms are able to provide levels of recall between 0.98 and 1, which means nearly no loss of relevant answers, except for the popularity ranking, which reduces recall to 0.54. The best ranking method for intersection and condition queries is the ranking by confidence at C@1. This ranking slightly increases precision with 0.03 points with respect to the baseline, keeping at 0.98 the level of recall. Finally, for this set of queries, the best combined approach is able to preserve the same precision and recall values as the baseline: 0.96/1. In other words, the effect of ranking measures on intersection queries is neutral, this was expected as for intersection and condition queries the filtering has already eliminated most (if not all) of the inaccurate answers.

In summary, we can say that the best ranking method for both subset of queries is the ranking by confidence at C@1 that is able to produce a 0.22 percentage increase of precision for union queries and a 0.03 for intersection ones. Semantic similarity depends on being able to calculate the semantic interpretation of each OT, but that's not the case if the OT entities are not covered in WordNet, or the taxonomical information is not significant enough to elicit the meaning of the entity in the ontology. The worst ranking method in both cases is ranking by popularity. It drops precision by 0.14 points for union queries and by 0.31 points for intersection and condition queries. This is because popularity at C@1 (answers obtained from at least two ontologies) is empty in the cases in which no answers were fused from different ontologies (empty set {} being equivalent to 0/0 for precision/recall). Interestingly, in the 25 cases where C@1 is not empty, this measure gives precision 1 in 22 cases. Therefore, precision would have been closer to 1 than with any other ranking if we would have chosen to C@1 all the answers with popularity 1, when there are not answers with popularity 2 or higher (empty set {} equivalent to 1/1 for precision/recall as all the answers are rank at the same level). The effect in recall is even worse, dropping to 0.31 for union queries and to 0.54 for the intersection and condition ones. At this early stage of the SW, PowerAqua's results are hampered by the knowledge sparseness and its low quality. We believe that any extension of the online ontologies and semantic data will result in direct improvements for both popularity and semantic similarity ranking measures.

Even with the different behavior of these ranking methods, the combined algorithm is over-performed by the confidence ranking in terms of precision but it is able to improve the precision and recall ratio. Contrary to what was expected, maximizing precision does not improve the precision value on the combined measure. This is because the average measure was affected by queries in which none of the answers ranked high enough (C@1={}).

These results confirm our initial hypothesis that the use of cross-ontological information to rank the retrieved answers helps to enhance the precision of the results, and therefore, to provide, from the wisdom of semantic crowds, better answers to users. An important remark is that, this increase of precision does not imply, in any case except for the popularity algorithm, a significant loss in recall.

Table 3. Test results for union queries

	Type /N° Ont.	Fil-ter	Baseline		P/R confidence		P/R Pop	P/R Sem. Sim.	P/R Combined	
			P	Er	C@1	C@1 U C@2			+R	+P
Q_1	U / 3	0.03	0.97	M	0.97/0.88	0.97/1	1/0.03	0.97/1	0.97/0.88	1/0.03
Q_2	U / 3	0.53	1	-	1/0.66	1/1	1/0.73	1/1	1/1	1/0.97
Q_5	U / 4	0	1	-	1/0.4	1/0.6	{}	1/0.97	1/0.97	1/0.97
Q_6	U / 4	0.27	0.77	R,I	0.66/0.4	0.5/0.4	1/0.4	0.77/1	0.77/1	{}
Q_7	U / 6	0.38	1	-	1/0.53	1/1	1/0.29	1/1	1/1	1/0.52
Q_{13}	U / 2	0.91	1	-	1/0.81	1/1	1/0.08	1/1	1/0.81	1/0.09
Q_{15}	U / 5	0.55	1	-	1/0.47	1/1	1/0.35	1/1	1/1	1/0.47
Q_{16}	U / 7	0.31	1	-	1/1	1/1	1/0.32	1/1	1/1	1/1
Q_{17}	U / 8	0.25	0.35	I,M	1/1	1/1	1/0.14	0.53/1	1/1	1/1
Q_{18}	U / 3	0.13	1	-	1/1	1/1	1/0.13	1/1	1/1	1/1
Q_{19}	U / 8	0.55	0.94	I	1/1	0.94/1	1/0.61	0.94/1	0.94/1	1/1
Q_{20}	U / 2	0.05	1	-	1/1	1/1	1/0.05	1/1	1/1	1/1
Q_{21}	U / 4	0.53	0.75	R	1/1	0.75/1	1/0.16	1/1	0.75/1	1/1
Q_{22}	U / 6	0.36	0.28	I	1/1	0.66/1	0/0	0.16/0.5	0.5/1	0.5/0.5
Q_{23}	U / 4	0.4	0.5	R	1/1	1/1	0.5/1	0/0	0.5/1	0.5/1
Q_{24}	U / 8	0.57	0.12	R	1/1	0.12/1	1/1	0/0	0.12/1	1/1
Q_{25}	U / 10	0.64	0.88	I,M	1/1	0.96/1	0.96/0.98	0.9/1	0.95/1	0.98/1
Q_{29}	U / 9	0.21	0.45	R	0.45/1	0.45/1	1/0.4	0.4/0.8	0.45/1	0.45/1
Q_{30}	U / 3	0.08	0.95	M	1/1	0.95/1	1/0.09	0.95/1	0.95/1	1/1
Q_{31}	U / 3	0	1	-	1/0.5	1/1	{}	{}	1/0.5	1/0.5
Q_{34}	U / 3	0.29	0.74	I	1/0.07	0.74/1	1/0.07	0.74/1	0.74/1	1/0.07
Q_{36}	U / 5	0.15	0.14	I	1/0.32	0.14/1	1/0.02	1/0.77	1/0.77	1/0.32
Q_{37}	U / 3	0.5	0.66	R	1/1	0.66/1	1/0.5	0.66/1	1/1	1/0.5
Q_{38}	U / 6	0.3	0.57	M,I	1/0.25	1/0.5	1/0.5	0.57/1	0.75/1	1/0.5
Q_{39}	U / 2	0	0.33	I	1/1	0.33/1	{}	0.33/1	1/2	{}
Avg	4.84	0.32	0.74		0.96/0.77	0.81/0.94	0.82/0.31	0.72/0.85	0.86/1	0.86/0.66

Table 4. Test results for intersection and condition queries

	Type /N° Ont.	Fil-ter	Baseline		P/R confidence		P/R Pop	P/R Sem. Sim.	P/R Combined	
			P	Er	@1	C@1 U C@2			+R	+P
Q_3	I / 3	0.96	1	-	-	-	-	-	-	-
Q_4	I / 3	0.93	1	-	-	-	-	-	-	-
Q_8	I / 3	0.97	1	-	1/1	1/1	{}	1/1	1/1	1/1
Q_9	C / 4	0.96	0.88	I	1/0.87	0.88/1	{}	0.88/1	0.88/1	1/0.87
Q_{10}	C / 3	0.92	1	-	1/1	1/1	{}	1/1	1/1	1/1
Q_{11}	C / 13	0.88	1	-	1/0.77	1/1	1/0.07	1/1	1/1	1/0.77
Q_{12}	C / 4	0.78	0.75	M	1/1	0.75/1	{}	0.75/1	0.75/1	1/1
Q_{14}	I / 8	0.98	1	-	1/1	1/1	1/1	1/1	1/1	1/1
Q_{26}	I / 2	0.72	0.97	I	0.97/1	0.97/1	1/0.03	0.97/1	0.97/1	0.97/1
Q_{27}	I / 6	0.99	0.83	R	0.83/1	0.83/1	0.83/1	0.83/1	0.83/1	0.83/1
Q_{28}	I / 1	0.99	1	-	-	-	-	-	-	-
Q_{32}	I / 1	0.99	1	-	-	-	-	-	-	-
Q_{33}	I / 2	0.84	1	-	1/1	1/1	{}	1/1	1/1	0/0
Q_{35}	I / 1	0.98	1	-	-	-	-	-	-	-
Q_{40}	C / 7	0.99	1	-	-	-	-	-	-	-
Avg	4.06	0.93	0.96		0.99/0.98	0.96/1	0.65/0.54	0.96/1	0.96/1	0.92/0.91

6 Related Work

The problem of retrieving information by means of the aggregation of data from different sources on the Web has been tackled in Sig.ma (http://sig.ma/). In this system the user enters a keyword and is able to explore all the aggregated data coming from the search engine Sindice. Their contribution at this early stage of the SW is to show that "the sum is really bigger that the single parts". The system uses large scale indexing, data aggregation heuristics, and ontology alignments for automatic semi-structured data discovery and consolidation. However, as opposed to our approach, sig.ma does not attempt to automatically disambiguate or rank between different interpretations.

More specifically, the problem of merging, or finding identical individuals, was mostly considered in the context of offline data fusion scenarios. Basic similarity metrics based on string comparison were developed in the database community (e.g., [3, 16]). These metrics are used as a basis for the majority of algorithms, which compare values of attributes of different data instances and aggregate them to make a decision about two instances referring to the same entity (see [6] for a survey). The main distinction of our work is that, in the PowerAqua scenario, the fusion of answers is done in real time.

The problem of ranking applied to semantic search results has been also addressed in the literature. Among these works we can highlight [10, 11, 13]. [10] provides a criterion for query result ranking in the SEAL Portal based on a similarity measure between query results and the original KB without axioms. [11] proposes the expansion of query results through arbitrary ontology relations starting from the initial query answer, where the distance to the initial results is used to compute a similarity measure for ranking. [13] proposes a sentence ranking scheme based on the number of times an instance appears as a term in a relation type, and the derivation tree by which a sentence is inferred. To our knowledge, none of these works is applied to results derived from different knowledge sources, therefore, they do not consider the so-called "wisdom of the crowds" paradigm within their ranking algorithms.

7 Conclusions and Future Work

In this work we present a set of merging and ranking algorithms that aim to integrate information derived from different knowledge sources in order to enhance the results obtained by semantic search systems. These algorithms have been integrated and tested in an open QA system, PowerAqua. The experiments are promising, showing that the ranking algorithms can exploit the increasing amount of collectively authored, highly heterogeneous, online semantic data, in order to obtain, more accurate answers to a questions. On the one hand, the merging algorithm is able to filter out a significant subset of irrelevant results. On the other hand, the ranking algorithms are able to increase the precision of the final set of, thus showing a deeper semantic "understanding" of the intent of the question.

The merging algorithm is able to filter out up to 91% (32% on average) for union-based queries, and up to 99% (93% on average) for intersection based queries.

The best ranking algorithm (ranking by confidence) is able to obtain an average of 96% precision for union queries and 99% for intersection queries. An interesting, observed, side effect of this approach is that, answers to some questions that are distributed across ontologies can only be obtained if the partial results are merged. In this case, the introduction of the merging algorithm provides PowerAqua with the capability to answer queries that cannot be answered when considering a single knowledge source.

The high precision values produced by the merging and ranking algorithms, that are responsible for amalgamating information from different sources, support the comparison with the idea of the *Wisdom of Crowds* that we suggested in the paper. We further observe that it is known that the Wisdom of crowds only works if the crowd is diverse and free to think independently[14], allowing it to converge on good solutions. Similarly PowerAqua works well where ontologies have different emphasis, to allow the assembly of composite answers, but also overlaps between ontologies exist, to allow mapping and identification of ranking criteria, such as popularity. Both too much homogeneity and isolated "silo" ontologies would weaken our approach.

Another interesting side-effect of this approach is that, apart from the obvious advantage to the final user, the filtering of negative results and the ranking capabilities of the retrieval system increase its adaptability for other tasks, e.g., query expansion using SW resources.

An issue remains nonetheless open: the use of our own dataset to perform the experiments. However, to our knowledge, the SW community has not yet proposed standardized benchmarks to evaluate semantic merging and/or ranking evaluation. Despite this fact we have tested our algorithms with a significant amount of queries and large amounts of distributed semantic metadata (around 4GB).

Finally, we are currently working on a trust propagation mechanism where the user can rank the answers as a way of giving feedback to the system. We believe that this mechanism will further improve the ranking so that answers replicated across many ontologies do not bias less frequently occurred facts generated from specialist knowledge from trusted ontologies.

References

1. Bartell, B.T.: Optimizing Ranking Functions: A Connectionist Approach to Adaptive Information Retrieval. PhD thesis, University of California, San Diego (1994)
2. Bernstein, A., Kaufmann, E.: GINO - A guided input natural language ontology editor. In: Cruz, I., Decker, S., Allemang, D., Preist, C., Schwabe, D., Mika, P., Uschold, M., Aroyo, L.M. (eds.) ISWC 2006. LNCS, vol. 4273, pp. 144–157. Springer, Heidelberg (2006)
3. Bilenko, M., Mooney, R.J.: Adaptive duplicate detection using learnable string similarity measures. In: KDD 2003, pp. 39–48 (2003)
4. Cimiano, P., Haase, P., Heizmann, J.: Porting Natural Language Interfaces between Domains- An Experimental User Study with the ORAKEL System. In: Proc. of the Int. Conf. on Intelligent User Interfaces (2007)
5. Cunningham, H., Maynard, D., Bontcheva, K., Tablan, V.: GATE: A Framework and Graphical Development Environment for Robust NLP Tools and Applications. In: Proc. of the 40th Anniversary Meeting of the Association for Computational Linguistics (2002)

6. Elmagarmid, A.K., Ipeirotis, P.G., Verykios, V.S.: Duplicate record detection: a survey. IEEE Transactions on Knowledge and Data Engineering 19(1), 1–16 (2007)
7. Fernandez, M., Lopez, V., Motta, E., Sabou, M., Uren, V., Vallet, D., Castells, P.: Using TREC for cross-comparison between classic IR and ontology-based search models at a Web scale. In: Semantic search workshop at WWW 2009 (2009)
8. Gracia, J., Lopez, V., d'Aquin, M., Sabou, M., Motta, E., Mena, E.: Solving Semantic Ambiguity to Improve Semantic Web based Ontology Matching. In: Aberer, K., Choi, K.-S., Noy, N., Allemang, D., Lee, K.-I., Nixon, L.J.B., Golbeck, J., Mika, P., Maynard, D., Mizoguchi, R., Schreiber, G., Cudré-Mauroux, P. (eds.) ASWC/ISWC 2007. LNCS, vol. 4825. Springer, Heidelberg (2007)
9. Lopez, V., Sabou, M., Uren, V., Motta, E.: Cross-Ontology Question Answering on the Semantic Web –an initial evaluation. In: Proc. of Knowledge Capture Conference (2009)
10. Maedche, A., Staab, S., Stojanovic, N., Studer, R., Sure, Y.: SEmantic portAL: The SEAL Approach. In: Fensel, D., Hendler, J.A., Lieberman, H., Wahlster, W. (eds.) Spinning the Semantic Web, pp. 317–359. MIT Press, Cambridge (2003)
11. Rocha, C., Schwabe, D., de Aragão, M.P.: A Hybrid Approach for Searching in the Semantic Web. In: Proc. of the 13th International World Wide Web Conference (WWW 2004), NY (2004)
12. Sais, F., Pernelle, N., Rousset, M.: L2R: a logical method for reference reconciliation. In: AAAI 2007, pp. 329–334 (2007)
13. Stojanovic, N., Studer, R., Stojanovic, L.: An Approach for the Ranking of Query Results in the Semantic Web. In: Fensel, D., Sycara, K., Mylopoulos, J. (eds.) ISWC 2003. LNCS, vol. 2870, pp. 500–516. Springer, Heidelberg (2003)
14. Suroweicki, J.: The Wisdom of Crowds. Doubleday, New York (2004)
15. Tablan, V., Damljanovic, D., Bontcheva, K.: A Natural Language Query Interface to Structured Information. In: Bechhofer, S., Hauswirth, M., Hoffmann, J., Koubarakis, M. (eds.) ESWC 2008. LNCS, vol. 5021, pp. 361–375. Springer, Heidelberg (2008)
16. Winkler, W.: The state of record linkage and current research problems. US Bureau of the Census Technical Report RR99/04 (1999)

LODE: Linking Open Descriptions of Events

Ryan Shaw[1], Raphaël Troncy[2,3], and Lynda Hardman[3]

[1] University of California, Berkeley, USA
ryanshaw@ischool.berkeley.edu
[2] EURECOM, Sophia Antipolis, France
raphael.troncy@eurecom.fr
[3] CWI, Amsterdam, The Netherlands
lynda.hardman@cwi.nl

Abstract. People conventionally refer to an action or occurrence taking place at a certain time at a specific location as an *event*. This notion is potentially useful for connecting individual facts recorded in the rapidly growing collection of linked data sets and for discovering more complex relationships between data. In this paper, we provide an overview and comparison of existing event models, looking at the different choices they make of how to represent events. We describe a model for publishing records of events as Linked Data. We present tools for populating this model and a prototype "event directory" web service, which can be used to locate stable URIs for events that have occurred, provide RDFS+OWL descriptions and link to related resources.

1   Introduction

Though their specific methods differ significantly, both historians and journalists work to produce narrative chains of events to explain phenomena in the past. The resulting historical records of events constitute valuable cultural heritage of interest to academics as well as the general public. The Linked Data[1] effort seeks to publish and connect RDF data sets on the Web using dereferenceable URIs for identifying web documents, real-world objects, links between them and/or other pieces of information. Yet, while standard and widely used vocabularies have emerged for representing people, places, and other types of entities as Linked Data, none has yet emerged specifically for events.

The term "event" has several meanings. It is used to mean both phenomena that have happened (e.g. things reported in news articles or explained by historians) and phenomena that are scheduled to happen (e.g. things put in calendars and datebooks). Various standards and formats have been proposed for representing the latter as structured data, usually for personal information management purposes. In this paper, we focus on the former category: phenomena that have happened in the past.

This paper makes two contributions. First, we compare existing models for representing historical events (Section 2). These models serve different communities and have different strengths. Our goal is not to propose yet another ontology

---

[1] http://linkeddata.org/

A. Gómez-Pérez, Y. Yu, and Y. Ding (Eds.): ASWC 2009, LNCS 5926, pp. 153–167, 2009.

per se, but rather to build an *interlingua* model that solves an interoperability problem by providing a set of axioms expressing mappings between existing event ontologies (Section 3). Second, we present tools for populating this model with data coming from existing sources, such as Wikipedia timelines. We describe a prototype of an "event directory"[2] web service which can be used to locate stable URIs for past events and to provide RDFS+OWL descriptions of those events and links to related resources (Section 4). Finally, we give our conclusions and outline future work in Section 5.

2   Comparison of Existing Event Models

A number of different RDFS+OWL ontologies providing classes and properties for modeling events and their relationships have been proposed (see Table 1). In this section, we present an analysis based on their main constituent properties: type (Section 2.2), time (Section 2.3), space (Section 2.4), participation (Section 2.5), causality (Section 2.6) and composition (Section 2.7). This builds upon previous work in which we examined a number of different non-RDFS+ OWL models for representing information about events [9].

2.1   Event Models Overview

Though all of the ontologies presented in Table 1 provide classes and properties suitable for representing events, they were created to serve different purposes. The CIDOC CRM [2] and ABC [6] ontologies aim at enabling interoperability among metadata standards for describing complex multimedia objects held by museums and libraries. The events they intend to describe include both historical events in the broad sense (e.g. wars, or births) as well as events in the histories of the objects being described (e.g. changes of ownership, or restoration).

Table 1. Ontologies for representing events

Event model	Ontology URL
CIDOC CRM	http://cidoc.ics.forth.gr/OWL/cidoc_v4.2.owl
ABC Ontology	http://metadata.net/harmony/ABC/ABC.owl
Event Ontology	http://purl.org/NET/c4dm/event.owl#
EventsML-G2	http://www.iptc.org/EventsML/
DOLCE+DnS Ultralite	http://www.loa-cnr.it/ontologies/DUL.owl
F	http://events.semantic-multimedia.org/ontology/2008/12/15/model.owl
OpenCYC Ontology	http://www.opencyc.org/

The Event Ontology (EO) [7] was developed by the Centre for Digital Music to be used in conjunction with music-related ontologies. Although intended to

---

[2] We provide an interface for searching and browsing linked descriptions of events at http://www.linkedevents.org

describe events such as performances or sound generation, there is nothing specific to the music domain. It is currently the most commonly used event ontology in the Linked Data community. EventsML-G2 has been developed by the International Press Telecommunications Council (IPTC) for exchanging structured information about events among news providers and their partners. It describes both planned, past or breaking events as reported in the news.

DOLCE+DnS Ultralite (DUL) is a lightweight "upper" ontology for grounding domain-specific ontologies in a set of well-analyzed basic concepts. It is a combination and simplification of the DOLCE foundational ontology and the Constructive Descriptions and Situations pattern for representing aspects of social reality [3]. The F Event Model is a formal model of events built on top of DUL. It provides additional properties and classes for modeling participation in events, as well as parthood relations, causal relations, and correlations between events. F also provides the ability to assert that multiple models represent views upon or interpretations of the same event [8]. OpenCYC is also an "upper" ontology, but at the other end of the spectrum from DUL: rather than being a lightweight set of core concepts it provides hundreds of thousands of concepts intended to model "all of human consensus reality".

2.2 Fundamental Types of Events: Aspect and Agentivity

Given their different intended applications, these ontologies define events in varying ways. Table 2 provides a comparison of the prose descriptions for the top-level event classes. Furthermore, all of these ontologies, with the exception of EO, make an attempt to distinguish among some fundamental types of events. The basis upon which these distinctions are made vary.

One way to distinguish types of events is their *aspect*, i.e. whether the event involved is an ongoing activity or process, or the completion of some activity or transition between states. For example, OpenCYC defines a concept called `Situation` and uses aspect to distinguish between two main specializations of this concept: `StaticSituation` and `Event`. The former denotes a situation in

Table 2. Definitions of top-level event-related classes

`cidoc:E2.Temporal-_Entity`	"[E2.Temporal_Entity] comprises all phenomena, such as the instances of E4.Periods, E5.Events and states, which happen over a limited extent in time."
`abc:Event`	"An `Event` marks a transition between `Situations`."
`eo:Event`	"An arbitrary classification of a space/time region, by a cognitive agent."
`eventsml:Event`	"...something that happens and is subject to news coverage."
`dul:Event`	"Any physical, social, or mental process, event, or state."
`f:Event`	"...perduring entities (or perdurants or occurants) that unfold over time, *i.e.*, they take up time.."
`cyc:Situation`	"...a state or event consisting of one or more objects having certain properties or bearing certain relations to each other."

which some state of affairs has persisted throughout the situation's interval of time, while the latter denotes a situation in which some change has occurred during the situation's interval of time.

CIDOC makes a similar but conceptually less clear distinction between two types of E2.Temporal_Entity: E3.Condition_State and E5.Event. It is less clear because CIDOC also introduces the concept E4.Period, a type of temporal entity that is not static, but does not necessarily involve a change of state. E3.Condition_State is also defined narrowly to denote only descriptions of "the prevailing physical condition of any material object or feature" which would seem to exclude descriptions of, for example, the relative state of two things. E3.Condition_State is similar to the ABC ontology's Situation concept, instances of which describe the states of tangible things at particular times. The ABC ontology then uses this Situation concept to narrowly define an Event concept as a transition between two different Situation instances. This makes it difficult to describe an event that is characterized by a change in the relationship between two things rather than a change in the state of a single object.

Another distinction is whether an *agent* is identified as having produced the event. Both OpenCyc and DUL distinguish an Action as a particular type of Event, and CIDOC distinguishes an E7.Activity as a particular type of E5.Event. The ABC ontology also distinguishes an Action concept as something performed by an agent, but rather than being a specialization of the Event concept, it is defined as disjoint with the Event concept, which can "contain" actions via a hasAction property. Thus the ABC ontology suggests that events are fully described as sets of actions taken by specific agents, which may be an issue for modeling events such as earthquakes.

One potential problem with building these types of classifications into an ontology for modeling things that happened is that they force a knowledge engineer to adopt a particular perspective on what happened. This is desirable for precise modeling in specific domains that share a descriptive paradigm, but it is undesirable if the goal is to enhance access to documents which may present different interpretations of the same events. Distinctions based on aspect or agentivity are not necessarily inherent to what happened, but instead are rooted in particular interpretations. Whether a historical event or a event reported in the news involves an identifiable change or not, or whether agency can be assigned, is often a matter of debate, and its resolution should not be a prerequisite for representing what happened using a concept from an ontology.

This desire to separate events from their interpretations is what drives the approach taken by DUL, which provides a Situation concept, instances of which may describe different views or interpretations of the same Event instance. Using the DUL ontology, the types of classifications discussed above would be applied to instances of Situation rather than to instances of Event[3].

---

[3] DUL does specialize its Event concept on the basis of agentivity, providing the Action concept for events that have at least one participating agent and the Process concept for events that are not recognized having participating agents.

2.3   Events and Temporal Intervals

Temporality is a major distinguishing feature of events as entities, requiring modeling spans of time and relating events to these. The relationship between events and chronological spans of time is analogous to the relationship between places and spatial coordinate systems. In each case, instances of the former have persistent, socially attributed meanings, while the latter are arbitrary systems for subdividing an abstract space. One approach to linking events to ranges of time uses datatype properties, directly relating event instances with RDF literals representing calendar dates (and thus typed using one of the date-related XML Schema datatypes such as xsd:date or xsd:dateTime). Another approach introduces a class for representing temporal intervals, and uses object properties to link event instances with instances of this class. Temporal interval instances can then be linked to calendar values using datatype properties.

ABC, CIDOC, and EO all take the second approach, with ABC and CIDOC introducing classes for temporal intervals, and EO using the TemporalEntity class from OWL-Time [5]. DUL allows both approaches: dates for an event can be directly asserted using the hasEventDate datatype property, or the temporal interval involved can be made explicit by instantiating the TimeInterval class and linking an event instance to it using the isObservableAt object property.

The advantage of associating dates directly with events is simplicity: there are fewer abstractions to deal with, and it is simple to filter or sort events using standard date parsing and comparison routines. This also makes it simple to export lists of events for visualization on a timeline. But the tradeoff for this simplicity is an inability to express more complex relationships to time, such as temporal intervals that do not coincide with date units, or uncertainty about when precisely an event took place within some bounded temporal interval. This is a problem for representing historical events.

By introducing classes for representing temporal intervals, one can use a temporal calculus for reasoning about these more complex relationships. For example, if the precise date of a historical event is not known but some boundaries can be established within which it must have occurred, the time between these boundaries can be represented as a temporal interval, and a containment relationship can be asserted between that interval and the (unknown) interval during which the event occurred. The drawback to such an approach is that it can be off-puttingly complex as it introduces a number of abstract entities. The problem also arises of how to either mint URIs to identify these entities or deal with the problems introduced by using blank nodes.

2.4   Events, Spaces and Places

Events can be linked to abstract temporal regions (Section 2.3) and to abstract spatial regions or to semantically significant places. ABC, CIDOC and EO only support linking to spatial regions. CIDOC provides a class (E53.Place) for "extent in space" to which events can be related via the P7.took_place_at property. Instances of E53.Place may have names (E44.Place_Appellation), but there is no way to link an event to a place name except through a specific spatial extent. ABC's `Place` class also emphasizes spatial location rather than meaningful place. EO's place property has a range of `wgs84:SpatialThing`, which is also defined in terms of spatial extent.

Only DUL makes an explicit place/space distinction between `Place` and `SpaceRegion`. An event instance can be related to a `Place` via the `hasLocation` property, or related to a `SpaceRegion` via the `hasRegion` property. This is the most flexible approach, as it allows one to make assertions about events that occurred in places not easily resolvable to geospatial coordinate systems. For example, scholars of ancient history may work with documents that do not distinguish between real and mythical events. These scholars may wish to indicate that some event is recorded as having occurred at a mythical place. Similar problems are posed by contemporary events which may occur at virtual places such as those found within massive multi-player online environments. In both cases it is convenient to be able to associate events to such places without having to specify geospatial coordinates for them. Furthermore, making a clear distinction between named places and spatial regions enables one to deal properly with the phenomenon of places changing their absolute spatial location over time.

2.5   Participation in Events

The event ontologies also provide properties for linking agents, such as people and organizations, and the things involved in them.

Object Involvement in Events. ABC defines two types of properties for relating an `Event` to a tangible thing (an `Actuality` in ABC parlance). The `involves` property does not imply anything beyond simple involvement. The `hasResult` property relates an `Event` to a tangible thing or attribute of a thing which exists as a result of that `Event`. ABC also defines various sub-properties of these two properties that further specialize these meanings. For example `destroys` is a specialization of `involves` implying that the involved `Actuality` ceased to exist as a result of its involvement in the `Event`.

CIDOC defines a property `P12.occurred_in_the_presence_of`, which like ABC's `involves` relates an `E5.Event` to a `E77.Persistent_Item` (endurant) without committing to any implied role for that item beyond simple involvement. `P12.occurred_in_the_presence_of` is the root of a hierarchy of properties expressing more specialized forms of involvement such as `P25.moved` and `P31.has_modified`. Unlike ABC's `Actuality`, CIDOC's `E77.Persistent_Item` encompasses not only tangible entities but also intangible concepts or ideas, making CIDOC's `P12.occurred_in_the_presence_of` a broader concept than ABC's `involves`. DUL defines a `hasParticipant` for relating an `Event` to an `Object`. Like CIDOC's `E77.Persistent_Item`, DUL's `Object` includes social and mental objects as well as physical ones. EO's `factor` property, having no range defined, is similarly broad. EO also defines a `product` property that, like ABC's `hasResult`, links an `Event` to some thing that exists as a result of that `Event`.

Agent Participation in Events. ABC defines a `hasPresence` property for weakly asserting that an agent was present at an event without implying that the agent took an active role. It is specialized by the `hasParticipant` property, which does imply an active or causal role for the agent. CIDOC's equivalent of ABC's `hasPresence` is `P11.had_participant`, and its equivalent of ABC's `hasParticipant` is `P14.carried_out_by`. DUL's `involvesAgent` property is a specialization of `hasParticipant` for relating an `Event` to an `Agent`. EO provides the `agent` property for the same purpose.

F stands apart from the other ontologies in what it offers for modeling participation. Using DUL, one can assert that a given object or agent participated in an event. F uses the *descriptions and situations* (DnS) pattern[3] to enable a further classification of this participation as an instance of some role-based class. For example, using DUL one might state that the agents Brian Boru and Máel Mórda mac Murchada participated in the Battle of Clontarf. Using F, one can further state that the Battle of Contarf is classified as a *battle*, that *battles* have *commanders*, and that Brian and Máel Mórda are classified as *commanders*.

CIDOC's `P14.1_in_the_role_of` property provides some support for classifying an agent's participation in an event as an instantiation of a particular role. However, since it is defined as a property of the `P14.carried_out_by` property, it requires the use of OWL Full. Furthermore, there does not seem to be a way to associate roles with generic event schemas in the manner described above.

2.6   Events, Influence, Purpose and Causality

Event models vary in their approaches to modeling relations of causality, purpose, or influence. Both EO and CIDOC provide properties for making broad assertions linking events to any relevant thing (tangible or not). CIDOC defines `P15.was_influenced_by`, while EO defines `factor`. EO does not distinguish between a thing's participation in an event and a thing's influence upon an event, using the same property for both relations. Likewise, it seems that the only difference between CIDOC's `P12.occurred_in_the_presence_of` and `P15.was_influenced_by` is whether the relevant thing was physically present (and, by implication, a `E77.Persistent_Item`). The only support that ABC offers for making assertions about causality is the `hasResult` property.

In historical discourse there is often a lack of consensus about causality, purpose, or influence. Thus simple properties like these are unlikely to be adequate for modeling assertions about such relations. Here the F model's DnS pattern provides a more powerful and flexible modeling tool. Unlike the other models, F takes the position that only other events can stand in causal relation to an event. Rather than directly linking events via a property expressing causality, events are included in an `EventCausalitySituation`. The `EventCausalitySituation` includes not only the events being classified as the cause and the effect, but also the theory under which causality is being asserted. Using the F model's interpretation pattern, one can assert that a given `EventCausalitySituation` is part of a specific interpretation of an event. Thus multiple, potentially conflicting causality relations can be asserted for the same set of events by specifying the interpretive context in which the relations are made.

2.7 Events, Parts and Composition

Often, it is desirable to model an event A as being part of some other event B. While an event A's being part of event B implies that event B's timespan contains event A's timespan, event parthood is more than temporal containment. One may get married during the Olympics, but that does not make one's marriage part of the Olympics. Thus, event ontologies must distinguish between mere temporal containment and mereological relationships between sub-events and some greater event. Ontologies that make a distinction between temporal spans and events can clearly distinguish between the two types of relationships.

CIDOC distinguishes between time-spans and periods/events, and provides the P86.falls_within property to express containment relations among time-spans, and the P9.consists_of property to express part-of relationships among events. EO defines a sub_event property, and ABC defines an isSubEventOf property for expressing mereological relationships among events. Since ABC conceptualizes events as sets of actions taken by specific agents, it also provides the hasAction property for linking events to the actions they contain.

DUL defines two properties for linking events to sub-events: hasPart and hasConstituent. hasPart can be used both for temporal containment relationships such as "the 20th century contains year 1923" and for semantic relationships such as "World War II included Pearl Harbour". dul:hasConstituent attempts to capture the notion that we sometimes model aspects of the world as consisting of layers at different levels of abstraction, which are not strictly parts of one another. Thus society is constituted of individual people, even though you might not want to say that people are "parts" of society because people and societies exist at different levels of abstraction. This distinction is useful for events as well, as it allows us to describe a large and complex event like the French Revolution as being constituted of many smaller events, even though these smaller events may not be "parts" of the larger event in the same sense that a set is part of a tennis match.

In keeping with its use of the DnS pattern, F enables one to define a high-level description of how an event can be composed of smaller events. Specific situations (i.e. specific groups of events) can then satisfy this description. This allows one to simply describe the conditions under which an event is considered to be part of another event, and infer parthood based on this description, rather than requiring parthood to be explicitly asserted every time. For large events that may contain large numbers of sub-events, this could be quite useful. And, of course, F's interpretation pattern allows for multiple, potentially conflicting decompositions of the same event.

3 Towards a Linked Data Event Model

We propose a minimal model that encapsulates the most useful properties of the models reviewed. Our goal is to enable interoperable modeling of the "factual" aspects of events, where these can be characterized in terms of the *four Ws*:

Table 3. Excerpt of approximate mappings between properties from various event models

ABC	CIDOC	DUL	EO	LODE
atTime	P4.has_time-span	isObservableAt	time	atTime
	P7.took_place_at		place	inSpace
inPlace		hasLocation		atPlace
involves	P12.occurred_in_the_presence_of	hasParticipant	factor	involved
hasPresence	P11.had_participant	involvesAgent	agent	involvedAgent

What happened, *Where* did it happen, *When* did it happen, and *Who* was involved. "Factual" relations within and among events are intended to represent intersubjective "consensus reality" and thus are not necessarily associated with a particular perspective or interpretation. Our model thus allows us to express characteristics about which a stable consensus has been reached, whether these are considered to be empirically given or rhetorically produced will depend on one's epistemological stance. We exclude properties for categorizing events or for relating them to other events through parthood or causal relations. We believe that these aspects belong to an interpretive dimension best handled through the DnS approach of the F event model.

Table 3 shows the main properties of our model, aligned with approximately equivalent properties from the models discussed above. For the actual equivalence relations, see the ontology itself at `http://linkedevents.org/model/`.

Agentivity. Our model is agnostic with regard to judgements of aspect or agentivity (see Section 2.2). Users are free to model historical or reported events without taking a position on what has changed or where agency lies. This agnosticism has consequences for mapping our `Event` class to those defined by other models. We consider our `Event` class to be directly equivalent to those defined by EO and DUL, as both of these are also agnostic with respect to aspect and agentivity. Our event class is not equivalent to the `E5.Event` class, since CIDOC defines `E5.Event` to exclude ongoing states, activities, or processes. Because we wish to support the modeling of such static entities as events, we define our `Event` class to be a subclass of CIDOC's `E2.TemporalEntity`, which is the superclass of `E5.Event` (via `E4.Period`) and `E3.Condition_State`. Our `Event` class is a subclass of `E2.TemporalEntity` because the latter is defined as "anything that happens over a limited extent in time", which is more general than the definition we wish to give. Specifically, we want to restrict our definition to only include those things happening over a limited extent in time that have been reported as events by some agent, e.g. a historian or journalist.

Time. We link events to ranges of time via instances of a temporal interval class. Like EO, we use `TemporalEntity` from OWL-Time as our temporal interval class, so our `atTime` property is directly equivalent to EO's `time` property. `atTime` is a subclass of DUL's `isObservableAt` property, as it restricts the domain of the latter to include only events. Likewise, `atTime` is a sub-property of CIDOC's `P4.has_time-span` because it restricts the domain of the latter to include only events (as we define them here) rather than any temporal entity (recall that our event class is a subclass of CIDOC's `E2.TemporalEntity`). We also define `atTime` to be an OWL `FunctionalProperty`, meaning that an event can be associated with at most one interval of time. Where there may be disagreement about the interval of time associated with an event, this disagreement should be modeled at an interpretive level beyond the scope of our model, and the value of `atTime` should either be specified as the shortest temporal interval that includes the conflicting interpretations, or left unspecified.

Space. We follow DUL in making an explicit distinction between abstract spatial regions and semantically significant places. Our `inSpace` property relates an event to some subjectively imposed spatial boundaries, i.e. a region of space. Like `atTime`, `inSpace` is a `FunctionalProperty`, so an event can be related to at most one such region of space. `inSpace` is a sub-property of DUL's `hasRegion` because it restricts its domain to include only events, not all entities, and because it restricts its range to include only spatial regions, not any dimensional space. In keeping with EO, we use `SpatialThing` from the Basic Geo (WGS84 lat/long) Vocabulary as our spatial region class, so our `inSpace` property is directly equivalent to EO's `place` property. Because our concept of an event is broader than the one defined by the CIDOC CRM, `inSpace` is a super-property of CIDOC's `P7.took_place_at`. While the range of `inSpace` is an abstract spatial extent, it is often desirable to express relationships to socially defined places. We define an `atPlace` property to associate an event with some meaningful place(s), whether or not it is possible to define spatial boundaries for those places. Unlike `inSpace`, `atPlace` is not a `FunctionalProperty`, so an event can be related to any number of places. `atPlace` is a sub-property of DUL's `hasLocation` property, because it restricts the latter such that the domain includes only events and the range includes only places (not any entity).

Participation. Like DUL, we define a property for linking events to arbitrary things (`involved`) and a single specialization of this property for linking events to agents (`involvedAgent`). These two properties are directly equivalent to DUL's `hasParticipant` and `involvesAgent`, respectively. They are roughly equivalent to CIDOC's `P12.occurred_in_the_presence_of` and `P11.had_participant` (though not directly equivalent given our broader event concept). The mapping to EO is more complicated. `involved` is more specific than EO's `factor` property because it restricts the range of the latter to include only objects and not, for example, "abstract causes." But it is also more general, because it does not imply (as `factor` does) a "passive" role for the involved object. Thus there is no formal equivalence relationship stated between the two. `involvedAgent` is a super-property of EO's `agent` property because it generalizes the latter to include all relations to agents, whether or not their role is "active" or "passive." Judgments of activity or passivity are higher-level interpretations that go beyond our goal of modeling only "factual" aspects.

Causality. Finally, as discussed above, our model contains no properties for expressing relations of influence, purpose, or causality. Therefore, there are no properties equivalent to CIDOC's P15.was_influenced_by or EO's factor. Similarly, we provide no properties for expressing parthood relations among events. We believe these higher-level interpretations are best handled via a layer of descriptions and situations over the basic statements expressible using our model. The F event model provides an exemplary blueprint.

4 Applications

For demonstrating the usefulness of our proposed model, we set up two experiments. First, we extract events from Wikipedia timelines in order to test whether we can represent these events accurately in the Web of Data (Section 4.1). Second, we load existing instances of events represented according to the various event models reviewed in this paper in order to test the interoperability we claim our model brings (Section 4.2). We provide an interface for searching, browsing and visualizing all these events at http://www.linkedevents.org.

4.1 Extracting Events from Wikipedia Timelines

The events found in Wikipedia timelines vary widely in scope and domain, making them a good challenge for modeling. We also demonstrate that Wikipedia timelines provide a source of structured data not yet tapped by projects such as DBpedia[4] and Freebase[5]. Since timelines on related topics are spread throughout Wikipedia, extracting their events and modeling them as linked data is useful for enabling aggregated views of these events and for exploring related topics.

Timelines appear in Wikipedia in two major forms. *Dedicated topic-specific* timeline articles, such as "Timeline of historic inventions", take the form of a list or table of events. As of October 2008, there were approximately 1000 such articles in Wikipedia. The list or table of events is usually divided into temporal groups (e.g. *September 1939* or *12th century*) by subheadings. Each event consists of (at a minimum) a date and a short description. The description generally contains words or phrases linked to other articles in the typical Wikipedia manner. The second form of timeline found in Wikipedia is *date-specific* timeline articles, such as "1996 in Ireland". In addition to short lists of events in the form described above, these articles usually also include some type-specific lists of events such as births, deaths, and sporting events that took place in that

---

[4] http://dbpedia.org/
[5] http://freebase.com/

year. The most general form of this type of article is the "Year" article (e.g. "1979"). Uses of a given year in any Wikipedia article are usually linked to the corresponding "Year" article. Similarly, uses of a given day of the month (e.g. "May 24") are usually linked to the corresponding "Month Day" article. These two types of article are highly mutually interlinked.

Date-specific timeline articles have a more standard format, making them more amenable to the extraction of structured data. But the events in date-specific timelines rarely have anything in common other than the year or day of the month with which they are associated. Since we were interested in linking events to one another via places, people, and other topics, we decided to focus on topic-specific timeline articles. Unfortunately, the formats for topic-specific timeline articles vary widely, making it difficult to create a generic parser and scraper. Many topic-specific timelines add additional fields for each event. For example, the "Timeline of Chinese history" includes a field for ruler or Emperor as well as the standard date and description. Other timelines group events in idiosyncratic ways, such as the "Timeline of punk rock" which categorizes the events of each year into "Bands formed", "Disbandments", "Albums [released]", and "Singles [released]". Furthermore, the timelines vary in the temporal granularity of their events: while some timelines specify specific days for their events, others only specify months or years. These variations illustrate how the structure of events can vary according to the topical context and the need for a flexible data model to accommodate them.

To populate instances of our event model, we wrote article-specific parsers for a number of the most active timeline articles. The parsers identify individual event entries within articles and from each entry extract the date and textual description. The parsers also extract the article subheading under which each entry appears for two reasons. First of all, the date specified in an entry is often given relative to the subheading. For example, events listed under the subheading *September 1939* may only specify a day of the month, with the month and year left implicit. Second, the subheadings provide a convenient means of linking back to the specific article section from which the event was extracted.

After the article-specific extraction, we use the extracted dates and descriptions to model our events. Dates are modeled using OWL-Time and linked to the event using the `atTime` property. Links to other Wikipedia articles found within the descriptions are used to identify other entities related to the event. We use type ontologies from DBpedia to determine what type of relation to create between an event and another entity. For example, if an event has the description "Canada declares war on Germany" and the word "Canada" is linked to the Wikipedia article of the same name, we then look up the corresponding resource in DBpedia (`http://dbpedia.org/resource/Canada`) and see what types have been assigned to it. `http://dbpedia.org/resource/Canada` has the type `http://dbpedia.org/ontology/Place` assigned to it, so we relate it to our event with the `atPlace` property. If DBpedia does not assign any usable types to the entity, we default to creating an `involves` relation.

Our initial set of events were extracted from four Wikipedia timelines:

- "Timeline of World War II" provides seven year-specific timelines of global events involving people at the granularity of single days.
- "Timeline of Irish History" provides events from a single geographic location spread over a wide temporal range, from the Stone Age to present day.
- "Timeline for the day of the September 11 attacks" provides a set of 147 very fine-grained events from a single day.
- "Timeline of evolution" tested our ability to model very coarse-grained events associated with times far in the past.

4.2   Interoperability with Legacy Event Collections

To evaluate the mappings between our model and other vocabularies, we combined our Wikipedia events with two collections of events modeled using other event vocabularies: the C4DM Event Ontology and the BIO[6] vocabulary for biographical information. The goal was to be able to browse and view event descriptions using Cliopatria, a generic semantic search web-server [12]. We defined views and facets only in terms of our event model but rely on our mappings to translate the legacy event collections to these views.

Congressional Biographies. The Biographical Directory of the U.S. Congress provides short biographical articles, as a series of statements describing life events, on every member of the United States legislature from 1774 to the present. The consistent structure allows simple extraction and modeling of events. In earlier work 69,228 events were modeled using the BIO vocabulary.

The Emma Goldman Chronology. The Emma Goldman Papers editors maintain a day-by-day chronology detailing where Emma Goldman and her associates were and what they were doing. This chronology serves as an internal reference tool, allowing the editors to make inferences about when or where documents may have been produced and to check for inconsistencies in historical accounts. Starting with a text document for the years 1910 through 1916, we produced an RDF data set by parsing dates, geocoding place names, and disambiguating personal names by linking them to DBpedia. These 1,041 Emma Goldman events were modeled using the C4DM Event Ontology.

Issues Mapping Between Vocabularies. To combine these legacy event collections with our Wikipedia events we used the mappings defined between our event model and the BIO and EO vocabularies. We found that our mappings were not sufficient to achieve our goal of using a single generic view to browse all three data sets, as there is not yet widespread support for the owl:equivalentClass and owl:equivalentProperty predicates, upon which our mappings rely. However, we were able to achieve our goal by making additional mapping statements using rdfs:subClass and rdfs:subProperty. These mappings enable us to work with multiple event collections as a unified whole without re-modeling.

---

[6] http://vocab.org/bio/0.1/

5   Conclusions and Future Work

There is a tremendous amount of timeline and chronology data on the web. There is also increasing interest in mining descriptions of historical events from narrative text, whether for temporal visualization of search results or for exploration of archival records. Historians and journalists are increasingly interested in presenting their work as structured data complementary to or in lieu of traditional narrative text. Yet, without some effort to bridge the various data models being developed and employed within these various applications, it will remain difficult to build the dense network of relations among them that could lead to new discoveries or novel modes of experiencing historical narrative. In this paper, we have presented a principled model for linking event-centric data that draws upon a close analysis of existing event ontologies. Our initial investigations show that it is useful for modeling a variety of timeline events and for mapping between events modeled using other vocabularies.

A number of questions remain to be answered. We have argued that a core event model should include only those relations about which a stable consensus has been reached, leaving more interpretive relations to a higher-level, application-specific models. But further application experience is needed before we can determine whether we have correctly identified those relations that are intersubjectively stable, or whether (for example) participation relations are interpretation-specific and ought to be moved outside the core model. A related problem is the question of event identification. In the applications discussed above, an event is identified with a single textual description. We have made no attempt to map multiple textual descriptions to the "same" event identifier. The reason for this is that it is not clear when (if ever) we should consider two textual descriptions to be of the "same" event. If we consider (as many contemporary philosophers of history do) events to be linguistic phenomena rather than objectively existing in the past, then there is no basis for arguing that two textual descriptions of an event refer to the same thing. At best we could say that they share a name, or that they refer to the same people, places, or spans of time. On the other hand, we clearly would like to say that two descriptions of past occurrences only differing in spelling or punctuation are the same event. These are deep philosophical questions about the nature of events that will likely only be answerable pragmatically, as we see which approaches are or are not useful for specific applications.

In future work, we plan on finding and working with more event collections modeled using the other ontologies discussed here, and putting these collections to use in a variety of applications. Current applications in development include event-centric searching and browsing of full-text historical scholarship, retrieval and display of historical context for documents by querying for related events, and interfaces for exploration, visualization, and comparison of events from a particular period or region.

Acknowledgments

The research leading to this paper was supported by the European Commission under contract FP6-027026, Knowledge Space of semantic inference for automatic annotation and retrieval of multimedia content – K-Space, and by the U.S. Institute of Museum and Library Services under a National Leadership Grant for Libraries (award number LG-06-06-0037-06).

References

1. Arndt, R., Troncy, R., Staab, S., Hardman, L., Vacura, M.: COMM: Designing a Well-Founded Multimedia Ontology for the Web. In: Aberer, K., Choi, K.-S., Noy, N., Allemang, D., Lee, K.-I., Nixon, L.J.B., Golbeck, J., Mika, P., Maynard, D., Mizoguchi, R., Schreiber, G., Cudré-Mauroux, P. (eds.) ASWC 2007 and ISWC 2007. LNCS, vol. 4825, pp. 30–43. Springer, Heidelberg (2007)
2. Doerr, M.: The CIDOC Conceptual Reference Module: An Ontological Approach to Semantic Interoperability of Metadata. AI Magazine 24(3), 75–92 (2003)
3. Gangemi, A., Mika, P.: Understanding the Semantic Web through Descriptions and Situations. In: Meersman, R., Tari, Z., Schmidt, D.C. (eds.) CoopIS 2003, DOA 2003, and ODBASE 2003. LNCS, vol. 2888, pp. 689–706. Springer, Heidelberg (2003)
4. Hildebrand, M., van Ossenbruggen, J., Hardman, L.: /facet: A Browser for Heterogeneous Semantic Web Repositories. In: Cruz, I., Decker, S., Allemang, D., Preist, C., Schwabe, D., Mika, P., Uschold, M., Aroyo, L.M. (eds.) ISWC 2006. LNCS, vol. 4273, pp. 272–285. Springer, Heidelberg (2006)
5. Hobbs, J., Pan, F.: Time Ontology in OWL. W3C Working Draft (2006), http://www.w3.org/TR/owl-time
6. Lagoze, C., Hunter, J.: The ABC Ontology and Model. Journal of Digital Information (JoDI) 2(2) (2001)
7. Raimond, Y., Abdallah, S., Sandler, M., Giasson, F.: The Music Ontology. In: 8th International Conference on Music Information Retrieval (ISMIR 2007), Vienna, Austria (2007)
8. Scherp, A., Franz, T., Saathoff, C., Staab, S.: F—A Model of Events based on the Foundational Ontology DOLCE+ Ultra Light. In: 5th International Conference on Knowledge Capture (K-CAP 2009), Redondo Beach, California, USA (2009)
9. Shaw, R., Larson, R.: Event Representation in Temporal and Geographic Context. In: Christensen-Dalsgaard, B., Castelli, D., Ammitzbøll Jurik, B., Lippincott, J. (eds.) ECDL 2008. LNCS, vol. 5173, pp. 415–418. Springer, Heidelberg (2008)
10. Troncy, R.: Bringing The IPTC News Architecture into the Semantic Web. In: Sheth, A.P., Staab, S., Dean, M., Paolucci, M., Maynard, D., Finin, T., Thirunarayan, K. (eds.) ISWC 2008. LNCS, vol. 5318, pp. 483–498. Springer, Heidelberg (2008)
11. van Hage, W., Malaisé, V., de Vries, G., Schreiber, G., van Someren, M.: Combining Ship Trajectories and Semantics with the Simple Event Model (SEM). In: 1st ACM International Workshop on Events in Multimedia (EiMM 2009), Beijing, China (2009)
12. Wielemaker, J., Hildebrand, M., van Ossenbruggen, J., Schreiber, G.: Thesaurus-based search in large heterogeneous collections. In: Sheth, A.P., Staab, S., Dean, M., Paolucci, M., Maynard, D., Finin, T., Thirunarayan, K. (eds.) ISWC 2008. LNCS, vol. 5318, pp. 695–708. Springer, Heidelberg (2008)

A Semantic Wiki Based Light-Weight Web Application Model

Jie Bao[1], Li Ding[1], Rui Huang[1], Paul R. Smart[2], Dave Braines[3], and Gareth Jones[3]

[1] Tetherless World Constellation, Rensselaer Polytechnic Institute, Troy, NY, USA
{baojie,dingl,huangr3}@cs.rpi.edu
[2] School of Electronics and Computer Science, University of Southampton, Southampton, UK
ps02v@ecs.soton.ac.uk
[3] Emerging Technology Services, IBM United Kingdom Ltd, Winchester, Hampshire, UK
{dave_braines,garethj}@uk.ibm.com

Abstract. Wiki is a well-known Web 2.0 content management platform. The recent advance of semantic wikis enriches the conventional wikis by allowing users to edit and query structured semantic annotations (e.g., categories and typed links) beyond plain wiki text. This new feature provided by semantic wikis, as shown in this paper, enables a novel, transparent, and light-weight social Web application model. This model let developers collectively build Web applications using semantic wikis, including for data modeling, data management, data processing and data presentation. The source scripts and data of such applications are transparent to Web users. Beyond a generic description for the Web application model, we show two proof-of-concept prototypes, namely RPI Map and CNL (Controlled Natural Language) Wiki, both of which are based on Semantic MediaWiki (SMW).

1 Introduction

The success of social Web applications (often called "Web 2.0" applications), such as Twitter, Wikipedia and Facebook, is evidenced by the fast growing, dynamic Web content contributed by millions of networked Web users. Unlike the conventional Web applications, where contents are primarily static and are exclusively contributed by the websites' owners, social Web applications grow contents by promoting Web users' collaborative contributions. These successful social Web applications share at least two common features:

- *Simple publishing*: a user can create, edit and publish a Web page without knowing much Web technologies, such as HTML and Web server configuration. For example, a Web form or a What-You-See-Is-What-You-Get (WYSIWYG) editor hides the details of HTML Web page editing; a click of "upload" button hides the details of uploading and publishing Web pages to a Web server.

- *Social interaction*: content publishing can be the result of user participation: users can collaboratively compose and improve one article on Wikipedia, or can update their status and opinions with their friends on Facebook. Such an interactive social content contribution mechanism promotes a network effect where the value of a service provided by a user increases as more people benefit from the service [5].

A. Gómez-Pérez, Y. Yu, and Y. Ding (Eds.): ASWC 2009, LNCS 5926, pp. 168–183, 2009.

Among successful Web 2.0 platforms, wikis, exemplified by Wikipedia, are known for promoting the convergence of collaborative writing. "The Wiki Way" [11] emphasizes the principle that the content of a wiki page should be collaboratively written using some *simple* markup languages in Web browsers, and collaborations are expected to grow and improve the content.

Social Web applications, including wikis, usually offer limited support to user contributed structured content. For example, a blog is usually submitted via a Web form with a fixed set of properties like title, content and tags. Although users may assert tags to annotate the semantics of a Web 2.0 page, they cannot declaratively publish the detailed structure or semantics of the content. Moreover, users have to follow the fixed user interaction design to access the structured annotation (e.g. author and date) of the published content. For example, the posts in the Craigs' list (www.craigslist.org) apparently contains latent structures (e.g., "2004 honda civic 2dr,..."), but users can only use text search to locate or filter their interested ones. This limitation leaves Web users limited means for preserving the structure of data to (i) avoid unnecessary overhead in natural language understanding and (ii) leverage smart services (such as semantic search and inference) that utilize the preserved semantics.

A number of efforts have been recently observed in addressing the above limitations with Semantic Web technologies. In particular, *semantic wiki* systems, such as Semantic Mediawiki (SMW) [9] and IkeWiki [13], have been developed to extend conventional wikis by additionally supporting simple semantic annotations on wiki pages, such as categories and typed links. A wiki page in semantic wikis may contain both conventional wiki text and structured data, and the structured data can be further accessed by customizable queries using some simple query languages.

By supporting both *annotation* and *query* of structured data on wiki pages, semantic wikis may serve as a platform for light-weight data modeling, computation and presentation tasks that are traditionally out of end users' control. Semantic wikis, therefore, promote a new application model with a couple of interesting characteristics:

- *Rich data modeling*: User contributed content may be a mixture of text and structured data, and semantic wikis can best preserve structured data without forcing structured representation of the free text part. With the structured data, a semantic wikis can function like a light-weight database or a knowledge base, and users can model data using several common modeling methods, e.g., relational modeling or rule modeling.

- *Transparent data processing*: as wiki allows simple computing logics (such as declaring an object and applying a data processing rule) to be published in forms of wiki, they are transparent to all wiki users and can be collaboratively authored and improved in Web browsers.

- *Social programming*: The transparency of data modeling and data processing and the convergence model of wiki itself opens up the development of Web applications to all interested users.

In this paper, we provide a generic description about the semantic wiki based Web application model, and then present two proof-of-concept prototypes, namely RPI Map and CNL (Controlled Natural Language) Wiki, both of which are based on SMW. The main contributions of the paper are the following:

- Identification of a light-weight Web application development model that possesses the aforementioned characteristics (Section 2);
- Working prototypes that embodies the identified model using the SMW platform (Section 3 and 4). In particular, we show that templates in SMW are useful in supporting some common data modeling and data processing tasks.

Although our description and demonstrations are limited to SMW-based implementations, the identified model is not necessarily limited to SMW or wiki-based implementations. We note that our demonstrations still carry some limitations which are mainly related to the SMW implementation. As more and more Web 2.0 applications are provided with semantic extensions (e.g., Drupal[1]), we believe results discussed in this paper can also be observed in other platforms.

2   Semantic Wiki Based Web Application Model

In this section, we introduce a semantic wiki based Web application model in the context of the evolution of Web application models.

2.1   Comparison of Web Application Models

The advance of Web technology drives the evolution of Web application models. Starting from just being able to browse Web pages, Web users are now able to control content publishing with the help of Web 2.0 technologies. Wikis, blog systems (e.g., Drupal and Wordpress) and similar online content management systems further provide extensible computing infrastructures that support scripting and/or customizable plugins to facilitate collective Web application development. Recent advance in social semantic Web, such as SMW, allows users to collaboratively control structured data management. In Figure 1, we compare several Web application models.

In the *Conventional Model*, a Web application is composed of three clearly-separated major components, namely the Web browser, the Web server and the backend storage system (e.g., a database or a file system). Users of such an application are provided with limited control for contents in the application, such as browsing or search. The representation, computation and presentation components are primarily hosted on the server side and are controlled by webmasters only.

Other models have extended the Conventional Model with extra client-side control of data or computation. The *AJAX Model* [3], which adds an AJAX engine to act as a mediator between the browser and the server, is getting increasing popularity due to its powerful client side computing ability. It improves user experience in both data transfer (e.g., asynchronous data retrieval from the server without interfering with page display) and data presentation. For example, a powerful word processing system (e.g., Google Docs) can be used within a browser where the data is actually stored on the Web. It is notable that users may also insert client side scripts into Web applications for customized processing.

The *Wiki-based Model* enables end users to directly control some data content and presentation on the server side. For example, Wikipedia articles are collaboratively

---

[1] http://drupal.org/project/rdf

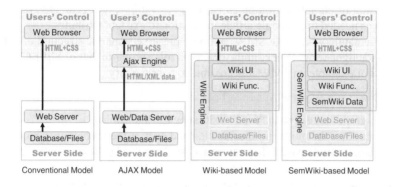

Fig. 1. A comparison of several Web application models

maintained by users and complex wiki templates are frequently used to enable advanced page layout (e.g., to render a calendar). A user may also call extensions of a wiki platform (e.g., "parser functions" in MediaWiki, the system used by Wikipedia) to perform certain computation tasks such as string processing, mathematical computation and visualization. It is also notable that a wiki page may embed external script languages (e.g., JavaScript) for advanced tasks.

Both the AJAX model and the Wiki-based model increase the user's control over data processing. The *SemWiki* (Semantic Wiki)-*based model* further grants users additional control on the management and consumption of structured data. For example, in Wikipedia, it is not yet possible to assert a structured, queryable annotation for a person's page, or to execute a query that "all European countries that have female government leaders". Semantic wikis address those limitations by extending wikis with the ability to create and query about structured annotations using a relatively simple modeling script and query language. As a result, users are now equipped with increasing ability to control data in the application. In particular, the SemWiki model enables a comprehensive in-browser scripting environment such that a light-weight Web applications can be built collectively with high transparency on computational logics (as computational scripts are included in wiki pages) and minimal required knowledge on Web server configuration. By light-weight we mean that the data structure, data processing logic and user interface of the application are relatively simple. In what follows, we elaborate the components of the SemWiki based model and several design patterns of this model.

2.2 Data Modeling

Semantic wikis are often built upon RDF triple stores for storing structured data. Thus, data in a semantic wiki does not required to be stored with a pre-defined schema (while it is also possible to do so) as an RDBMS will require. This conforms to the open nature of the Web and enables significant flexibility and extensibility in data modeling. Please also note that the semantic wiki model allows the hybrid modeling with predefined "schema", schema-free user added metadata, and unstructured data, thus makes the extension of an application much easier. For example, users can always add new attributes as needed to a specific article in addition to existent attributes.

Some semantic wikis (like SMW) not only preserve semantic structure of data, but also provide light-weight query ability (with its role similar to that of SELECT queries in SQL). For example, in SMW it is possible to pose a query

```
{{#show [[Category:Article]][[tag::<q>Category:food</q>]] }}
```

to find all articles tagged with "food" or its subtags (like "donut").

Note that since the modeling specification and queries themselves are also presented as some semantic wiki pages, they can be accessed, updated or deleted in the same fashion as for other wiki pages in the browser. Thus, semantic wikis function as a virtual abstraction layer over the Web server and database/file systems, such that programmers are not required to directly access the layers hidden below semantic wikis. This characteristic naturally enables collective construction of an application.

2.3   Data Processing

Several MediaWiki extensions provide scripting functionalities similar to that of the basic constructs of a programming language. When combined with templates, semantic annotations and semantic queries, these extensions can be used to support a wide range of light-weight data processing abilities. Some most useful extensions include[2]:

• **Variables:** General variables are supported by the Variable Extension so that users can name a long expression as a variable, and then reuse it later on the same wiki page. A special type of wiki pages called "template" pages also allow the use of variables as input parameters.

• **Datatypes:** The String Functions extension provides some common string functions such as string length and concatenation; the Array Extension provides array operations (e.g., search and sort) and set operations (e.g., union and intersect) on arrays.

• **Control Flow:** The Parser Functions Extension offers: (i) expression calculation that evaluates, e.g., mathematical expression like "(1+2)", and logical expressions like "(true and false)"; and (ii) conditional statements such as a IF-THEN-ELSE conditional flow. The Loop Extension supports loop structures such as WHILE and DO-WHILE.

2.4   User Interface

In SMW, many elements of a user interface (UI) in an application can be constructed using scripts. For example, the Semantic Forms[3] extension offers a form-based editing interface for users to edit template-based data. Utilizing templates and queries also allow us to control the look-and-feel of the user interface and present the data with various visual elements (e.g., table, picture and tree). Since templates and forms are also wiki pages and can be edited in browsers, the design and improvement of UI are also supported by collective scripting enabled by the semantic wiki model.

---

[2] See http://tw.rpi.edu/wiki/ASWC2009Bao#Links for their URLs
[3] http://www.mediawiki.org/wiki/Extension:Semantic_Forms

In addition, in MediaWiki (thus, also in SMW) users can also inject JavaScript code into a wiki page, either by including server-side scripts or code in some client-editable special wiki pages. Some SMW-based applications (e.g., wikicafe.metacafe.com and metavid.org) have developed advanced UIs such as video browsing and annotation. By aggregating the data management and data processing features with JavaScript, we are able to design interactive, visualized interfaces for the manipulation of semantically enriched data.

2.5   Strength and Limitations

By allowing data modeling, processing and presentation (via a user interface) abilities, semantic wikis provide a transparent platform for light-weight Web application development. In particular, such a development model enjoys several advantages:

- **Flexibility:** Because contents and scripts are both stored as wiki pages, users can always read and update them directly through browsers. Thus, the improvement to both contents and the application (as constructed with scripts) becomes a dynamic, highly portable, and easily accessible process.
- **Socialization:** Semantic wikis inherits the inherent collaborative nature of wikis, in particular the support of social user participation, e.g., user login, collaborative editing, and revision history. This may encourage large-scale, collaborative interactions between users.
- **Inference Ability:** The availability of semantically enriched content in semantic wikis makes it possible to do some inference with data, thus allows potentially better means in the consumption of data (e.g., search and query).

Nevertheless, it should be noted that the semantic wiki based model may carry some limitations:

- **Efficiency:** Semantic wikis often use a triple store for data storage. The state-of-the-art of triple stores has not yet reached the same level of maturity and scalability as that of relational databases. This may present some efficiency problem for applications with very large number of wiki pages. In addition, overhead of parsing and rendering structured data in semantic wiki pages often results in delays in response. Performance tuning for commercial deployment is thus often crucial.
- **Modeling Ability:** The native modeling support of semantic wikis is usually limited to a subset of RDF or OWL. The page-centric structure of knowledge organization in semantic wikis also makes the modeling of complex knowledge structure and data structure difficult. Thus, building an application that requires very complex data structure or logic with the semantic wiki based model can be challenging. This will be further discussed in Section 5.
- **Safety:** As wikis in general are designed to be an open collaborative environment, safety control is usually not natively supported, or with only limited realization. For Web applications requiring stronger access control to avoid malicious changes to the

content of the application, additional efforts are required to ensure data and application safety[4].

It should be noted that these limitations are mainly related to the current implementation of semantic wiki systems (like SMW), not to the general semantic wiki based model we have presented. We believe that many of these limitations will be overcome or alleviated with the advance in the semantic wiki development community.

In the next two sections, we will introduce two concrete examples of Web applications based on SMW, namely RPI Map and CNL Wiki. They illustrate, with emphasis of different usage patterns, how SMW enables light-weight data modeling, data processing and user interface building with an open, extensible architecture.

3   Case Study: RPI Map

This section introduces **RPI Map** (http://map.rpi.edu), a SMW-based Web application that exemplifies the general methodology we described in the previous section.

RPI Map is a campus map application for the Rensselaer Polytechnic Institute (RPI) community. It integrates and visualizes location based information, such as buildings, events and classes, on an interactive map using the Google Map API[5]. At its core is a Semantic MediaWiki along with a set of mediators that perform data mash-up from multiple external data sources. In what follows, we will describe in details how SMW help build RPI Map.

3.1   Data Modeling

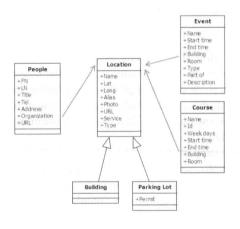

Fig. 2. Data Schema of RPI Map

Template as Schema. In RPI Map, templates play an important role for data organization as they serve as a "virtual" schema for data involved. Main types of data on RPI Map include locations (and its subtypes such as buildings and parking lots), people, events, courses, campus shuttle routes and real time shuttle positions. Many of these data are published by various individual entities across RPI. For example, event information is published as an RSS feed of the institutional calendar, course information is available from the RPI catalog as a text table, and people information is provided as downloadable vCard files from the RPI directory. To integrate these data in RPI Map, for each external data source there is a

---

[4] We noticed some recent advance in SMW access control, e.g., the HaloACL extension from Ontoprise.

[5] http://code.google.com/apis/maps/

mediator (implemented as server-side scripts[6]) to transform the original data into a form that can be consumed by the wiki platform. These data will then be linked by various semantic queries based on the location information (e.g., the building name) inside of them.

We use templates as the general output format of these mediators. Each template corresponds to a type of data in the system and describes a set of attributes that one such data instance must possess. For example, `Template:LocationInfo` defines a template with parameters of a location, e.g., its name, latitude, longitude and aliases. Together, these templates define a "schema" to organize data in the application, which is shown in Fig. 2.

Please note that while RPI Map uses schema-like templates for data modeling, these templates should not be understood as a relational schema in the database domain. These templates provide, on the basis of the triple-based data representation infrastructure of SMW, a higher level abstraction of some related triples. It is not required in RPI Map to have all data fits in a rigidly defined schema, or an instance of a template not having extra data that beyond what the template describes. Such an ability brings additional flexibility in accommodating data from heterogenous data sources.

3.2 Data Computation

Stored Query. Many queries are repeatedly used in many different components of RPI Map. For example, one commonly used query is to map a location based on its aliases. Such a query is stored as a template page `Template:Alias`:

```
{{#ask: [[Has alias::{{PAGENAME}}]] |link=none|limit=1}}
```

It may be embedded in other pages that need such a query.

Thus, templates can play a role similar to that of stored procedures in a relational database. As those templates can be edited in the browser by users (with some necessary protection mechanisms), it is more transparent and easier to access than stored procedures (which are normally hidden behind a server-side DBMS).

Data Cleansing. SMW can also help clean up corrupt or inaccurate data in the course of integrating data into RPI Map. For example, in transforming people information (in the vCard format) into wiki, the same location (e.g., a person's office address) may be called in a couple of different names across different branches of the university. In addition, new variations of a location's name may be discovered when new data is added (e.g., from the event RSS feed on daily basis). A special name recognition template was designed, partly leveraged by a fuzzy string similarity comparison parser function, to identify the closest known location or its aliases.

3.3 User Interface

Query-based Map Generation. Each of the map pages on RPI Map is based on some semantic queries. For example, the "Today Event" page relies on a query in the form[7]:

---

[6] It is also possible to use client-side script to do data importing, thus users may add other types of data to the system.

[7] For ease of presentation, the query is simplified from the actual query.

```
{{#vardefine: eventLocations|
   {{#ask: [[Category:Event]]
     [[has end time::> {{LOCALMONTHNAME}}
        {{LOCALDAY}},{{LOCALYEAR}} 00:00]]
     |?has location=|mainlabel=-|link=none}} }}
{{#map_objects:
   {{#ask: [[Has alias::
     {{#var:eventLocations}}]]
   | ?Has LatLong
   ...
   |limit=200|link=none }} }}
```

where map_objects is a function that will automatically generate a map via Google Map API from the result of the "ask" semantic query. The query asks for "all the locations (potentially in their alias forms) of today's events and their latitude/longitude (some other attributes omitted)". Please note that the example also demonstrates the use of variables in constructing complex queries.

Fig. 3. RPI Map Main Interface

Integration with JavaScript. In RPI Map, JavaScript is intensively used together with wiki scripts. Some examples are:

• Popping up a new window with additional information of a location, e.g., its full name, picture and services;

• Generating labeled markers and customizing icons;

• Validating of user-input location information;

• Displaying geographic information in an extant data format, e.g., Keyhole Markup Language (KML).

The main page of RPI Map is shown in Fig. 3.

Part of RPI Map source code for data representation and navigation has been released as the Tetherless Map Extension to Mediawiki[8].

4 Case Study: CNL Wiki

CNL Wiki (http://tw.rpi.edu/proj/cnl) is another application we developed that conforms to the architecture described in Section 2. The CNL Wiki is motivated at providing an end-user friendly interface for collaborative ontology building. We use a SMW

---

[8] http://www.mediawiki.org/wiki/Extension:Tetherless_Map

as the application platform exploiting its inherent collaborative nature, high portability and accessability. In addition, we utilizes Controlled Natural Language (CNL) to provide some support for ontology development in OWL with the intention to improve the comprehensibility of generated knowledge statements to end users. In this section, we will introduce in details how SMW enables the representation of strongly structured data (i.e., OWL knowledge bases), data computation (e.g., CNL sentence generation), and user interface generation. Additional details about CNL Wiki can be found in [1].

4.1 Data Modeling

In this subsection, we introduce how SMW templates can be used in modeling structured data and generating semantic data.

Modeling Structured Data. In order to accommodate ontology construction in OWL within SMW, we need to address a number of expressivity constraints associated with SMW. Currently, SMW does not provide full native support for OWL modeling. In order to address this limitation, we developed a meta-model extension to SMW, called SMW-mOWL (where "m" stands for meta model). Please refer [1] for a complete description of SMW-mOWL.

SMW-mOWL represents an OWL ontology using a set of wiki pages, each of which encodes some ontology elements (i.e., classes, properties, individuals and axioms) as wiki template instances. For example, suppose we have an OWL statement saying that "every father is a person that has a child who is also a person", which can be given in the OWL Abstract Syntax (OWL-AS) as:

```
Class(Father partial Person restriction
    (hasChild someValuesFrom(Person)))
```

This statement can be broken down into several template instances and represented as SMW pages. For example, on the page "Category:Father", the above statement in OWL-AS is represented with three template instances:

```
{{NamedClass          |label=Father      |plural=Fathers  }}
{{NamesClassRelation  |type=subClassOf   |class=Person  }}
{{someValuesFrom      |on property=hasChild |on class=Person  }}
```

Thus, each category page represents a single class in OWL along with some axioms about the class. The "`Template:NamedClass`" describes annotations to this class, such as comments and natural language labels. "`Template:NamesClassRelation`" describes relationship between two classes (here a class inclusion relationship). "`Template:someValuesFrom`" represents a restriction that the class in question must satisfy.

Semantic Data Generation. The use of a template-based mechanism for SMW-mOWL also allows us to store the knowledge model in the SMW database (tuple store). For example, an instance of `Template: someValuesFrom` will be persisted as an instance of the ternary property `owl:someValuesFrom` in the wiki of which the first element is the class where the template instance resides, the second element is the "on property" parameter, and the third element is the the "on class" parameter. Such persisted data in database can be further consumed by other scripts, e.g. for CNL generation (will be described in the next subsection), or external tools, e.g., a SPARQL query engine.

4.2 Data Computation

Once the SMW-mOWL meta-model is persisted in the database, the query language for SMW (SMW-QL) will be used to retrieve specific information from the model, which can be consumed by other wiki scripts. In this subsection, we describe two such usage patterns.

Templates as Functions. To query information stored in the SMW tuple store, we use a set of templates to implement query and some additional processing. In that sense, templates are used in a role similar to that of functions in a usual programming language. For example, the `Template:CNL.Rabbit.getLabel` takes input of a page's name (denoted as {{page}}), and do the following queries:

- query if the page is an anonymous class using

```
{{#ask: [[:{{page}}]][[Category:Anon]]
    |format=list|limit=1|link=none }}
```

The query result will be stored as a variable: it is empty (false) iff the class is an anonymous class.
- If it is an anonymous class, call `Template:CNL. Rabbit.Anon` to construct its label in the Rabbit CNL, otherwise return its label by calling a SMW query:

```
{{#ask: [[:{{page}}]]   |?CnlLabel= |mainlabel=-
                        |format=list|limit=1|link=none }}
```

CNL Generation. Queried results from SMW database will be further parsed and processed by a set of CNL generation templates. Currently, we support two CNLs in English, namely Rabbit [4] and Attempto Controlled English (ACE) [7].

For example, the `Template:CNL.Rabbit.getSomeRestrictionAssertion` template generates Rabbit CNL sentences about "someValuesFrom" restrictions of a class "{{page}}". It preforms the following tasks:

- Use `Template:CNL.Rabbit.getLabel` to get the natural language label of the class in question (i.e., the input parameter "{{page}}").
- Use a query to fetch all "someValuesFrom" restrictions related to "{{page}}":

```
{{ask: [[{{page}}]]   |?owl:someValuesFrom
    |mainlabel=-|format=list|link=none}}
```

- For each such a restriction, parse its "on property" and "on class" values, use `Template:CNL. Rabbit.getLabel` to get their natural language labels, and generate an Rabbit sentence using the Rabbit grammar. For instance, for the example in the last subsection, we will have "Every Father has child Person.".

A meta-model template like `Template:NamedClass` may call a CNL generation template, such as `Template:CNL.Ace.Concept` (which in turn calls other templates to construct all CNL sentences about a specific class). Thus, users will get CNL description of a knowledge statement whenever the statement is constructed by form-base editing or by importing from an external ontology. Fig. 4 shows such a CNL generation result about a property in an ontology in the Rabbit CNL.

Property:Eat [Edit]	
Type	String
Name	eats
Passive Name	is eaten by
Comments:	This property defines the relationship of **eat**
In Ontology	Rabbit Ontology

"Property:Eat" in the "Rabbit" controlled natural language

"**eats**" is a relationship

- The "**eats**" relationship can only have a "Life" as a subject
- The "**eats**" relationship can only have a "Consumable Thing" as an object
- The relationship "**eats**" is a special type of the relationship "consumes"
- The relationships "**eats**" and "takes as food" are equivalent
- The relationship "**eats**" is the complement of "is eaten by"
- The relationship "**eats**" can only refer to one thing

Fig. 4. A property represented in the Rabbit CNL

4.3 User Interface

Furthermore, structured data representation in SMW also allows user interface construction. This is again facilitated by (semantic) templates.

Controlling Page Layout. Similar to conventional wikis, templates in semantic wikis play important roles in controlling page layout. For example, `Template:Property` controls look-and-feel of a property, such as

- Content organization (e.g., as tables), color schema, font size and other visual elements of a page;
- Linking to the editing interface;
- CNL statements in selected CNL languages, each in a separate table section.

Different from conventional templates, a template in SMW is able to use semantic queries so that content from other pages can also be displayed on the page in question. In addition, by separating text content and semantic content of a page, SMW is able to partially reuse a page's content, and does not need to keep the original layout of the content of other pages. Those features are not available by conventional page inclusion in MediaWiki.

Light-weight GUI. Using semantic queries, structured information across multiple pages can be aggregated on one page with graphical representation. One such practice on the CNL Wiki is query-based class hierarchy tree. The template `Template:GUI.Tree` defines a recursive query that fetches class inclusion relations from a root class in a specified ontology. For example, the following script will create a tree presentation of all subclasses of Animal in the "Rabbit Ontology".

```
{{GUI.Tree |root=Category:Animal |ontology=Rabbit Ontology }}
```

We believe this approach can be extended to support displaying other types of GUI elements, such as toolbar, list and menu bar.

Form Generation. By utilizing the "Semantic Forms" extension of SMW, some template instances can be edited using a form-based interface. Generation of such forms can be automated from the template definition. Thus, having the template-based OWL meta-model immediately provides us with a light-weight OWL ontology editor within the SMW environment. Each form comprises some controls (textboxes, checkboxes, radio buttons, and so on) that support various editing operations. Auto-completion (which may in turn involves some queries) in semantic forms allows sentence editing using existing entities in the ontology.

5   Discussion and Related Work

5.1   Collaborative Web Application Development

A recent review [6] on the trends of Web application development has analyzed several popular Web development models and showed how they benefit from the advance of technologies and the evolution of user behavior. It showed that collective intelligence can benefit not only content creation but also application development. Our semantic wiki based model clearly exemplifies this trend by its emphasis on general-purposed, in-browser scripting that enables users to contribute to the representational, computational and visualization capabilities of the target system. Thus, an application could be extended in a collaborative fashion as the result of activities of multiple individuals, e.g., by adding new information sources via the creation of client-side mediators, creating new datatypes and associated templates, and making data available to other systems by the creation of new export formats.

A number of semantic wikis (e.g., AceWiki [10] and IkeWiki [13]) and Semantic Web platforms (e.g., HyperDE [12,14] and social semantic desktop [2]) have been used to support Web application development following the similar approach as we adopted in the proposed application model. These efforts share common characteristics in that they all allow social publishing of semantically enriched data. Our approach, different from these efforts, provides that users can be allowed to contribute not only semantically enriched data but also some data consumption scripts, both using a simplified, easy-to-use, browser-based publishing process, to collectively build Web applications.

5.2   Users Participation

The semantic wiki based model we proposed has shown some advantages to the extant Web collaborative programming approaches, including off-browser approaches like Concurrent Versions System (CVS), and in-browser approaches like Bespin[9]. The built-in support for Semantic Web features makes our approach easier to build knowledge-intensive applications. In addition, since the components of an application (e.g. data structure and UI) can be all presented as wiki pages, they can be edited via the usual wiki editing interface without requiring a special client devolvement software. This may encourage the social participation of users in improving the application, therefore better exploiting the network effect.

---

[9] http://labs.mozilla.com/projects/bespin/

We may observe in the proposed model, as in Wikipedia, a "long tail" effect [8] that, while a large portion of edits is done by a small, core group of "elite" users, the aggregation of the small numbers of edits from the majority of "common" users also contributes a significant portion of contributions. It has be shown that although Wikipedia was driven by the influence of elite users early on, there has been a shift in increasing contribution from the common users [8]. We expect the same social participation pattern to occur in the semantic wiki based application model: there may be an elite group that intensively involved in designing the templates, data models and UI of the application, while the majority of users interact mainly with forms-based entry and prebuilt queries, but also occasionally contribute to the improvement of the application. Yet, it is notable that this is different from the traditional programming paradigm where developers and users are two distinctive groups. In the semantic wiki based model, the boundary of the two groups is not absolute and the change of roles is easy.

5.3 Data Modeling

The data modeling ability in semantic wikis is a tradeoff between the complexity and flexibility of data models. Semantic wikis extend traditional wikis with the ability to add and query metadata thus also additional complexity in the scripting language (e.g., the syntax for "ask" query). However, the core extension to conventional wiki scripts is quite small thus is relatively easy to learn. Compared with the conventional techniques for developing Semantic Web applications, the semantic wiki approach is limited in providing the ability to model heavy-weight semantic structure, e.g., the page-centric representation of wiki modeling makes it sometimes hard to freely create knowledge statements. On the other hand, this simplicity also makes the semantic wiki model easier to learn and to use, therefore lowers the threshold for mass user participation.

It should be noted that while the demonstrated examples in the paper only use form-based data entry from users, the semantic wiki based model does not exclude other forms of editing, e.g., by using an ontology browser in the Halo extension[10] or a video stream editor of the MetaVidWiki extension[11].

When compared with the conventional Web model where data is organized with a pre-defined database schema, the semantic wiki based model shows a clear advantage from its inherent support for the RDF graph model. This is evident in applications where database schemas tend to be too rigid and too slow to evolve, as both requirement, user expectations and data structures are often consistently changing. In semantic wiki based modeling, relationships between data elements can be represented explicitly and transparently as an RDF graph, thus making it easier for creating, extending and combining of data, and for an application to utilize unanticipated new data source (e.g., to add a new GeoRSS source to RPI Map), and vice versa[12].

---

[10] http://semanticweb.org/wiki/Halo_Extension

[11] http://metavid.org/wiki/MetaVidWiki

[12] This paragraph is influence by a presentation "Drupal and the Semantic Web" by Jamie Taylor on Jun 6th, 2009. http://bit.ly/2H5Pil

6 Conclusions

In this paper, we present a light weight Web application model based on semantic wikis. The model utilizes the data modeling, processing and presentation abilities of semantic wikis, which enable better flexibility, socialization and inference ability in building a Web application compared with conventional models. We illustrate our approach with two proof-of-concept applications, RPI Map and CNL Wiki, based on Semantic MediaWiki (SMW). Using the two examples, we show that semantic queries and templates are useful building components in realizing many of the data modeling, processing and presentation abilities of semantic wikis.

Our future work will focus on the enhancing of the mentioned prototype systems. The extensible architecture of the two applications allows them to evolve with user contributed scripts. For example, in CNL Wiki, we plan to add additional CNL verbalization support by new sets of CNL templates, and the ontology repository management ability by using a set of ontology templates. The ultimate goal is to better demonstrate how to create and update an application using semantic wiki thus to encourage the adoption of the proposed semantic wiki based development model.

Acknowledgement. This research was sponsored by the U.S. Army Research Laboratory and the U.K. Ministry of Defence and was accomplished under Agreement Number W911NF-06-3-0001. The views and conclusions contained in this document are those of the author(s) and should not be interpreted as representing the official policies, either expressed or implied, of the U.S. Army Research Laboratory, the U.S. Government, the U.K. Ministry of Defence or the U.K. Government. The U.S. and U.K. Governments are authorized to reproduce and distribute reprints for Government purposes notwithstanding any copyright notation hereon.

We thank Jin Guang Zheng for part of the RPI Map implementation and Zhenning Shangguan for part of the CNL Wiki implementation.

References

1. Bao, J., Smart, P.R., Braines, D., Jones, G., Shadbolt, N.R.: A controlled natural language interface for semantic media wiki. In: 2nd Annual Conference of the International Technology Alliance (2009), http://tw.rpi.edu/wiki/TW-2009-05
2. Decker, S., Frank, M.R.: The networked semantic desktop. In: Bussler, C., Decker, S., Schwabe, D., Pastor, O. (eds.) WWW Workshop on Application Design, Development and Implementation Issues in the Semantic Web. CEUR Workshop Proceedings, vol. 105. CEUR-WS.org (2004)
3. Garrett, J.J.: Ajax: A new approach to web applications. adaptivepath.com, [Online; Stand 18.03.2008] (February 2005)
4. Hart, G., Johnson, M., Dolbear, C.: Rabbit: Developing a control natural language for authoring ontologies. In: Bechhofer, S., Hauswirth, M., Hoffmann, J., Koubarakis, M. (eds.) ESWC 2008. LNCS, vol. 5021, pp. 348–360. Springer, Heidelberg (2008)
5. Hendler, J.A., Golbeck, J.: Metcalfe's law, web 2.0, and the semantic web. J. Web Sem. 6(1), 14–20 (2008)
6. Jazayeri, M.: Some trends in web application development. In: FOSE, pp. 199–213 (2007)

7. Kaljurand, K., Fuchs, N.E.: Bidirectional mapping between OWL DL and attempto controlled english. In: Alferes, J.J., Bailey, J., May, W., Schwertel, U. (eds.) PPSWR 2006. LNCS, vol. 4187, pp. 179–189. Springer, Heidelberg (2006)
8. Kittur, A., Chi, E., Pendleton, B.A., Suh, B., Mytkowicz, T.: Power of the few vs. wisdom of the crowd: Wikipedia and the rise of the bourgeoisie. In: Presentation at 25th Annual ACM Conference on Human Factors in Computing Systems, CHI 2007 (2007)
9. Krötzsch, M., Vrandecic, D., Völkel, M., Haller, H., Studer, R.: Semantic wikipedia. J. Web Sem. 5(4), 251–261 (2007)
10. Kuhn, T.: Acewiki: A natural and expressive semantic wiki. In: Semantic Web User Interaction Workshop at CHI 2008 (2008)
11. Leuf, B., Cunningham, W.: The Wiki way: quick collaboration on the Web. Addison-Wesley Longman Publishing Co., Inc., Boston (2001)
12. Nunes, D.A., Schwabe, D.: Rapid prototyping of web applications combining domain specific languages and model driven design. In: ICWE, pp. 153–160 (2006)
13. Schaffert, S.: Ikewiki: A semantic wiki for collaborative knowledge management. In: WET-ICE, pp. 388–396 (2006)
14. Schwabe, D., da Silva, M.R.: Unifying semantic wikis and semantic web applications. In: International Semantic Web Conference (Posters & Demos) (2008)

Guidelines for the Specification and Design of Large-Scale Semantic Applications

Óscar Muñoz-García and Raúl García-Castro

Ontology Engineering Group, Departamento de Inteligencia Artificial
Facultad de Informática, Universidad Politécnica de Madrid, Spain
{omunoz,rgarcia}@fi.upm.es

Abstract. This paper presents a set of guidelines to help software engineers with the specification and design of large-scale semantic applications by defining new processes for *Requirements Engineering* and *Design* for semantic applications. To facilitate its use to software engineers not experts in semantic technologies, several techniques are provided, namely, a characterization of large-scale semantic applications, common use cases that appear when developing this type of application, and a set of architectural patterns that can be used for modelling the architecture of semantic applications. The paper also presents an example of how these guidelines can be used and an evaluation of our contributions using the W3C Semantic Web use cases.

1 Introduction

A large-scale semantic application is an application that makes use of semantic technologies and that manipulates huge quantities of heterogeneous decentralized knowledge and semantic data presenting different degrees of quality. The application produces and consumes its own and external data and retrieves knowledge automatically by exploring different sources.

As a particular domain for large-scale semantic applications, the Semantic Web is a large-scale source of knowledge that requires to design a new generation of Semantic Web applications, which are very different from classic knowledge-based systems (KBS) [1]. In classic KBS the ontologies (usually one) and instances are bound to a particular domain. On the other hand, the next generation of Semantic Web applications permits the execution of the applications in multiple domains, integrates heterogeneous proprietary and legacy solutions, and makes use of big networks of ontologies. In addition, the next generation of Semantic Web applications needs to deal with significant problems associated with the scale, heterogeneity, interoperability and distribution of the information processed, such as the need for searching, accessing and integrating the appropriate knowledge according to the task at hand [1]. These problems do not appear in the Semantic Web but also in other knowledge management or data interpretation systems.

On the other hand, software engineers without expertise in the development or use of semantic applications do not know how to define or implement the

A. Gómez-Pérez, Y. Yu, and Y. Ding (Eds.): ASWC 2009, LNCS 5926, pp. 184–198, 2009.

semantic functionalities of applications, and therefore, it is difficult for them to carry out the development process of these types of applications.

Software development methodologies are broadly used in Software Engineering and Knowledge Engineering. Nevertheless, while there are methodologies that support the development of data models (i.e., ontologies) for semantic applications [2,3], there are no methodologies that support the development of such applications. Since semantic applications are a subset of software applications, they could be built by applying any general-purpose software development methodology. However, a set of guidelines that specifically deals with large-scale semantic applications will lead to a more efficient development of these types of applications.

Our goal in the present paper is to provide guidelines for the requirements analysis and architectural design of a new generation of practical, large-scale semantic applications that draw on contextualized networked ontologies, heterogeneous data and other knowledge-level resources. The guidelines can be easily adapted and integrated in existing development processes by application developers whose aim is to design the architecture of semantic applications from scratch or to include semantic components into traditional information systems.

To do so, we have extended the work presented in [4] by refining the process and techniques presented there, and by defining a new process and technique for designing the architecture of a large-scale semantic application.

The main research results here described are the definition of the *Requirements Engineering* and the *Design* processes for semantic application development and the associated techniques for carrying out these processes, namely, a set of questionnaires for identifying the semantic requirements of the application being developed, catalogues of common use cases that appear when developing these types of applications, system models for understanding the context of the application, and patterns used for modeling the architecture of the application.

This paper is structured as follows. Section 2 presents previous work from which the guidelines are based. Section 3 provides an overview of the guidelines, while sections 4 and 5 detail the activities covered by the guidelines. Section 6 illustrates the guidelines with an example application, and Section 7 shows the results obtained after evaluating the guidelines. Finally, Section 8 presents the conclusions of this work and future lines of research.

2   Related Work

In order to elicit and analyse the requirements of a semantic application, it is necessary to understand the characteristics that commonly appear in such applications and the different scenarios where semantic solutions are applied. Besides, to obtain the architectural design of large-scale semantic applications it is also necessary to define, among others, a set of independent components commonly used in semantic applications.

We have extracted a **characterization of semantic applications** regarding the characteristics of this type of applications presented in [1,5,6,7]. Figure 1

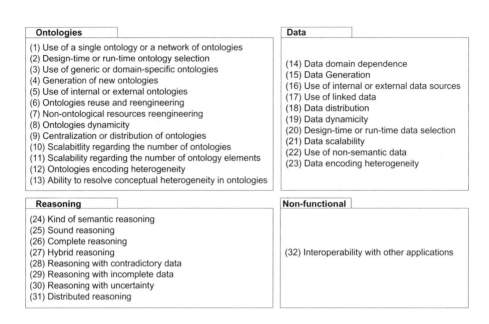

Fig. 1. Characteristics of Large-scale Semantic Applications

shows the result of the analysis made. As can be observed, we have clustered the characteristics according to the nature of the ontologies used, the data produced and consumed, the kind of reasoning applied, and other non-functional characteristics.

In [8], the following set of **scenarios for applying ontologies** to applications is presented: (1) *Neutral Authoring*, where an information artefact is authored in a single language and converted into a different form so that it can be used in multiple target systems; (2) *Ontology as Specification*, where an ontology of a given domain is created and used as a basis for the specification and development of some software; (3) *Common Access to Information*, where information is required by one or more persons or by computer applications; this information, however, is expressed in unfamiliar vocabulary or in an inaccessible format; and (4) *Ontology-based Search*, where an ontology is used for searching an information repository for desired resources.

In [9] there is a classification of the type of ontology usage in Semantic Web applications from where several scenarios can be derived: (1) *Usage as a Common Vocabulary*, (2) *Usage for Search*, (3) *Usage as an Index*, (4) *Usage as a Data Schema*, (5) *Usage as a Media for Knowledge Sharing*, (6) *Usage for a Semantic Analysis*, (7) *Usage for Information Extraction*, (8) *Usage as a Rule Set for Knowledge Models*, and (9) *Usage for Systematizing Knowledge*. The work presented in [10] adds the scenario of *Collaborative Construction of Knowledge* to those here presented.

The **Semantic Web Framework** (SWF) [11] is a component-based framework from which Semantic Web applications can be organized and developed;

this framework provides the skeleton for the specification of the independent components needed for the component-based engineering of Semantic Web applications. The SWF describes the functionalities that the components of Semantic Web applications provide and require, classifies these components, and identifies the main dependences between them. The SWF components are defined at the conceptual level and are decoupled of the technology that implements such components.

3   Overview of the Processes Described

The main objective of the guidelines here presented is to lead application developers from the elicitation of semantic application requirements to the description of the architecture of pure large-scale semantic applications, as well as to the description of the semantic part of applications that include semantic components. To achieve such a goal, we have described the *Requirements Engineering* and *Design* processes bearing in mind the development processes defined for Component Based Software Engineering [12] and the agile methods employed in software development. Figure 2 shows an overview of the overall process.

During the *Requirements Engineering* process, the requirements of the application must be analysed, agreed and documented. On the other hand, *Design* is the process of describing the structure of the software to be implemented and the interfaces between system components [12]. These processes cover different

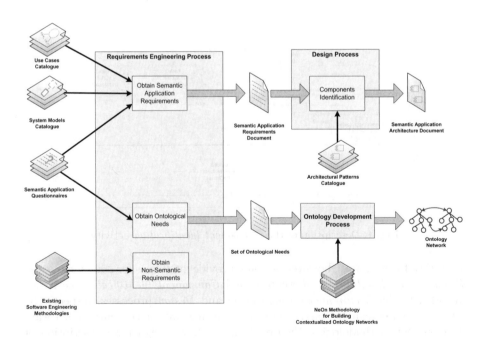

Fig. 2. Overview of the *Requirements Engineering* and *Design* processes

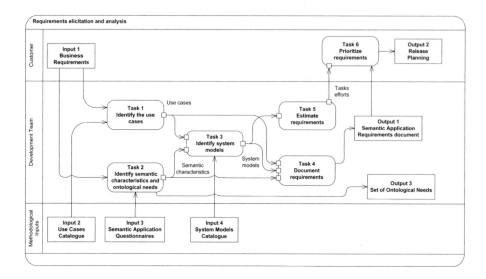

Fig. 3. Description of the *Requirements Elicitation and Analysis* activity

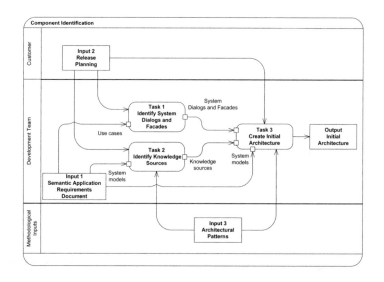

Fig. 4. Description of the *Component Identification* activity

activities. In this paper, however, we only provide guidelines for carrying out the
Requirements Elicitation and Analysis and *Component Identification* activities
included in the *Requirements Engineering* and *Design* processes, respectively.
Other activities can be carried out by following any of the current software
development methodologies. Figures 3 and 4 show a high level description of
the activities commented above. Next, we summarize each of these activities. A
complete description of the guidelines can be found in [13].

4   Guidelines for Requirements Elicitation and Analysis

To facilitate the requirement analysis, our guidelines propose that the requirements be divided into three different types: (1) the **Non-semantic Requirements** gather the application requirements not related to semantic functionalities; (2) the **Semantic Application Requirements** bring together the software requirements that tackle the semantic functionalities of the application; (3) finally, the **Set of Ontological Needs** reflects the ontological needs to be taken into account when developing the ontologies required by the semantic application. Such ontologies can be constructed following the guidelines given by any ontology development methodology, as, for example, the one described in [14]. Since any software engineering methodology supports the discovery of non-semantic requirements, our guidelines only provide techniques for obtaining the last two groups of requirements.

Semantic Application Questionnaires. Accompanying the guidelines, we provide a set of questionnaires that can be used by application developers for identifying the semantic characteristics of a given application (see Figure 1). In these questionnaires, each characteristic is covered with one question. The questionnaires also serve to identify the set of ontological needs and the data sets used by application.

Use Cases Catalogue. A catalogue of use case templates is also supplied. The catalogue describes the scenarios commonly appearing in semantic applications, such as the performance of a search based on ontologies or the semantically browsing of resources. Each template is graphically represented using UML 2.0 use case diagrams and includes detailed textual descriptions. The use case templates have been abstracted from the scenarios analysed in Section 2 (see Table 1). For identifying the use cases the following guidelines are provided: (1) to select the appropriate template from the catalogue; (2) to adapt the selected by modifying the use case information fields; and (3) to append the use case to the application requirements.

System Models Catalogue. We also provide a catalogue of system models. System models are graphical representations commonly used in Software Engineering that describe business processes, the problem to be solved, and the

Table 1. Mapping between the use cases and the scenarios analysed

	State of the art scenarios	
Use case	**Scenario in [8]**	**Scenario in [9]**
1. Query Information	3. Common access to Information	4. Usage as a Data Schema 5. Usage as a Media for Knowledge Sharing
2. Search Resources	4. Ontology-Based Search	2. Usage for Search
3. Browse Resources		3. Usage as an Index
4. Extract Information		7. Usage for Information Extraction 6. Usage for a Semantic Analysis
5. Manage Knowledge	1. Neutral Authoring 2. Ontology as Specification	1. Usage as a Common Vocabulary 8. Usage as a Rule Set for Knowledge Models 9. Usage for Systematizing Knowledge

system that is to be developed [12]. In our case, the system models let application developers to preliminarily specify the system from (1) an external perspective, where the context or environment of the application is modelled by showing the limits of the application and the external systems or applications that will interoperate with the application, and (2) a structural perspective, where the structure of the ontologies and the data processed by the application are modelled.

The system models catalogue contains a set of basic symbols (e.g., ontological and non ontological resources, applications) and the relationships between these symbols, which reflect the aforementioned structural perspective of the system. The system models will reflect the scenarios identified during the use case identification task, which is constrained by the application characteristics.

Figure 5 shows an example of a system model template that represents multiple data sources expressed according to several ontologies or non-ontological schemas and aligned with a shared vocabulary. The template has been obtained from the different approaches to ontology-based integration of information described in [15]. Figure 6 illustrates an example of an instantiation of the template shown in Figure 5, where several data sources that conform to an ontology or to a non-ontological schema are integrated through a shared ontology. Also in Figure 6, there are an ontology and a set of instances that are discovered at run-time (e.g., in the Semantic Web).

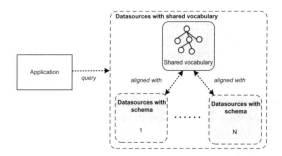

Fig. 5. Template example: *Query Information with a Hybrid Ontology approach*

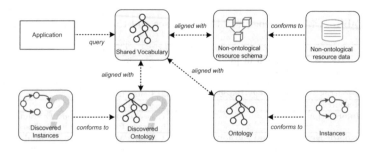

Fig. 6. Example of an instantiation of the system model template in Figure 5

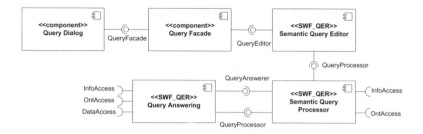

Fig. 7. Pattern example

For elaborating the system model we provide the following guidelines: (1) to associate a basic symbol to each of the resources that the application will deal with; (2) to identify the existing relationships between the basic symbols; (3) to identify the system model templates associated to the identified use cases and application characteristics; (4) to combine the symbols, relationships, and system model templates in order to conform a unique system model.

5 Guidelines for Component Identification

Our guidelines are focused on obtaining the structure (i.e., the architecture) of the semantic application using the technique explained next.

Architectural Patterns Catalogue. This catalogue provides 28 architectural patterns that reflect common organizations of semantic-related software components in large-scale semantic applications; the patterns were obtained from the analysis of the architecture of existing applications such the ones described in [11]. The components in the architectural patterns are those described in the SWF [11]. The patterns are represented as UML 2.0 component diagrams.

During the *Component Identification* activity, the patterns are selected regarding each of the symbols, relationships and templates used to depict the application system model.

Figure 7 shows an example of an architectural pattern used for solving the scenario depicted in Figure 5. In the pattern, the *Semantic Query Editor* component takes care of all issues related to the user interface. The *Semantic Query Processor* component is in charge of all the issues related to the physical processing of a query, while the *Query answering* component is responsible for all the issues related to the logical processing of a query. The *Query Dialog* component implements the *Query Information* use case logic, and the *Query Facade* component provides the operations to meet the use case responsibilities.

6 Example

This section presents an example of how to carry out the *Requirements Elicitation and Analysis* and *Component Identification* activities; the example is drawn from a fictitious case study whose *Business Requirements* are explained next:

A logistics company has proved that setting dynamic shipment routes will decrease their shipment risks and delivery time, while it will increase its income due to factors such as weather, transport companies availability and fares, etc.

The company wants to upgrade its system to enable intelligent search of optimal routes. To do this, the system will take into account weather information coming from different Internet providers and information owned by transport companies, for example, delivery times, transportation costs, and availability of service for a certain route stretch. The candidate routes are obtained from maps available on the Web. Besides searching for the most adequate routes and transport companies, the logistics company wants to make use of the aforementioned integrated information to provide its clients with real time tracking of their shipments.

The information that the new application will use is encoded according to different formats: the weather information providers expose their information as instances expressed according to a given ontology; the transport companies provide a set of XML resources to facilitate the interoperability with the logistics companies; and the maps are published in the Semantic Web formats. Additionally, the logistics company will also use information stored in a relational database included in its own information system.

The logistics company works with several known transport companies. The information about weather previsions will be discovered at run-time and integrated with the rest of the information.

6.1 Requirements Elicitation and Analysis Activity

This subsection presents how to carry out the three first tasks of a *Requirements Elicitation and Analysis* episode starting from the *Business Requirements*.

Task 1. To Identify the Use Cases. The development team starts by identifying the use cases and then finding the two use cases that are shown in Figure 8.

The purpose of first use case, *Obtain Optimum Route*, is to identify the interactions between the logistics company and the different external systems when an optimum route is obtained, whereas the purpose of the second use case, *Track Shipment*, is to show the interactions between the customer of the logistics company with the system and the interactions of the system with the external information provider systems. Both use cases can be seen as realizations of the use

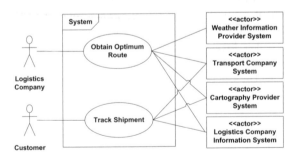

Fig. 8. Use cases identified for the sample case study

case template *Query Information*, contained in the catalogue. The template in the catalogue has to be instantiated by identifying the primary actor and the stakeholders, including the external systems, and by modifying the flow specified in the template.

Task 2. To Identify Application Characteristics and Ontological Needs. As previously seen, the set of characteristics that commonly appear on semantic applications is intended to help developers to identify the semantic requirements of the application under development. To answer the questionnaires, developers are not required to master semantic technologies, but, at least they should have a minimum knowledge of such technologies.

Next, we provide part of the responses to the questionnaires and the values obtained for some characteristics.

- Will the ontologies be identified by developers at design-time or located by the application at run-time?
    Response: "Mixed (some at design time and some at run-time)".
    Characteristic: Design-time or run-time ontology selection. Value: "Mixed"
- Will the application aggregate non-semantic data?
    Response: "Yes (transport companies data and corporate database)".
    Characteristic: Use of non-semantic data. Value: "Yes"
- Will the application deal with contradictory data?
    Response: "Yes (e.g. contradictory weather previsions)".
    Characteristic: Dealing with contradictory data. Value: "Yes"

Task 3. To Identify System Models. Table 2 shows the resources identified with their associated basic symbols. Table 3 depicts the relationships identified, obtained from the catalogue of system model templates.

As previously in *Task 1*, both use cases are associated to the use case template *Query Information*. Therefore, the development team has chosen the system model template *Query Information with a Hybrid Ontology Approach* (see Figure 5) because of the characteristics previously discovered.

Table 2. Symbols associated to the resources used by the example application

Resource Identifier	Resource Description	Basic Symbol
Cartography Ontology	Ontology of the cartography provider	Static Ontology
Cartography Instances	Instances of the cartography provider	Static Instances
Transport Schema	XML schema of the transport company provider	Static Non-ontological Resource Schema
Transport Data	XML data of the transport company provider	Static Non-Ontological Resource Content
Weather Ontologies	Ontologies of the weather information providers	Dynamic Ontology
Weather Instances	Instances of the weather information providers	Dynamic Instances
Logistics DB	Corporate database of the logistics company	Non-ontological Resource Content that Conforms to a Given Schema Abbreviation

Table 3. Relationships between the resources used in the example application

Resource 1	Resource 2	Relationship
Cartography Instances	Cartography Ontology	1. Instances that Conform to a Given Ontology
Transport Data	Transport Schema	2. Non-ontological Resource Content that Conforms to a Given Schema
Weather Instances	Weather Ontologies	1. Instances that Conform to a Given Ontology

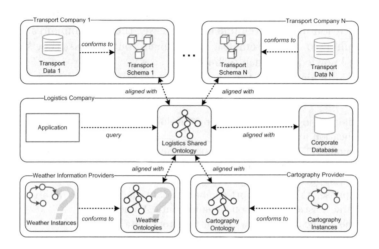

Fig. 9. System model identified for the sample case study

As the template chosen indicates, it is necessary to create and incorporate another ontology, having a shared vocabulary, that will be aligned with the rest of the ontologies and schemas to facilitate information integration. Therefore, several *Aligned With* relationships must be included in the system model.

By integrating the basic symbols and their relationships with the *Query Information with a Hybrid Ontology Approach*, the system model we obtain is the one shown in Figure 9.

6.2   Component Identification Activity

This subsection presents how to carry out the three tasks of the *Component Identification* activity, considering the use cases and system model obtained in the previous subsection.

Task 1. To Identify Dialogs and System Facades. Within this task the development team introduces in the architecture a system dialog and a facade for each use case identified as specified in [16]. The *dialog* components implement the logic of each use case, that is, the software that handles the dialog between the actors of a given use case and the system. The *facade* components provide operations for every step specified in the use case flow definition and are used by the *dialog* components.

Table 4. Patterns associated to the repositories used by the sample case study

Resource	System	Pattern
Transport Data	Transport Companies	2. Data Repository
Transport Schema	Transport Companies	2. Data Repository
Logistics Shared Ontology	Logistics Company	1. Ontology Repository
Corporate Database	Logistics Company	2. Data Repository
Weather Ontology	Weather Information Prov.	3. Dynamic Ontological Resource Access
Weather Instances	Weather Information Prov.	3. Dynamic Ontological Resource Access
Cartography Ontology	Cartography Providers	1. Ontology Repository
Cartography Instances	Cartography Providers	1. Ontology Repository

Task 2. To Identify Interfaces to Knowledge Sources. Within this task, the developers catalogue the repositories containing the ontological and non-ontological data that the application will use. For each ontological and non-ontological resource reflected in the system model, its containing repository is identified. Table 4 shows the system and patterns associated to each resource.

Task 3. To Create the Initial Architecture. The architecture shown in Figure 10 is obtained directly by integrating all the components and patterns.

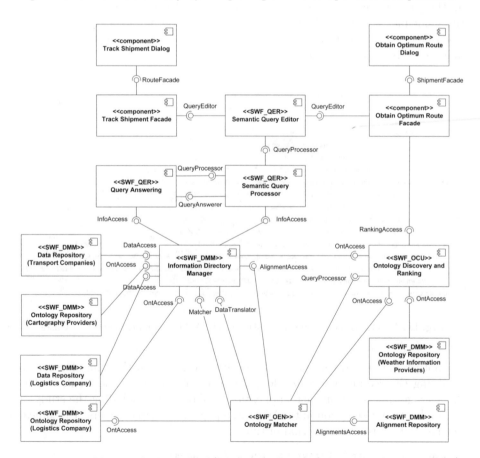

Fig. 10. Architecture identified for the sample case study

7   Evaluation

For evaluating our work, we have analyzed ten use cases, described in the *W3C Semantic Web Case Studies and Use Cases* web page[1]. These test cases are not related to the applications analysed for developing the guidelines described in this paper. For each use case we have applied the activities and techniques proposed in this paper to obtain (1) the characteristics of each application, (2) the use cases that the application covers, (3) the system models, and (4) the architecture of the application.

Figure 11 summarizes the values obtained for some characteristics of the applications analysed; it also shows how many times these values appear in the whole set of applications. With regard to the scenarios that the applications cover, all the use cases templates provided by the guidelines address those scenarios. Since all the use cases described by the guidelines appear in the applications analysed, almost all the system model templates provided by the guidelines can be used to model the structure of some of these applications. However, with respect to the patterns applied to build the architecture, it should be explained that not all the patterns have been used. The reason is that some patterns described in the guidelines are used when ontological and/or non-ontological resources are discovered by the application at run-time, and such dynamic behaviour is not present in the analysed applications.

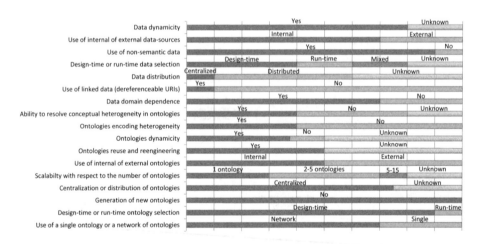

Fig. 11. Values for some characteristics of the W3C Semantic Web Use Cases

Another measure taken during the evaluation has been the time devoted to analysing the requirements and designing each application. The result is that the average time spent in each application is of one day (including the study of the application description in the W3C web page and related papers or technical reports), which is a short period of time.

---

[1] http://www.w3.org/2001/sw/sweo/public/UseCases/

8   Conclusions

Large-scale semantic applications require different software development methods and techniques from those for classic knowledge-based systems because they manipulate huge quantities of heterogeneous decentralized information, integrate semantic and non-semantic data, and explore different sources at runtime. Therefore, software engineers without expertise should be provided with methodological guidelines for the development of semantic applications.

For this purpose, we have adapted the *Requirements Engineering* and *Design* processes from methodologies widely accepted in Software Engineering. This adaptation allows to design the architecture of semantic applications from scratch and to include semantic components into traditional information systems, by integrating the activities and techniques here described into existing application development processes. The techniques described are novel and especially oriented to the specification and design of the semantic functionalities of an application.

The architectural patterns dealt with are not bound to a particular implementation. Therefore, after using the guidelines here presented, the application architecture will remain independent of concrete component implementations. Architecture realizations in particular settings are out of the scope of this paper.

The catalogues and patterns presented can be extended, and for this purpose a collaborative space (e.g., a wiki) will be enabled to facilitate community feedback, extension and enrichment. The immediate lines of work include to continue defining the rest of the development processes (i.e., *Implementation*, *Integration* and *Testing*). Other future line of work is to specialize the guidelines in order to deal with particular settings, for example, the Open Linked Data initiative.

Another extension should be to give software support to the guidelines by building or adapting an existing CASE tool and by formalizing the processes, activities, methods, catalogues and patterns of the guidelines with ontologies. The purpose here is twofold: to automatically document the large-scale semantic application development process and to support the application code generation. For this last issue, it is necessary to define the rest of the processes and to provide interoperable implementations of the components involved in the semantic application. Finally, the questionnaires will be used to characterize and categorize existing semantic applications and then to carry out an analysis of the current panorama of semantic applications.

Acknowledgements

This work has been partially supported by the NeOn project (IST-2005-027595).

References

1. Aquin, M., Motta, E., Sabou, M., Angeletou, S., Gridinoc, L., Lopez, V., Guidi, D.: Towards a New Generation of Semantic Web Applications. IEEE Intelligent Systems 23 (2008)

2. Fernández-López, M., Gómez-Pérez, A., Juristo, N.: METHONTOLOGY: From Ontological Art Towards Ontological Engineering. In: Ontological Engineering on Spring Symposium Series, Stanford (1997)
3. Staab, S., Schnurr, H.P., Studer, R., Sure, Y.: Knowledge Processes and Ontologies. IEEE Intelligent Systems (2001)
4. Muñoz-García, O., García-Castro, R., Gómez-Pérez, A.: Facilitating Requirements Engineering of Semantic Applications. In: Proceedings of the 5th Workshop on Semantic Web Applications and Perspectives (SWAP 2008), Rome, Italy, CEUR Workshop Proceedings (2008)
5. Motta, E., Sabou, M.: Next Generation Semantic Web Applications. In: Mizoguchi, R., Shi, Z.-Z., Giunchiglia, F. (eds.) ASWC 2006. LNCS, vol. 4185, pp. 24–29. Springer, Heidelberg (2006)
6. Domingue, J., Fensel, D.: Towards a Service Web: Integrating the Semantic Web and Service Orientation. IEEE Intelligent Systems (2008)
7. Krummenacher, R., Simperl, E., Fensel, D.: Scalability in Semantic Computing: Semantic Middleware. In: Proceedings of the IEEE Conference on Semantic Computing, pp. 538–544 (2008)
8. Jasper, R., Uschold, M.: A Framework for Understanding and Classifying Ontology Applications. In: Twelfth Workshop on Knowledge Acquisition Modeling and Management, KAW 1999 (1999)
9. Kozaki, K., Hayashi, Y., Sasajima, M., Tarumi, S., Mizoguchi, R.: Understanding Semantic Web Applications. In: Domingue, J., Anutariya, C. (eds.) ASWC 2008. LNCS, vol. 5367, pp. 524–539. Springer, Heidelberg (2008)
10. Coskun, G., Heese, R., Luczak-Rösh, M., Oldakowski, R., Schäfermeier, R., Streibel, O.: Towards Corporate Semantic Web: Requirements and Use Cases. Technical report, Freie Universität Berlin (2008)
11. García-Castro, R., Gómez-Pérez, A., Muñoz-García, O., Nixon, L.J.: Towards a Component-Based Framework for Developing Semantic Web Applications. In: Domingue, J., Anutariya, C. (eds.) ASWC 2008. LNCS, vol. 5367, pp. 197–211. Springer, Heidelberg (2008)
12. Sommerville, I.: Software Engineering, 8th edn. International Computer Science Series. Addison-Wesley, Reading (2007)
13. Muñoz-García, O., García-Castro, R., Gómez-Pérez, A., Sini, M.: D5.5.1. NeOn Methodology for the development of large-scale semantic applications. Technical report, NeOn Project (2009)
14. Gómez-Pérez, A., Suárez-Figueroa, M.: NeOn Methodology: Scenarios for Building Networks of Ontologies. In: 16th International Conference on Knowledge Engineering and Knowledge Management Knowledge Patterns (EKAW 2008), Conference Poster, Italy (2008)
15. Wache, H., Vögele, T., Visser, U., Stuckenschmidt, H., Schuster, G., Neumann, H., Hübner, S.: Ontology-based integration of information - a survey of existing approaches. In: IJCAI workshop on Ontologies and Information Sharing, pp. 108–117 (2001)
16. Cheesman, J., Daniels, J.: UML Components. A Simple Process for Specifying Component-Based Software. Component Software Series. Addison-Wesley, Reading (2001)

Semantic-Linguistic Feature Vectors for Search: Unsupervised Construction and Experimental Validation

Stein L. Tomassen[1] and Darijus Strasunskas[2]

[1] Dept. of Computer and Information Science
stein.l.tomassen@idi.ntnu.no
[2] Dept. of Industrial Economics and Technology Management,
Norwegian University of Science and Technology, Norway
darijuss@gmail.com

Abstract. In this paper, we elaborate on an approach to construction of semantic-linguistic feature vectors (FV) that are used in search. These FVs are built
based on domain semantics encoded in an ontology and enhanced by a relevant
terminology from Web documents. The value of this approach is twofold. First,
it captures relevant semantics from an ontology, and second, it accounts for statistically significant collocations of other terms and phrases in relation to the
ontology entities. The contribution of this paper is the FV construction process
and its evaluation. Recommendations and lessons learnt are laid down.

1 Introduction

Search is among the most frequent activities on the Web. However, the search activity
still requires extra efforts in order to get satisfactory results. One of the reasons is
heterogeneous information resources and exponential growth of information. There
are many different approaches proposing a solution for this problem. Some approaches are relying on semantic annotations (e.g., [2, 19]) by adding additional
metadata; some are enhancing clustering of retrieved documents according to topic
(e.g. [13]); some are developing powerful querying languages (e.g. [4]). Therefore,
many efforts are devoted to research on improvement of information retrieval (IR) by
the help of ontologies that encode domain knowledge (e.g. [5, 17]).

The objective of this paper is to discuss our approach to semantic search that builds
on a concept of *feature vector* (FV) and elaborate on the FV construction (FVC) process. The approach is based on pragmatic use of ontologies by relating the concepts
(domain semantics) with the actual terminology used in a text corpus, i.e. the Web.
We propose to associate every entity (classes and individuals) of the ontologies with a
FV to tailor them to the terminology in a text corpus. First, these FVs are created offline and later used on-line to filter, and hence disambiguate search, and re-rank the
search results from the underlying search system. The proposal is based on a non-
supervised solution that is applicable to any ontology as long as there is some correlation between the ontology and the text corpus. Moreover, the approach is independent
from a collection of relevant documents. Possibility to use a diverse corpus (the Web)
is the main advantage of the approach since the approach builds on word sense disambiguation by utilizing the relationships between the entities. Nevertheless, the FV

A. Gómez-Pérez, Y. Yu, and Y. Ding (Eds.): ASWC 2009, LNCS 5926, pp. 199–215, 2009.
© Springer-Verlag Berlin Heidelberg 2009

quality will be highly depended on both the quality of the ontology and the correlation of terminologies in the ontology and the text collection.

In [15], we focused on FVs used to disambiguate search that was evaluated with real users. While in [17], the FVC algorithm used in Strasunskas and Tomassen [15] was presented. Therefore, in this paper we focus on the aspects of the components of FV construction algorithm that affect the feature vector quality. Furthermore, in the evaluation we analyse the effect of alternative techniques on the FVs.

Moreover, many approaches build on similar artefacts as our FVs, although they target various application areas (e.g., ontology alignment, ontology mapping, semantic search, ontological filtering), cf. [7, 9, 14, 16]. Despite they are differently built, this paper provides useful insights on how the process of FVC can be evaluated and the FV quality assessed.

This paper is organized as follows. In section 2, related work is discussed. In section 3, the algorithm of how the FVs are constructed and a small example of the process are presented. In Section 4, we present the conducted experiments and explain the evaluation. Then in section 5, the results will be analyzed. Finally, in section 6, we conclude this paper.

2 Related Work

The focus of this paper is the construction of feature vectors (FV). Therefore, scope of related work synopsis provided here is limited correspondingly. In general, FVs can be classified in three groups, numerical, textual, and a mixture of both. Numerical FVs are typically used in machine learning (e.g. [10]) and are not relevant here, which neither is the case for approaches using mixed FVs. Textual FVs on the other hand, are typically based on a lexical resource like WordNet (e.g. [9]) or extracted from a set of documents (e.g. [1, 14, 16, 20]). The latter form of FVs is most relevant and will be reviewed in more details.

There are approaches that depend on highly relevant document collections (e.g. [14, 16]) as distinct from our approach. Approaches that are more interesting are based on topic signatures. A topic signature is a list of topically related words [1]. There are many topic signature approaches (e.g., [1, 20]. Zhou et al. [20] propose a Topic Signature Language Model that is used to perform semantic smoothing to increase the retrieval performance. They create topic signatures for each concept defined in domain specific ontology using a highly relevant document collection. The topic signature terms are found by collocation. They assume that the concepts are unique and consequently circumvent the problem of word disambiguation. For general domains where no ontology exists, they propose to use multiword expressions as topic signatures. The multiword expressions contains context in nature and are consequently mostly unambiguous.

While Agirre et al. [1] propose enriching WordNet with topic signatures using the Web. A concept in WordNet can contain several senses. Nevertheless, for each sense a set of cue-words (hyponyms, hypernyms, etc.) is used to create a highly specific query that is submitted to the search engine. The top 100 documents are retrieved and keywords are extracted. They experienced formulating the queries being the weakest point of their approach. The quality of the queries highly affected the quality of the retrieved documents. In contrast to our approach that is not depended on a high quality query but uses clustering and domain identification, based on neighbouring entities, to find relevant documents from a set of diverse documents.

3 Feature Vector Construction

Every ontology entity has an associated feature vector with a set of relevant terms extracted from the text corpora. An ontology entity can be either a class or an individual. In this approach, we use the term *entity* instead of *concept* because a concept is often a synonym for a class when it comes to ontologies. Our approach associates feature vectors to both classes and individuals that hereinafter are referred to as entities. In this section, we will describe the process of how these FVs are constructed, but first a definition of a feature vector is provided. At the end of this section, an example of the construction process is presented.

3.1 Definition of a Feature Vector

The development of the approach is inspired by a linguistics method for describing the meaning of objects - the semiotic triangle [11]. In our approach, a feature vector "connects" a concept (entity) to a document collection, i.e., the FV is tailored to the specific terminology used in a particular collection of the documents. FVs are built considering both semantics encoded in an ontology and a dominant lexical terminology surrounding the entities in a text corpus. Therefore, a FV constitutes a rich representation of the entities and is related to actual terminology used in the text corpus. Correspondingly, a FV of an entity e is represented as a two-tuple (see Definition 1):

Definition 1: *Feature Vector (FV)*

$$FV_e = \langle S_e, L_e \rangle \mid S_e \in O_d, L_e \in D_d \tag{1}$$

$$S_e = \left(e_i, DR_{e_i} \right)$$

$$DR_{e_i} = Parents_{S_{e_i}} \cup Children_{S_{e_i}} \cup Others_{e_i} = \left\{ \langle e_i, e_k \rangle \right\} \subseteq E \times E$$

$$L_{e_i} = collocated \left(S_{e_i}, L_{e_{Dd}} \right)$$

where S_e is a semantic enrichment part of FV_e that represents a set of neighbourhood entities and properties in an ontology O of a domain d. L_e is a linguistic enrichment of a entity that is a set of terms (from document collection D of a particular domain d) with a significant proximity to an entity and its semantic neighbourhood.

3.2 Feature Vector Construction

The Feature Vector Construction (FVC) process is visualized in Figure 1. The algorithm constitutes two phases (main steps). The first phase aims to extract and group candidate terms being potentially relevant to each entity. However, the candidate terms are not necessarily relevant to the domain defined by the ontology (terms can be ambiguous). Consequently, the aim of the last phase is to identify those groups of candidate terms being relevant to the entities w.r.t. the ontology. Finally, an FV for

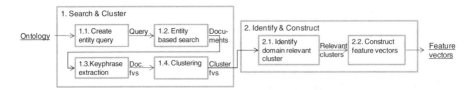

Fig. 1. The Feature Vector construction process

each entity is created based on the most prominent group of candidate terms for each entity. The result of this algorithm is a list of entities with corresponding FVs, which consist of terms associated to both the entities and the domain terminology (Eq. 1).

The FVC algorithm is designed to be flexible in the sense that it can be tailored to the intended usage of the FVs as well as the different quality of the ontologies. Consequently, the algorithm provides several options at each step. The effect of some of these options is evaluated in section 4 and 5, while detail description follows below.

Step 1: Search and cluster
This step constitutes four sub-steps where the aim is to extract candidate terms that are relevant to each entity. The candidate terms are grouped and then, in Step 2, further processed to identify which of the candidate groups being most relevant to the domain of interest defined by the ontology.

Step 1.1: Compose entity query
In this step, a search query is prepared for each entity while the actual search is performed in Step 1.2. The query is based on the entity label with an option to include relevant neighbouring entities and/or keyword(s) (more of this Section 4.2). Here we aim at creating a query that reflects on the ontology by considering closest neighbours of a particular entity.

A *parent* of a class is defined to be its super class, while a parent of an individual is the class the individual being an instance of. A *child* of a class is defined to be its sub class or individual, the latter if it does not have a sub class. An individual does not have a child. Finally, *other* neighbouring entities are any other object property defined in OWL. The motivation behind expanding the initial query with neighbouring entities is to create a query that reflects both the ontology and the relationship of each entity to other neighbouring entities.

Larger ontologies tend to include several minor domains. By experimentation we found that for diverse ontologies, like the Wine[1] ontology that also imports the Food[2] ontology, it can be beneficial to add keyword(s) that represents the overall subject domain. The result of using keyword(s) is less unique and more homogeneous FVs while omitting keywords would create FVs that are more unique and more true to the local variances in the ontology.

Step 1.2: Entity based search
The query for each entity created in Step 1.1 is used to retrieve candidate documents for each entity. Any search engine can be used in this step. Currently, Yahoo! and

---

[1] Wine, http://www.w3.org/2001/sw/WebOnt/guide-src/wine.owl
[2] Food, http://www.w3.org/2001/sw/WebOnt/guide-src/food.owl

Google (for searching in Web documents) and Nutch[3] (for searching in local documents) are supported. In the experiments described in Section 4 Yahoo! is used. The retrieval session is keyword-based.

Step 1.3: Contextual key-phrase extraction

For each document a set of key-phrases and keywords is extracted, hereinafter referred to as key-phrases. First, a part of speech (POS) tagger is used to tag the retrieved documents (snippet or full text). In the experiments described in Section 4 we have selected to use FastTag[4], because it is fast and by experiments found it to perform adequate on Web documents and snippets with diverse quality.

Then a set of tagging rules (39 rules), inspired by Justeson and Katz [8], is applied. Based on these rules a set of candidate noun key-phrases are extracted. However, only those key-phrases within what we call a contextual window are extracted. A contextual window is a frame of a specified size surrounding a keyword (in the experiments described in section 4 a window of size 50 is used). If a keyword appears several places in the document then more windows are created. Each key-phrase is stemmed to remove duplicates by finding their common root. If a duplicate is found then the frequencies are summed up and the duplicate removed. Finally, those candidate key-phrases above a specified frequency threshold (dependent on the document length) are kept and stored in the *document feature vector* (DFV) of the corresponding documents.

Step 1.4: Cluster search results

In order to identify (discriminate) different subject domains within the documents found for each entity, clustering techniques are used. Recall that the retrieval session is keyword-based (Step 1.2) consequently the terms (entities) can be part of many different domains. Clustering allows us to find these different domains. Currently the Lingo [12] algorithm is used since it performs well for both snippets and full-text documents. The result of this step is a set of clusters for each entity. In addition, for each cluster a *cluster feature vector* (CLFV) is created. A CLFV is a combination of all the DFVs of a cluster. In the following step, we deal with selecting the relevant cluster w.r.t. the domain of interest.

Step 2: Identify and construct

This step is constituted of two sub-steps and aims at identifying the most relevant clusters w.r.t. the ontology and create the final feature vectors.

Step 2.1: Identify domain relevant clusters

A problem at this stage is to identify the correct subject domain, that is, the most relevant clusters found in Step 1.4 w.r.t. the ontology. Therefore, we compute the similarity between the cluster feature vectors of an entity with the CLFVs of the selected neighbouring entities. In order to find the most prominent cluster, an entity must have at least one neighbour otherwise this check would fail. The neighbouring entities are grouped according to their relation type, as in Step 1.1, i.e., *parents, children*, and *other* entities.

---

[3] Nutch, http://lucene.apache.org/nutch
[4] FastTag, http://www.markwatson.com/opensource/

Commonality (i.e. high similarity) here identifies the document sets (clusters) being relevant to the domain of our interest. The hypothesis is that individual clusters having high similarity with neighbouring entities are with high probability of the same domain defined by the ontology. This hypothesis is backed up with observed patterns of collocated terms within the same domain, and consequently different domains will have different collocation pattern of terms. However, the similarity of clusters depends a lot on the quality of the ontology, especially on semantic distance between the different entities.

The result of this step is a Domain Relevance Measure score for each cluster of an entity. The relations of each entity are given different weighting according to Definition 2.

Definition 2: *Domain* Relevance *Measure (DRM)*
Let $S = \{S_1, S_2, S_3\} = \{parents[e], children[e], other[e]\}$, $c_i \in \{clusters[e]\}$, and $c_k \in \{clusters[S_j]\}$

$$DRM(e,c_i) = \sum_{j=1}^{3} \frac{1}{n_j} \sum_{k=1}^{n_j} w_j S_j sim(c_i, c_k)$$ (2)

where $DRM(e, c_i)$ is the Domain Relevance Measure for entity e and cluster c_i of e. w_j is a weight factor set to a default value of 1, and S_j is either 1 if S_j is true or 0 if S_j is false. Further, n_j is the number of clusters of each neighbouring entity defined in S.

Step 2.2: Construct feature vector
The cluster with the highest DRM score, calculated in Step 2.1, is selected for each entity. The step of creating the final FV for the selected cluster can either be based on the already created CLFV of that cluster (Step 1.4) or a deeper analysis of the documents of the selected cluster can be done. In the experiments described in section 4, the CLFVs were used.

3.3 Feature Vector Construction Example

In this section, a small example is presented to illustrate the steps of the Feature Vector Construction algorithm described in Section 3.2.

Step 1.1: Create entity query
In order to better illustrate the purpose of the clustering (step 1.4) and the identification of the domain relevant clusters in step 2.1, the illustrative query for the entity Jaguar, seen in Figure 3, is: <jaguar>

Step 1.2: Entity based search
The query created in Step 1.1 is submitted to Yahoo! Search and the three top ranked documents (of 30 used in this example), as of 18th of April 2009, are shown in Table 1. Not surprisingly was Jaguar the car brand most popular for the moment (23 of 30 top ranked), then panther (5/30), perfume (1/30), and vodka (1/30).

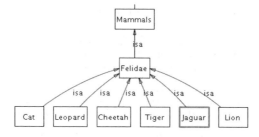

Fig. 2. A small fragment of the Animals[5] ontology, where the Jaguar entity is highlighted and used in this example

Table 1. Top three search results for jaguar

1. Jaguar Official site of Jaguar featuring new models and local dealer information. *http://www.jaguar.com* **2. Jaguar – Home** Jaguar USA official website ... Build Your XK. Find Your XK. Locate a Dealer. Build Your Jaguar. Find Your Jaguar. Request Brochure ... *http://www.jaguarusa.com* **3. Jaguar - Wikipedia** The jaguar, Panthera onca, is a big cat, a feline in the Panthera genus. It is the only Panthera found in the Americas. The jaguar is the third-largest feline after the tiger and the lion, and the largest and most power-ful... *http://en.wikipedia.org/wiki/Jaguar* ...

Table 2. A text fragment from the third search result (Table 1) is shown at the top and a set of corresponding extracted key-phrases for the whole document is seen at the bottom

Text fragment of the third search result (Table 1)
"The jaguar, Panthera onca, is a big cat, a feline in the Panthera genus. It is the only Panthera species found in the Americas. The jaguar is the third-largest feline after the tiger and the lion, and the largest and most powerful feline in the Western Hemisphere...."
Extracted key-phrases from the whole page
cat (17), culture (11), habitat (13), jaguar (136), panthera (11), population (11), prey (19), range (20), species (27), state (11)

Step 1.3: Contextual key-phrase extraction

For illustration purposes, only a small text fragment is shown in Table 2 to illustrate the contextual key-phrase extraction process. The contextual window was of size 50. Typical noise in the documents, like menus, is removed. For instance, Wikipedia documents got start content and end content tags, which are utilized, and hence only the text between theses tags is processed.

Step 1.4: Cluster search results

We used the Lingo clustering algorithm. The full text documents were used. Four clusters were created for the jaguar entity as shown in Table 3.

---

[5] Animals, http://nlp.shef.ac.uk/abraxas/ontologies/animals.owl

Table 3. FVs for clusters found for the jaguar entity

Cluster#1={advice car (4), auto insurance (4), auto show (2), calculators true cost (1), car (94), chevrolet (4), compact awd sport sedan (2), company (13), detailed jaguar (6), drivetrain engine (2), econ msrp (2), engine (12), engine trans (2), flagship 4-door sedan line (2), ford (10), information jaguar (2), information pictures (6), invoice (14), invoice price (4), jaguar (197), land rover (7), line (3), low dealer price (4), market value (2), midsize sport sedan (2), model (23), model name (6), motor (7), motor company (6), msrp (16), price (8), quotes inside line (1), review (17), saloon (8), search sitemap company privacy (1), sedan (17), select (4), series jaguar (5), sports (8), stars (14), style (5), system premium sound system (1), terms (1), tips advice (3), trans fuel (2), truck (29), trucks tips advice (1), xj-series (3), yahoo autos (8), zip (4)}
Cluster#2={accolades (2), conditions (2), contact (2), dealer (1), disclaimer international sites (2), features (2), gallery (2), gtr company (2), international sites faq (2), jaguar (2), ownership quality highlights (2), pre-owned (2), privacy policy (2), profile site (2), request brochure (1), site (2), sites faq gtr (2), site map (2), specs (2), terms (2)}
Cluster#3={cat (26), culture (11), habitat (13), jaguar (164), panthera (11), population (11), prey (19), range (20), species (27), state (11)}
Cluster#4={accessories (5), blue grass (1), blvd louisville (2), brake (7), car (10), careers (1), contact info links (1), deal (1), department (1), exterior jaguar (3), fax (1), genuine (10), genuine parts order (1), inventory (1), inventory pre-owned inventory (1), jaguar (204), jaguar blue (1), jaguar brake (5), jaguar fuel (5), jaguar jaguar (71), land rover (2), news (2), order parts service (1), part (24), parts catalogaccessories catalogjaguar (1), part number (13), parts service schedule (1), phone (1), pre-owned (1), pre-owned inventory events (1), rotor part number (2), rover jaguar (1), saab land (1), service (1), service contact (1), service schedule service contact (1), serviceservice (1), shop jaguar (3), specials events news (1), specialsparts (1), specialsservice (1), specialsservice department (1), system (12), technivision (1), tool (4), type (10), upcoming events news (1), vehicle (10), wagner (1), wagner jaguar (1)}

Table 4. Cluster DRM for the entity Jaguar

Cluster #	3	1	2	4
DRM	0,070	0,011	0,000	0,000

Step 2.1: Identify domain relevant clusters
By calculating the similarity with the clusters of the neighbouring entities of Jaguar, which are Felidae (super-class) we can identify the relevant cluster for this domain. In this case, Cluster#3 had the highest similarity (see Table 4) with a DRM score of 0,070. This cluster is therefore selected as the candidate cluster for the construction of the feature vector to be done in the next step.

Step 2.2: Construct feature vector
The last step for the Jaguar entity is to create the final entity feature vector, which in this example will be the same as the CLFVs for Cluster#3 as seen in Table 5. At this stage, we could do a more thorough analysis of the cluster documents to improve the quality of the feature vector even further.

Table 5. The final fv created for the Jaguar entity

Jaguar={cat (26), culture (11), habitat (13), jaguar (164), panthera (11), population (11), prey (19), range (20), species (27), state (11)}

4 Experiments

We have conducted a set of experiments (described in Section 4.2) to validate the feature vector construction algorithm discussed in Section 3. The goal of the experiments is to measure the sensitivity both w.r.t. some of the components of the approach and some ontologies of different granularity (presented in Section 4.1). We are using Normalized Google Distance (NGD) (described in Section 4.3) and two additional measures to get a representative value of the feature vector quality. In Section 5, we will present and discuss the results of the experiments.

4.1 Ontologies

FVs' construction is semantics based and heavily relies on ontologies. Consequently, we would like to measure the effect of ontologies of different granularity. We have chosen three ontologies that have been used in our earlier experiments [18]. All the ontologies are formalized in OWL DL. Next, short descriptions of the ontologies[6] are provided:

Animals ontology: this little ontology classifies some species, does not contain any individuals, and has only hierarchical properties. The original ontology is adapted to be more correct w.r.t. biological classification. The ontology was selected to see the effect of applying the approach on a typical *taxonomy*.

Travel ontology: A bit more advanced compared to the Animals ontology by having individuals and some object properties. This ontology is classified in this work as a *lightweight* ontology.

Wine ontology: Even more advance than the Travel ontology with more individuals than classes and many relations. This ontology was originally constructed to test reasoning capabilities. Maybe as a result, the ontology contains some entity labels that are not found elsewhere (e.g. the entity McGuinnesso is according to the ontology a winery; however a search with Google provides no results). Consequently, there will be several entities that will not be populated with this ontology. This ontology can indicate the robustness of this approach and is classified in this work as *advanced*.

We have selected not to include any large or *heavyweight* ontologies in this experiment since we believe that larger ontologies will not provide any significant new insight except of processing time, which is not the focus of this evaluation.

The key characteristics of the ontologies are displayed in Table 6. The evaluation has restrictions as follows:

- All OWL object properties are treated as other relations.
- Disjoint classes as a feature are ignored since we do not consider siblings in this evaluation.
- The following equality features are ignored: equivalentClass, sameAs, and differentFrom.
- No reasoner is used. A reasoner can be used to extract more relationships between the entities than are available without using a reasoner. These additional relationships can be utilized to improve the FV quality.
- The maximum length of the FVs has been set to 30. In earlier experiments [18], the average FV length was 24±3.

---

[6] The ontologies used can be found here: http://folk.ntnu.no/steint/ontologies/

Table 6. Ontology key characteristics

Ontology	Classes	Individuals	Properties
Animals	51	0	0
Travel	34	14	6
Wine	82	155	10

- For query expansion, there have been set a limitation of maximum 3 entities from each of the possible neighbour relation types (parents, children, and others), that implies query expansion by maximum 9 entities in total.

4.2 Experimental Configurations

In this section, we will describe the experiments and the motivation behind them. The conducted experiments are summarised in Table 7. Next, we briefly describe each of the experiments.

Baseline (Bl#1, 2): A baseline was created in order to compare the results. For the domain identification component (Step 2.1), we selected to use *parent* entities for comparison since it must compare with at least one neighbouring entity. The baseline was conducted twice: at the beginning and at the end of the experiments. This was done in order to isolate influence of time span (see Section 5). The experiments were conducted in a period of one week.

Query expansion - neighbours (Ex#2-8): We test what kind of neighbouring entities (parent, child, other) are optimal to include.

Query expansion - keywords (Ex#12, 13): By populating an ontology with global keywords it is expected that all the FVs will have higher similarity and be less unique compared to omitting the global keywords. However, is this the case?

Number of search results (Ex#14, 15): 30 search results have been set for the baseline. Is this an optimal number and what implication has it on the FV quality? We test if 100 or even 200 are more optimal. We expect that more search results will have a positive effect on the FV quality.

Content (Ex#9): It is expected that using full text documents will provide better feature vector quality than using snippets.

Clustering - input (Ex#10, 11): The clustering algorithms used are optimized for processing snippets. As a result, it is assumed that using document feature vectors will be a better candidate than using raw full text documents. However, for snippets it might be better to use the raw text than creating document feature vectors since snippets do in general provide little information and if only some of the key-phrases are extracted then even less information will be available to the clustering algorithm.

Domain identification (Ex#16-21): It is expected that comparing with neighbouring entities by relation type filtering will have a major effect on the feature vector quality. Utilizing parents are assumed in general to have the most positive effect.

Best practice (Ex#22): As the experiment proceeded we started to get some indications of what components and parameters that had a positive effect on the feature vector quality or not. Consequently, we would also like to test if a combination of these findings would yield the same positive effect or not. Therefore, we have combined some of these findings to assess the effect.

Table 7. Summary of the experiments conducted

	Bl#1	Bl#2	Ex#2	Ex#3	Ex#4	Ex#5	Ex#6	Ex#7	Ex#8	Ex#9	Ex#10	Ex#11	Ex#12	Ex#13	Ex#14	Ex#15	Ex#16	Ex#17	Ex#18	Ex#19	Ex#20	Ex#21	Ex#22
0. Ontology																							
Animals	X	X	X	X	X	X	X	X	X	X	X	X	X	X	X	X	X	X	X	X	X	X	X
Travel	X	X	X	X	X	X	X	X	X	X	X	X	X	X	X	X	X	X	X	X	X	X	X
Wine	X	X	X	X	X	X	X	X	X	X	X	X	X	X	X	X	X	X	X	X	X	X	X
1. Query expansion																							
neighbors																							
parents			X		X	X			X					X									X
children				X		X		X	X														X
others					X		X	X	X														X
keywords											x[1]	x[1]											
2. Search results																							
content																							
snippet	X	X	X	X	X	X	X	X	X		X		X	X	X	X	X	X	X	X	X	X	X
full text										X		X											
nbr of results	30	30	30	30	30	30	30	30	30	30	30	30	30	30	100	200	30	30	30	30	30	30	100
3. Clustering																							
input																							
document fv												X	X										X
text	X	X	X	X	X	X	X	X	X	X				X	X	X	X	X	X	X	X	X	
4. Domain identification																							
neighbors																							
parents	X	X	X	X	X	X	X	X	X	X	X	X	X	X	X	X			X	X		X	X
children																	X			X		X	X
others																		X		X	X		X

[1] Animals ontology: 'animals'; Travel ontology: 'travel'; Wine ontology: 'wine'

4.3 Evaluation Measures

In this section, we will define the similarity measures used. First, we define the Average Fv Similarity (AFvS) as follows.

Definition 3. *Average Fv Similarity (AFvS) gives an indication of the uniqueness of the FVs.*

$$AFvS(o) = \frac{2}{n^2 - n} \sum_{i=1}^{n} \sum_{j=i+1}^{n} sim\left(fv_i, fv_j\right) \qquad (3)$$

where n is the number of fvs in the ontology o and $sim(fv_i, fv_j)$ is the traditional cosine similarity measure between the two vectors. A score of zero would indicate that all FVs are unique. However, this is hardly possible since the approach is based on similarity between the entities to be able to populate the ontology. In general we would like this score to be as low as possible, in order to discriminate the entity FVs. However, this depends a lot on ontology.

Next similarity score is the Average Fv Neighbourhood Similarity (AFvNS) defined as follows.

Definition 4. *Average Fv Neighbourhood Similarity (AFvNS) indicates the degree of overlap with neighbouring entities.*

$$AFvNS(o) = \frac{1}{n} \sum_{i=1}^{n} \frac{1}{m} \sum_{j=1}^{m} sim\left(f v_i, f v_j\right) \qquad (4)$$

where n is the number of fvs in the ontology o and m is the number of neighbouring entities with fvs of entity i with fv_i. In this experiment, we have selected to use all the neighbours of an entity and do not differentiate the neighbours by weighting. As for AFvS this score will be highly depended on the ontology quality. Nevertheless, the ideal score would depend on the intended usage of the populated ontology (e.g. when used in search, for a comprehensive search we would like this value to be higher than for a fact-finding kind of search).

We have chosen to use the Normalized Google Distance (NGD) [6] as a measure to evaluate the quality of each feature vector. NGD can be used to compute the semantic distance between two terms. The NGD equation [6] is provided below for the clarity:

$$NGD(x,y) = \frac{\max\{\log f(x), \log f(y)\} - \log f(x,y)}{\log N - \min\{\log f(x), \log f(y)\}} \tag{5}$$

where $f(x)$ denotes the number of pages containing x and $f(y)$ for y, and $f(x, y)$ denotes the number of pages containing both x and y. N denotes the "total number" of pages assumed index by Google, which in this experiment was set to 20 billion (at this magnitude the precise amount of pages is not significant). The range of NGD is between 0 and ∞, where 0 denotes best match. However, in practice most values are in the range from 0 to 1. Consequently, for the special case where $NGD(x, y) > 1$ we set $NGD(x, y) = 1$. The motivation behind this is that the distance is too large to be of any interest anyway. Note, for this assumption to be valid the constant N must be set to a representative value. NGD is symmetric by definition, however searches with Google are not (e.g. a search for "x y" often yield different results than "y x"). We tackle this issue by ordering the search term (for instance, always putting the parent entity before a child entity).

NGD will be used in the next similarity scores as follows.

Definition 5. *Average Fv NGD (AFvNGD)* indicates the semantic distance between the entities and their FVs.

$$AFvNGD(o) = \frac{1}{n}\sum_{i=1}^{n}\frac{1}{m}\sum_{j=1}^{m}NGD\left(fvn_i, kp_j\right) \tag{6}$$

where n is the number of fvs in the ontology o and m is the length of the fv_i and fvn_i is the name of the fv_i, the entity name, and kp_j are the key-phrases of fv_i. Note if an entity got a parent then the name of the parent is also included to provide a more specific similarity distance (adapted from Bouquet et al. [3] that in our case is limited to the closest parent). *FvNGD(fv)* will have a score in the same range as NGD.

Once we have found the AFvS, AFvNS and the AFvNGD measures for an ontology the total score can be calculated. The total score is an aggregated score of the three measures. The total feature vector quality score is defined as follows.

Definition 6. *Fv Quality Score (FvQS)* provides the overall quality of the FVs.

$$FvQS(o) = \alpha(1 - AFvS) + \beta AFvNS + \gamma(1 - AFvNGD) \tag{7}$$

where $\alpha + \beta + \gamma = 1$ are weight factors (defaults are 1/3). The total FV quality score for an ontology will be in the range (0-1), where 1 indicates the best score.

5 Results and Analysis

In this section, the results of the experiments are presented and analysed. Note, because of constant change of the Web corpora and update of search engines the results of this evaluation may vary in time. Therefore, the evaluation was conducted in a week to minimize the issue of results changing due to changes provided by the search engine providers.

Table 8. Experimental results

	Animals ontology				Travel ontology				Wine ontology			
	AFvS	AFvNS	AFvNGDS	FvQS	AFvS	AFvNS	AFvNGDS	FvQS	AFvS	AFvNS	AFvNGDS	FvQS
Bl#1	0,019	0,154	0,266	**0,623**	0,019	0,186	0,253	**0,638**	0,040	0,286	0,163	**0,694**
Bl#2	0,020	0,168	0,255	**0,631**	0,023	0,147	0,253	**0,624**	0,041	0,286	0,180	**0,688**
Ex#2	0,048	0,304	0,194	**0,687**	0,042	0,326	0,227	**0,686**	0,079	0,412	0,149	**0,728**
Ex#3	0,021	0,288	0,277	**0,663**	0,021	0,313	0,254	**0,679**	0,046	0,322	0,155	**0,707**
Ex#4	0,021	0,178	0,265	**0,631**	0,020	0,139	0,241	**0,626**	0,041	0,304	0,152	**0,704**
Ex#5	0,040	0,404	0,214	**0,717**	0,035	0,243	0,231	**0,659**	0,075	0,403	0,150	**0,726**
Ex#6	0,048	0,288	0,200	**0,680**	0,041	0,334	0,231	**0,687**	0,079	0,409	0,149	**0,727**
Ex#7	0,020	0,278	0,276	**0,661**	0,021	0,259	0,258	**0,660**	0,043	0,316	0,158	**0,705**
Ex#8	0,039	0,406	0,215	**0,717**	0,034	0,272	0,233	**0,668**	0,073	0,412	0,149	**0,730**
Ex#9	0,015	0,211	0,261	**0,645**	0,049	0,239	0,246	**0,648**	0,102	0,458	0,192	**0,722**
Ex#10	0,019	0,130	0,270	**0,613**	0,019	0,092	0,241	**0,611**	0,042	0,277	0,177	**0,686**
Ex#11	0,014	0,224	0,249	**0,654**	0,049	0,225	0,229	**0,649**	0,099	0,446	0,196	**0,717**
Ex#12	0,182	0,280	0,262	**0,612**	0,161	0,253	0,243	**0,616**	0,286	0,452	0,182	**0,661**
Ex#13	0,133	0,358	0,197	**0,676**	0,098	0,261	0,220	**0,648**	0,218	0,467	0,170	**0,693**
Ex#14	0,017	0,210	0,259	**0,644**	0,022	0,201	0,241	**0,646**	0,045	0,386	0,184	**0,719**
Ex#15	0,015	0,221	0,268	**0,646**	0,026	0,233	0,249	**0,652**	0,054	0,397	0,185	**0,720**
Ex#16	0,022	0,070	0,149	**0,633**	0,029	0,177	0,249	**0,633**	0,079	0,345	0,208	**0,686**
Ex#17	-	-	-	-	0,014	0,195	0,192	**0,663**	0,048	0,293	0,268	**0,659**
Ex#18	0,018	0,181	0,260	**0,635**	0,026	0,136	0,236	**0,625**	0,045	0,308	0,156	**0,703**
Ex#19	0,019	0,180	0,259	**0,634**	0,025	0,132	0,226	**0,627**	0,043	0,307	0,152	**0,704**
Ex#20	0,010	0,030	0,150	**0,623**	0,023	0,151	0,258	**0,623**	0,059	0,337	0,241	**0,679**
Ex#21	0,018	0,172	0,231	**0,641**	0,023	0,137	0,234	**0,627**	0,042	0,311	0,180	**0,696**
Ex#22	0,044	0,487	0,198	**0,749**	0,043	0,343	0,237	**0,687**	0,101	0,553	0,174	**0,759**

5.1 Results and Analysis

Table 9 summarises the test results where the evaluation measures described in Section 4.3 were used. In total 23 experiments were conducted. The first experiment conducted was Bl#1 while the last was Bl#2 that are used as the baseline for the other experiments. In the next section, we will analyse the results.

Table 9. Experimental analysis

	Bl#1	Ex#2	Ex#3	Ex#4	Ex#5	Ex#6	Ex#7	Ex#8	Ex#9	Ex#10	Ex#11
Animals ontology	0,0%	9,1%	5,3%	0,0%	13,8%	7,9%	4,8%	13,9%	2,3%	-2,8%	3,7%
Travel ontology	0,0%	9,7%	8,7%	0,4%	5,5%	10,0%	5,7%	7,0%	3,8%	-2,0%	4,0%
Wine ontology	0,0%	5,7%	2,7%	2,2%	5,5%	5,6%	2,4%	6,0%	4,8%	-0,3%	4,1%
Average	0,0%	8,2%	5,6%	0,9%	8,3%	7,8%	4,3%	9,0%	3,6%	-1,7%	4,0%
Standard deviation	0,0%	2,2%	3,0%	1,2%	4,8%	2,2%	1,7%	4,3%	1,2%	1,3%	0,2%

	Ex#12	Ex#13	Ex#14	Ex#15	Ex#16	Ex#17	Ex#18	Ex#19	Ex#20	Ex#21	Ex#22
Animals ontology	-3,0%	7,3%	2,2%	2,4%	0,4%		0,6%	0,6%	-1,2%	1,7%	18,9%
Travel ontology	-1,1%	3,8%	3,5%	4,5%	1,5%	6,2%	0,2%	0,5%	0,0%	0,5%	10,0%
Wine ontology	-3,9%	0,7%	4,4%	4,5%	-0,3%	-4,2%	2,1%	2,2%	-1,3%	1,1%	10,2%
Average	-2,7%	3,9%	3,4%	3,8%	0,5%	1,0%	1,0%	1,1%	-0,9%	1,1%	13,1%
Standard deviation	1,4%	3,3%	1,1%	1,2%	0,9%	7,3%	1,0%	1,0%	0,7%	0,6%	5,1%

An overview of the experiments and their percentage difference relative to the baseline is shown in Table 9. Since we used Bl#1 as the baseline the values for this experiment is set to 0. Further, since we are using the Web and depends on search results from a commercial search engine, where we have little control of potential changes that might affect the search results, we conducted the same baseline test as the final test of these experiments. This new baseline test is denoted as Bl#2. Consequently, Bl#2 serves as deviation value and therefore subtracted from the results shown in Table 9. Next, we will provide some comments about the findings of the experiments:

Query expansion - neighbours (Ex#2-8): Ex#8 provided in average the best results, and also the best results for the Animals and the Wine ontologies. However, for the Travel ontology Ex#8 provided the fourth best results while Ex#6 gave the best results for this ontology. It was assumed that Ex#2 in average would provide the best results however it turned out that it provided the third best results. If we look at both the standard deviation and mean results then Ex#2 yields the best results. This could indicate that independent of the quality of the ontology Ex#2 would be the best choice.

Query expansion - keywords (Ex#12, 13): The results from Ex12# indicate that adding global keywords is not beneficial w.r.t. the overall FV quality score. The AFvS score is high for both Ex#12 and Ex#13. Ex#13 indicates an increase but compared to Ex#2 it is a decrease. However, as discussed in Section 3 Step 1.1, homogeneous FVs can be a feature that is beneficial depending on the intended usage.

Number of search results (Ex#14, 15): In Ex#14 and Ex#15 we tested if the number of search results retrieved and processed would affect the FV quality, which provide to be the case with 3.4% and 3.8% respectively. More clusters are more expensive to compute. In Ex#2, the Animals, Travel and Wine ontologies took 3, 3, and 16 minutes to process respectively while Ex#14 took in average 3 times as long to process and Ex#15 took 7 times as long. In this experiment we have not tried to find the optimal number of results to process, but just by looking at the increase of FV quality from 30 to 100 results versus 200 results indicate that 100 is the best candidate in this test w.r.t. both the FV quality and processing time.

Content (Ex#9): The results of Ex#9 show a slight improvement with an average of 3.0% compared to the baseline. It is uncertain if this result is optimal since we have experienced some difficulties using full text documents. Many sites do not allow direct download of Web pages for other purposes than browsing. Consequently, some of the documents became unavailable which would influence the quality of the FVs. Nevertheless, Ex#9 showed an improvement compared to the baseline.

Clustering - input (Ex#10, 11): In Ex#11 we tested if it is more beneficial to use document FVs, key-phrases extracted from the full text documents, as input to the clustering algorithms or snippets. Ex#11 showed some improvement of using document FVs compared to Ex#9 with only 0.4%, probably because the document FVs are more focused by extracting only those parts of the documents considered most relevant to the search. However, when creating document FVs for the snippets, Ex#10 showed a decrease in performance by 1.7% indicating that the snippets are best used as is.

Domain identification (Ex#16-21): Not surprisingly we got more or less the same results as for the query expansion experiments (Ex#2-Ex#8) where using *parents, children*, and *other* neighbouring entities provided the best results (Ex#21). Ex#19 got the same results as Ex#21 but with higher standard deviation indicating that Ex#21 provides better results independent of the ontology quality. Ex#18 and Ex#19 provides more or less same results. For Ex#16, Ex#17, and Ex#20 the algorithm failed to populate most of the entities (see Table 10). In fact, for Ex#17 no entities were populated for the Animals ontology since the ontology only got super- and sub-class relationships and hence no *other* relations. Consequently, the results from Ex#16, Ex#17, and Ex#20 can be disregarded.

Best practice (Ex#22): These experiments were conducted to test the combination of some of the best results from the other experiments. Both Ex#22 performed considerable better that the other experiments with an increase of 13.1% and 10.6% respectively.

5.2 Key Findings

Based on the findings in the conducted experiments we conclude the following:

(1) Query expansion: Query expansion increases the quality of the search results and hence the quality of the FV quality. Including the parents, children, and other related entities provide the best results.

(2) Search results and (3) Clustering: Using full text documents in combination with extraction of the most relevant key-phrases seems to provide the best positive effect on the FV quality. However, this increases the processing time considerably compared to using just snippets (assumes this is mainly due to download of each page).

(4) Domain identification: Including the parents, children, and other related entities seem to provide the best results when identifying the most prominent cluster candidates.

However, these are general conclusions independent of ontology quality. The most important component with respect to the FV quality is the query expansion component (Step 1.1). The parent entities are the most important neighbouring entities both for query expansion (Step 1.1) and when identifying the most prominent candidate cluster (Step 2.1). Further, utilizing the neighbouring entities when expanding the query yields better FV quality than using scope keywords. A high number of search results minimises the difference between the search engines and probably the change in ranking they provide over time.

6 Conclusions and Future Work

In this study, we have described and evaluated an unsupervised approach to feature vector construction. These feature vectors typically contain terms that are associated with the concepts reflected by the actual text corpora, i.e. the Web. We have focused on the aspects of the components w.r.t. both the FV quality and the ontologies used. Ontologies with different granularity where used, we have shown how this affect the quality of the feature vectors. In total 23 experiments were conducted. Based on the findings a set of recommendations for the construction of ontology based feature vectors are proposed.

We have also done some minor experiments with the NGD measure to assess the semantic distance between the entities of the ontologies used in this experiment. Preliminary results indicate that there is a connection between the findings and characteristics of each ontology used in this experiment and the NGD ontology score. This needs to be explored further. Therefore, one of the future tasks is to conduct a similar experiment with a broader set of ontologies. We need to categorize the ontologies according to different key characteristics to find trends relevant to the categories.

References

[1] Agirre, E., Ansa, O., Hovy, E.H., Martínez, D.: Enriching very large ontologies using the WWW. In: ECAI Workshop on Ontology Learning. CEUR-WS.org, vol. 31 (2000)

[2] Bergamaschi, S., Bouquet, P., Giazomuzzi, D., Guerra, F., Po, L., Vincini, M.: An Incremental Method for the Lexical Annotation of Domain Ontologies. Int. J. on Semantic Web and Information Systems 3(3), 57–80 (2007)

[3] Bouquet, P., Serafini, L., Zanobini, S.: Semantic Coordination: A New Approach and an Application. In: Fensel, D., Sycara, K., Mylopoulos, J. (eds.) ISWC 2003. LNCS, vol. 2870, pp. 130–145. Springer, Heidelberg (2003)

[4] Bry, F., Koch, C., Furche, T., Schaffert, S., Badea, L., Berger, S.: Querying the Web Reconsidered: Design Principles for Versatile Web Query Languages. Int. J. on Semantic Web and Information Systems 1(2), 1–21 (2005)

[5] Castells, P., Fernandez, M., Vallet, D.: An adaptation of the vector-space model for ontology-based information retrieval. IEEE TKDE 19(2), 261–272 (2007)

[6] Cilibrasi, R., Vitanyi, P.: The Google Similarity Distance. IEEE Transactions on Knowledge and Data Engineering 19(3), 370–383

[7] Formica, A., Missikoff, M., Pourabbas, E., Taglino, F.: Weighted Ontology for Semantic Search. In: Meersman, R., Tari, Z. (eds.) OTM 2008, Part II. LNCS, vol. 5332, pp. 1289–1303. Springer, Heidelberg (2008)

[8] Justeson, J.S., Katz, S.M.: Technical terminology: some linguistic properties and an algorithm for identification in text. Natural Language Engineering, vol. 1, pp. 9–27. Cambridge University Press, Cambridge (1995)

[9] Lopez, V., Sabou, M., Motta, E.: PowerMap: Mapping the Real Semantic Web on the Fly. In: Cruz, I., Decker, S., Allemang, D., Preist, C., Schwabe, D., Mika, P., Uschold, M., Aroyo, L.M. (eds.) ISWC 2006. LNCS, vol. 4273, pp. 414–427. Springer, Heidelberg (2006)

[10] Mitchell, T.M.: Machine Learning. McGraw-Hill, New York (1997)

[11] Ogden, C.K., Richards, I.A.: The meaning of meaning: a study of the influence of language upon thought and of the science of symbolism. Kegan Paul, Trench, Trubner & Co., London (1930)

[12] Osinski, S., Weiss, D.: A Concept-Driven Algorithm for Clustering Search Results. IEEE Intelligent Systems 20, 48–54 (2005)

[13] Panagis, Y., Sakkopoulos, E., Garofalakis, J., Tsakalidis, A.: Optimisation mechanism for web search results using topic knowledge. Int. J. Knowledge and Learning 2(1/2), 140–153 (2006)

[14] Solskinnsbakk, G., Gulla, J.: Ontological Profiles in Enterprise Search. Knowledge Engineering: Practice and Patterns, 302–317 (2008)

[15] Strasunskas, D., Tomassen, S.L.: The role of ontology in enhancing semantic searches: the EvOQS framework and its initial validation. Int. J. Knowledge and Learning 4, 398–414 (2008)

[16] Su, X., Gulla, J.A.: An information retrieval approach to ontology mapping. Data & Knowledge Engineering 58, 47–69 (2006)

[17] Suomela, S., Kekalainen, J.: Ontology as a search-tool: A study of real user's query formulation with and without conceptual support. In: Losada, D.E., Fernández-Luna, J.M. (eds.) ECIR 2005. LNCS, vol. 3408, pp. 315–329. Springer, Heidelberg (2005)

[18] Tomassen, S.L., Strasunskas, D.: Construction of Ontology based Semantic-Linguistic Feature Vectors for Searching: the Process and Effect. In: WI-IAT 2009. IEEE Computer Society, Milano (2009)

[19] Yang, H.-C.: A method for automatic construction of learning contents in semantic web by a text mining approach. Int. J. Knowledge and Learning 2(1/2), 89–105 (2006)

[20] Zhou, X., Hu, X., Zhang, X.: Topic Signature Language Models for Ad hoc Retrieval. IEEE Transactions on Knowledge and Data Engineering 19, 1276–1287 (2007)

Utilising Task-Patterns in Organisational Process Knowledge Sharing

Bo Hu[1], Ying Du[1], Liming Chen[2], Uwe V. Riss[1], and Hans-Friedrich Witschel[1]

[1] SAP Research
{bo01.hu,ying.du,uwe.riss,hans-friedrich.witschel}@sap.com
[2] University of Ulster, UK
l.chen@ulster.ac.uk

Abstract. Pattern based task management has been proposed as a promising approach to work experience reuse in knowledge intensive work environments. This paper inspects the need of organisational work experience sharing and reuse in the context of a real-life scenario based on use case studies. We developed a task pattern management system that supports process knowledge externalisation-internalisation. The system brings together task management related concepts and semantic technologies that materialise the former through a variety of semantic enhanced measures. Case studies were carried out for evaluating the proposed approach and also for drawing inspiration for future development.

1 Introduction

Recently *agility* has become an important requirement facilitating businesses of large as well as small sizes to reach their goals with reduced cost and increased efficiency. A major challenge presents in such a vision. The employees should effectively share their best practice in the form of process knowledge, also referred to as "know-how". Process knowledge, in many cases, manifests itself as the so-called tacit knowledge that is difficult to capture and denies easy reuse and sharing. In real-life, we acquire "know-how" through observation and gardened participation. Apprentices watched and learnt from their masters while working and achieving goals together. In the modern society, although apprenticeship still exists, its importance diminishes due to formal education system. There is, however, an evident shortcoming of modern education. That is while one can systematically acquire "know-what" in classroom or from online e-learning materials, "know-how" is more situated that is bound to particular problems and contexts and somehow proprietary to the individuals. It is not uncommon that when an experienced employee leaves an organisation, so does the "know-how" possessed and demonstrated by this employee. One way to ensure the sustainability of organisational "know-how" is to capture such knowledge and make it sharable and reusable. Moreover, by sharing knowledge of problem solving, we are able to extract and migrate such knowledge from individuals to teams and communities, and eventually to a stage that it becomes organisational knowlege. Externalising and formalising "know-how" is by no means a

A. Gómez-Pérez, Y. Yu, and Y. Ding (Eds.): ASWC 2009, LNCS 5926, pp. 216–230, 2009.

new research topic. Relevant research includes business process modelling, case-based reasoning, etc. Yet, the not-so flourishing results of current approaches are attributed to the rigidity of process models and the significant effort required in constructing such models. This daunting fact immediately prompted us to find a less formal approach facilitating process knowledge sharing. We propose to mimic what happens in real-life. That is demonstrating process knowledge through performing tasks (as *process knowledge externalisation*) and obtaining process knowledge through copying "actions" from accomplished tasks (referred to as *process knowledge internalisation*). During the externalisation phase, tacit process knowledge is concretised by attaching each task with pertinent information such as documents used, URLs visited, people contacted, etc. Accumulated data are abstracted and classified into repeatable patterns. Externalised process knowledge is not truly shared unless it is internalised again by others when similar tasks are to be fulfilled.

1.1 Motivating Scenario: SAP ByDesign™ Support

Nowadays, Digital Divide [12] manifests itself in a totally different form. Apart from being separated by our accessibility to the digital world (differed by how frequent and how well we use digital devices), we are more and more divided by how effectively we reuse and situate past experience into new problems and how proficiently we attach to it the relevant information. This is particularly evident when we studied the SAP ByDesign™ Support Team where work efficiency is not solely determined by whether one has access to product documentation, manuals, internal Wiki pages, customer discussion forum—all the information is readily available to every employee and some is even open to public—rather by whether she knows how to leverage such information and project it upon the problem at hand. It is our contention that the difference in information handling capabilities leads to potentially significant variations in productivity, creativity, and work efficiency. When interviewing the support team in SAP, Galway, Ireland, we found significant difference between numbers of queries (referred to as incidents) handled by individual staff. An important issue raised was the lack of mechanism supporting reuse and sharing of past experiences, successful ones especially. All the interviewees were well trained and aware of the high-level formal procedures of dealing with incident reports. They, however, adopted different strategies when proceeding following the formal procedures. The difference in strategies, consequently, impinges significantly on the outcomes of work. For instance, some experts keep a separate record of key points when attending incidents (using notepad or software tools); some organise pertinent information of similar cases together in one file folder; some link together apparently different cases and emphasise on their commonality. Such good practice is not normally transferred from one expert to another and thus seldom go beyond individuals into team knowledge.

The need of sharing and reusing experience (in terms of established process knowledge) has been addressed previously in various contexts. This study, however, unveiled some technical barriers which, once overcome, might indeed change the terms of engagement. These barriers are: 1) the lack of low-cost formalisation of past experiences, 2) the lack of tools to systematically collect relevant information against a task, and 3) the lack of more sophisticated mechanisms for retrieving previous experiences.

1.2   Challenges and Considerations

Our user studies raised many challenges that can be, to a great extent, generalised to other situations where sharing and reuse of past problem solving experience is a major concern. Typically, good practices are formally modelled in workflow systems and made available to an organisation. However, workflow systems depend on predefined models, which are expensive and rigid. This is against our philosophy in searching for an answer to facilitating agility by cutting short the normal process formalisation cycle. We, therefore, focus on one easy-to-start approach to process knowledge sharing in terms of past problem solving experience. Process knowledge *internalisation* through task pattern reification provides an answer to our quest. Instead of formal process models, users adopt a task pattern and situate it with information specific to a problem's context leading to the instantiation and creation of new tasks from the task pattern. In summary, a task pattern is a carrier and externalisation of the user's past experiences. When the task pattern is applied as a template of creating new tasks, a knowledge worker internalises the past experiences embedded therein and contextualises them against her problems at hand (Figure 1).

Fig. 1. Process knowledge externalisation and internalisation

An onus of this process knowledge sharing duality is the additional effort and heavy user interaction in the task pattern lifecycle which have raised the major challenge of enabling the task pattern based approach. This is because knowledge workers are usually focusing on the tasks at hand and are likely to reject the extra work of abstracting existing tasks, e.g., contribute to task patterns, as part of their activities. Such activities merely add more administrative overhead for which the immediate benefit is not always clear. Hence, costs of performing such extra work must be reduced to as low as possible. A system enhanced with semantic technologies helped us to make a step forward in this direction.

In the follow, we first explain the pattern-based process knowledge sharing in Section 2 and how semantic technologies could be leveraged. In Section 3, we propose an architecture that facilitates the synergy of defined task patterns and bottom-up style task sharing. We detail what semantic technologies were employed. User study was carried out whose results are summarised in Section 4. Finally we conclude the paper in Section 5.

2   Sharing Process Knowledge as Task Patterns

The task pattern approach has been suggested in [9] as a low cost means to capture process knowledge in knowledge-intensive work. At the heart of this approach is the concept of dissecting process models on the level of individual tasks and use them to record and abstract activities necessary to fulfill the tasks. By doing so, the sharing of process knowledge becomes well focused and grounded as sharable and reusable patterns of tasks.

2.1   Task Patterns

Hereinafter, we differentiate the concepts of *tasks*, *task patterns*, and *processes* as follows: A *process* is a collection of structured activities (tasks) with a precise goal to be achieved over a period of time. The activities (tasks) of a process are partially ordered and can be further divided into finer-grained sub-tasks. A *task* is an action requiring completion. A *task pattern* is an abstraction of tasks replacing specific resources with *abstractors*. *Task patterns* can be instantiated by assigning concrete instances to task resource abstractor as the abstraction of artefacts associated with tasks.

The Task Pattern approach bottom-uply involves users in the process management without implicating them in actual process management activities. This is done through task-oriented experience reuse or task copy transferring users' past experiences by copying details of accomplished and successful tasks. With task copying in mind, the next natural question is how such a copy operation will be realised. There are two options. First, entire task structures and details are duplicated, with the assumption that everything is implicitly relevant to the next task context; second, the user is responsible for explicitly selecting every detail to be copied. The former is likely to be useful for a small set of tasks. This is because it will likely result in information overload for a user. It results in a situation which requires the user to spend potentially more effort customising the duplicated task than to start a new one. The second option could also overwhelm users as it requires them to consider too many details from previous tasks, thus leading to a situation similar to that of the first. This reflection informs us that a more helpful position to consider the reuse of past experience lies somewhere in between these two extremes. To that end, we propose the concept of *task journal* as the basis on which previous experience can be shared (Figure 1). Task journals are the records of previous task activities and information artefacts, which are harvested by continually monitoring the interaction between users and a task management system, and collecting valuable information of events during task execution. The task history actually provides an explicit view on how the task is completed with critical information artefacts attached to it. The transition between task, task journal, and task pattern is fully supported in the task pattern management system (TPMS).

2.2 Supporting Task Pattern with Semantic Technologies

We investigated how existing semantic technologies are utilised in the context of a TPMS. We would argue that although semantic technologies are not the only solution to the challenges presenting in a TPMS, they offer unprecedent advantages over technologies that are not semantically enhanced. This can be seen from the followings aspects:

Firstly, semantic technologies offer machine processable meanings through ontologies. Even though obtaining shared domain knowledge increases the overhead, its value is evident in that i) a controlled vocabulary rooted in the ontology helps to regulate user interface reducing random inputs from users and thus increases system efficiency; ii) a common reference based on the ontology would serve as the foundation for aligning heterogeneous data; and iii) machine processable formalisation has a knock-on effect on automated reasoning. We are aware of arguments in the community against the formality of predefined global ontology due to its rigidity and the significant efforts involved therein. We, however, would like to emphasise that these would not be a major barrier in our application domain. Nowadays, organisations from different sectors, being both large enterprises and SMEs (Small and Medium Enterprises), have strict regulations in place and exercise an organisational common vocabulary to some extent as an effort towards organisational standards. Even though this is still far away from a formal ontology, it is already an embryonic form with which ontologies are considered as a natural subsequent step.

A formal ontology also demonstrates the capability of explicitly and implicitly linking apparently isolated data "islands". Semantic technologies increase data linkage through well-formed logic formulae. Linking data together is important in process knowledge capturing. One goal of process knowledge sharing is to allow re-execution of the processes when and where it becomes necessary. In order to do so, it is crucial to not only share the skeleton problem solving steps but also "beef" each step up with necessary supporting evidences. For instance, when sharing the "flight ticket booking" process, backing each individual step with necessary organisational regulations, white papers, airline web pages, contacts of internal people in charge of payment, contacts of travel agencies, etc., would allow others to instantly pick up the correct process as well as *how* each step of it can be achieved with *what* means. These connections are not always evident and can only be established through instantiating the domain ontology with linkages among data concretising the properties defined in the ontology.

Tasks are introduced as instances of ontology concepts and properties, coded in RDF triples. In an organisation, archiving and easy retrieval of historical data is important for quality assurance purposes. This practical consideration leads to design requirements on the reliability and scalability of data repositories. Thus far, semantic data (mainly RDF triples) storage has been intensively studied. Efficient and scalable platforms include those of industrial strength (e.g. Oracle 11g, Jena SDB[1]) and those rooted in academic "proof-of-a-concept" prototypes (e.g. 3Store[2], Sesame[3]). The advantage of a purpose-built RDF triple store is its reasoning capability and the native query language facilitating easy access to apparently complex, intertwined RDF triples. In our scenario, RDF triple store is the backend for task instances, resource metadata, and semantic annotations.

RDF triples in its native representation suffer from poor readability. An intuitive user interface is preferable for better comprehension. In the meantime, the user interface should support team work, as in organisations collaboration sometimes is the key to fulfill tasks. We propose to materialise the frontend of data repository through Semantic Wikis[4]. Wiki is widely used as a platform for collecting and exchanging knowledge within communities. Collaboration is natural in Wikis in that registered users can jointly contribute to the contents of a topic while changes are managed with versioning tools. Enhanced with machine-processable markups, semantic wikis lend themselves to better organisation of information held therein. In this way the semantic structure is preserved while user interactions and information provision are better adapted to users' needs.

Finally, established semantic technologies can be of great help in addressing data interoperability. The so-called *semantic similarity* algorithms were recognised as an enabling technology for aligning heterogeneous perspectives over the same domain. Even though an organisation-wide ontology is reinforced, we still see the needs of attending data interoperability when one compares her goals against existing task pattern repository, retrieves established good practice and identifies candidate resources to be associated with / instantiate task patterns.

3   The Task Pattern Management System

The proposed TPMS (Figure 2) is underpinned by a task management ontology coded using OWL [2] as the ontology representation language. Having considered many existing approaches, we opted for the semantically enhanced wiki, Semantic Mediawiki (SMW) [5], for its improved content management, intuitive collaborative user interface, and smooth learning curve. Finally, we experimented and leveraged a variety of similarity algorithms to identify suitable tasks, task patterns, and information artefacts.

The system also consists of an shared repository for existing task patterns. Towards the aim of reusing past experience and work structures, knowledge workers are unlikely to consider abstracting previous tasks to task patterns as an initial step. Therefore, we hypothesise that knowledge workers will refer to

---

[1] http://jena.sourceforge.net/SDB/
[2] http://threestore.sourceforge.net/
[3] http://www.openrdf.org/
[4] http://semantic-mediawiki.org

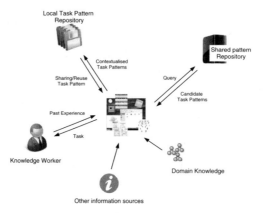

Fig. 2. System architecture

similar ongoing or completed tasks whose goals or contexts match those of her tasks at hand based on certain similarity measures [8], e.g. those reviewed in [7].

Since every task will differ in its context and details, the user is unlikely to be able to use the details directly from a task pattern. Instead, we expect that the user will be required to adapt existing task patterns to suit the context of the new task using domain knowledge encoded in the ontology and other useful information artefacts. For example, the user may select a document template as the basis for a report in the new task. Or the user may involve a person for the role suggested by the template task. This step of user adaptation through adding contextualised information and ontological domain knowledge is therefore necessary to enable the task patterns to be situated in the new work context. The user can then decide which adapted task patterns go into her private pattern repository and which are open to be shared with others. The adaption is effectively a classification and instantiation process during which the ontology and technologies built on top of ontologies play an important role.

3.1   Task Management Ontology, TMO

Ontology is currently considered the carrier of semantics. Its implication on our work is two-fold. On the one hand, ontology provides the necessary formalisation to increase data interoperability. When creating a task instance, one would impose some sort of structure on the information through ontologically regulated properties. For instance, one would normally annotate a person with names, affiliation, contact details, work places, etc. Such information facilitates both the readability of human users and accessibility of software tools. On the other hand, ontology offers inferencing capability. For instance, annotating artefacts based on the ontology enables us to perform (semi-)automated classification.

The central challenge of task management is providing effective task-related information support to knowledge workers. To this end, the TMO must be highly expressive and yet extensible to cater for ill-defined and continuously changing knowledge-intensive work situations. Consequently, the TMO is structured in two layers: (1) a set of concepts and resources which describe task-oriented information and work activities and (2) an underlying set of concepts which support the elaboration or concretisation of the more generic domain knowledge [10]. The first layer is centred around concept Task which is restricted with various properties, e.g. title, initiator, delegated_to (people), subtask, etc. The second layer includes such concepts as People, Document, Device, etc. These concepts are further refined with finer details. For instance, Document has WhitePaper, StaffManual, WebPage, etc, as sub-concepts. TMO is coded in OWL [2] due to its expressiveness, standard status, wide acceptance, and the Description Logic (DL [1]) ready feature for reasoning.

TMO was constructed by closely working together with target users, i.e. knowledge workers from various organisations. The evaluation of TMO was performed through expert review in an ethnographic study carried in the context of MATURE[5] and as part of the system evaluation to investigate whether the ontology facilitate a smooth integration of different functionalities of the TPMS.

3.2   Task Pattern Abstraction and Classification

Task journals play the role of experience carriers which convey various information during task execution. The decisive advantage of task journals is that they provide a chronological structure to task activities. This helps users to better understand which activities come first and which resources where used in the context of these activities. We utilise TMO to annotate and then classify and refine the information recorded in task journals. Whenever resources are attached to a task during its execution, a dialog box (Figure 3) is displayed to prompt users for annotating the resources with a list of concepts drawn from TMO. Annotated data are then processed with DL-based classification. For instance, when performing a task, one might make contact with a variety of individuals (introduced as instances of People) within different departments. Based on annotations detailing their positions and roles, these contacts are classified and abstracted as, for instance, "line manager", "cashier", "accountant", etc. Hence, who contributed to the fulfillment of previous tasks become less important as long as the correct types of people are contacted when one needs to carry out a similar task. Classifying information associated with tasks facilitates the creation of task patterns. Obtaining task patterns from task journals minimises users' effort to provide reusable process information to others and thus lowers one of the most important barriers in process knowledge sharing.

---

[5] www.mature-ip.eu

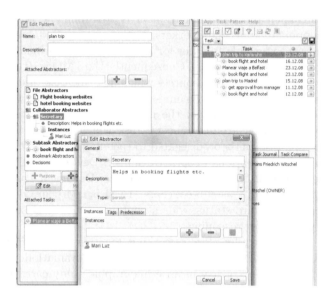

Fig. 3. Task pattern editing environment

3.3   Semantic WiKi for Content Management

Our TPMS utilises SMW for a collaborative working environment. SMW extends the widely used MediaWiki[6] by adding machine-readable meta-data to a wiki article. A direct result of such an extension is improved search and classification, as well as better interoperability with other applications through importing existing ontologies and exporting contents in the standard OWL/RDF format, in addition to retaining the ease of collaboration that any wikis offer. Meanwhile, with an RDF interface, it is possible to query the wiki's content using popular query languages such as SPARQL. The advantage of such a combination is evident in the smooth learning curve for ordinary users while being sophisticated enough to create and store machine processable semantics.

Viewing in the light of task management, the decisive advantage of applying SMW is the combination of metadata and content handling with respect to task patterns. The metadata provided with a task pattern are ideal for automatic processing but less friendly for users to really understand what a task pattern is about. The SMW approach allows users to augment task patterns with sufficient textual descriptions that help users understand the goal and proceeding in a task pattern while at the same time support automatic manipulation.

In the SMW, a task or task pattern is represented as an individual page with embedded inter-page links corresponding to the properties from an ontology (in our case the TMO). We leverage the SMW templates for nicely-formatted task patterns (shown below).

---

[6] semantic-mediawiki.org/wiki/MediaWiki

```
{{Task
   |Description=                      |Document abstractors=
   |People abstractors=               |Subtask abstractors=
   |Problems and solutions=          |Rating=
   |Rating Comment=
}}
```

An exemplary "prepare case" task pattern is shown in Figure 4 which links the task pattern with two file abstractors, one person abstractor, one subtask, textual descriptions, etc. The person abstractor in turn restricts that only one's colleague can and should be contacted when carrying out case preparation. And finally, when one instantiates a task pattern, she links a person instance to the `PersonAbstractor` via `Colleagues`. The page `Roger Smith` instantiates `Person` concept in the ontology with concrete values for defined properties.

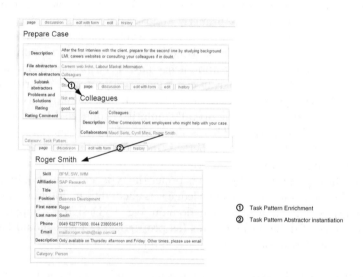

Fig. 4. Instantiating task patterns in Semantic Wiki

As wikis have been widely used in different working environments, wiki-based TPMS can easily blend into the daily working environment of knowledge workers.

3.4   Similarity Measures

One of the fundamental operations in our approach is to retrieve similar tasks that have been successfully carried out before. In our system, we utilise different similarity measures depending on the characteristics of the data.

String similarity. The initial task patterns are normally retrieved by comparing task pattern descriptions with users' request, normally as a small set of keywords, e.g. "travel booking". Existing task patterns are summarised with plain text detailing the goal of the task pattern, features of its various abstractor services, outcomes of the tasks, and open issues. This is done utilising established techniques from Information Retrieval that the similarity of two pieces of textual descriptions as numeric values. The simplest form of text similarity is Edit Distance. Free Java based string distance library, SecondString [3], is used in our implementation.

Ontology-based similarity. Pure string similarity algorithms ignore the structure of data which in many cases provide important information. They, therefore, are less attractive when the data in question are well structured against an ontology. For instance, when enriching a task pattern, users are prompted with candidate information artefacts that could instantiate abstractors. These artefacts are instances of People, WebPage, Document, etc, from the TMO. Hence, the way we understand artefacts is constrained by the ontology commitment. When information artefacts are properly annotated using the ontology, the comparison is tantamount to computing instance-level semantic similarity.

Instance-level semantic similarity is addressed in ontology mapping which is still an ongoing research. Many approaches have been proposed, implemented, and evaluated [7]. When comparing candidates, we reuse the algorithms developed in previous projects (e.g. CROSI [4]) that compute similarities of individual properties of task instances. The overall similarity is obtained from the similarities of individual properties. This is done by utilising weighted-average with human inputs indicating which property is more important than the others. For instance, one might emphasise on origins, dates, or places and thus give higher weights to these properties while others prefer the type and format of information artefacts. The domain heuristic knowledge was elicited from domain experts with an assumption that for a particular domain, e.g. ByDesign^{TM} Support Team, the importance of different attributes (as properties) of tasks should have been and can be clarified prior to running the similarity algorithm.

Graph-based similarity. Similarity measure is needed for comparing tasks. A task consists of sub-tasks and is associated with artefacts, being human experts and information resources. This naturally becomes a labelled and directed graph, $G = (V, E)$ with nodes V corresponding to (sub-)tasks and individual artefacts and edges E corresponding to either part-whole relationships or associations. The easy conversion of task graphs inspires us to consider graph similarity measures. Graph similarity has been extensively studied (c.f. [6]).

In our TPMS, a task graph is forced into a task tree. The root of a task tree is the task itself. Children of the root are the first level sub-tasks which in turn have their sub-tasks as child nodes. The leaves of a *task-tree* are artefacts supporting the fulfillment of the task. We duplicate a piece of artefact when it is referred to by more than one sub-task. Focusing on task trees, we can largely ignore how a sub-task node is labelled. When two sub-tasks are supported by the same set of evidences, we can assume, based on the closed world assumption, they have overlapping instance data and require the same knowledge to proceed.

This leads to a further assumption that sub-task requiring the same knowledge can be considered as similar tasks even though they are labelled differently. The closed world assumption is supported by two observations of our problem domain, i.e. process knowledge sharing in organisations. First, all the supporting evidences are shared and frequently used by a large number of employees and thus a common understanding can be easily negotiated. Second, such a set of artefacts is relatively stable. The creation or introduction of new information artefacts in a mature organisation is constrained by protocols and regulations. Hence, information artefacts that are not present are considered to be excluded from the organisational knowledge space and from our similarity computation. The algorithm is formalised as:

$$\mathtt{sim}(t, t') = \frac{|\,\gamma(t) \cap \gamma(t')\,|}{|\,\gamma(t) \cup \gamma(t')\,|}$$

where $\gamma(x)$ gives the set of supporting artefacts of x and t and t' are two bottom-level subtasks.

For pairs of inner nodes from different task trees (T and T'), the similarity is computed from those of their children using tree edit distance [11]. Tree edit distance is an approximate measure computing the difference of two trees as a numeric value between 0 and 1. Based on the TreeDiff algorithm, task tree edit distance is defined as the minimum number of node deletes and inserts when one task tree is transformed into another.

$$\mathtt{sim}(T, T') = 1 - \mathtt{diff}(T, T')$$
$$\mathtt{diff}(T, T') = \min\{\epsilon(S) \mid S \text{ is a sequence of edit operations } T \to T'\}$$

where $\epsilon(\cdot)$ is the cost function mapping an edit operation to a numeric value based on users' preference. The initial alignment among nodes stems from the similarity among bottom-level subtask nodes computed as above.

4   Use Case Studies

The TPMS was studied in two different use scenarios. Thus far we mainly focused on i) the investigation of task-pattern as a general approach to process knowledge sharing and ii) the applicability of semantic technologies in task pattern management. In order to achieve the first goal, use case studies were carefully designed so that different types of organisations were presented. More specifically, one scenario was carried out in an organisation from the public sector while the other scenario focused on the support department of a large enterprise which regularly interacts with other departments within the same enterprise as well as external customers. In the meantime, the two scenarios differ in that the public sector organisation is normally in contact with individual clients with different types of requests while the support team's customers are relatively stable and predictable. The preliminary results are promising although further detailed studies are still necessary. The second goal was achieved through observing user behaviours so as to detect whether the use of semantic technologies distracts users from their ordinary working activities and whether extra work incurs.

Procedure. These studies were performed as follows:

- Presentation of the prototype, its general functionality (including a very brief live demo) and the concept of task patterns.
- Pre-interview to determine how participants currently manage their tasks and how they deal with recurring tasks
- Performance phase during which participants had to work in two groups: the first group was asked to create a task, re-using information available in the system and at the same time refining that existing information into a task pattern. The second group was asked to create and populate another task by finding and using the task pattern provided by the first group.
- After the performance, participants were given a post-survey questionnaire to leave feedback and comments.
- Final discussion of the potentials of the task pattern concept in general.

Results. We first examined the acceptance of the pattern-based approach to experience and process knowledge sharing. It was evident that all the interviewees acknowledge the significance of having a formal procedure to help them collect past experience when starting with a new task, while they pointed out that having the flexibility of attacking the problems in different manners is equally important. The interviewees in general saw task pattern as a low cost approach to facilitate process knowledge sharing, even though they have different working ethic, different regulations and are from different industrial sectors. Able to break down an apparent complex process into manageable task patterns was well received in both studies. During their daily work, they have both formal and less formal process prescribing how queries should be dealt with. It would be useful to deploy task pattern management system and then code some of these repeating queries as task patterns, thus making the sharing and reuse more flexible, allowing for individualised execution through process knowledge *internalisation*. Similar task patterns were actually identified across different industrial sectors, such as "report writing" and "travel booking", suggesting a wide applicability of repeatable patterns in organisational process knowledge.

Enriching a task pattern by associating with it different resources is an important functionality that was identified as a welcoming feature of the prototype. A particular emphasis was made on the possibilities of using arbitrarily *any* artefacts to enrich a task pattern, as all of the interviewees frequently search internal competence management system for experts on particular subjects and subsequently talk to them over the phone or communicate via instant messaging. It is important to classify human experts in the same way as other information artefacts. To this end, semantic technologies (e.g. ontologies) were credited for making connections among data easier and more intuitive.

Usability was the most important thing raised unanimously by all the participants. During their fast-pace daily work, a software tool should be less interruptive and blend in with the other tools used. Wikis were accepted as a more user-friendly alternative as oppose to pop-up windows and dialog boxes. Some of the interviewees from the public sector were particular keen on a Wiki-based collaborative framework as such systems had already been embedded into their everyday working environment.

When asked for potential barriers, the participants would prefer the system to "speak" perfectly their "language" suggesting a certain terminology gap between the current version of TMO and some domain specific vocabulary. We, however, would argue that instead of denying the importance of having a task management ontology, this feedback emphasises the needs of mapping across different industrial sectors. Due to the relatively small size of TMO and the expressiveness of OWL, such a request can be easily accommodated by bridging new "jargons" with existing TMO concepts or properties, using either the *equivalent* construct from OWL or alternatively using SKOS[7] constructs.

Another problem raised in the post-interview discussion was that people have very different approaches to handle tasks (e.g. with respect to the granularity of modelling the task) and thus it might be hard to share patterns in some domains. Therefore, the application should be able to detect differences in granularity and react to it. On the other hand it is said that even if some abstractors were not re-usable in a particular context, the pattern (with its subtasks) could be seen as a checklist and would always be useful in that respect for sharing useful experience in organisations.

5   Discussion and Conclusions

In this paper, we described a semantic-enriched, task pattern based approach to organisational process knowledge sharing and reuse. We view process knowledge as the "know-how" that are normally not transparent to others. In order to pass on to others, it goes through first a series of dissection and abstraction phases to externalise knowledge possessed by individuals into "tangible" and repeatable task patterns. We then employ a series of internalisation phases situating task patterns into the context of new tasks transferring others' experience to one's own. This operation is supported by tools and methods leveraging semantic technologies.

The presented user studies are the initial step of evaluating the applicability of task patterns. Though showing promising results, we acknowledge the major issues raised by users. One of such issues is the *incentive measure*. Like other systems relying on user inputs, our approach is based on the assumption that individuals are willing to record their problem solving activities and share such knowledge with others. Although this is still subject to further investigation, we would argue that motivating individuals within an organisation is less challenging. Employee performance evaluation is widely adopted in organisations from different sectors. The proposed framework is expected to increase the productivity of work and also help individual workers to identify gaps of her knowledge. It, therefore, presents good motivation for more active involvement in pattern-based good practice sharing and reuse.

---

[7] http://www.w3.org/2004/02/skos/

While, obviously, there are many important issues to address, the crux of our immediate future work lies in the optimisation of the current TPMS and further improvement of the user interface. Thus far, our evaluation work focused on qualitative aspects investigating the underpinning theory of pattern-based process knowledge sharing and its impact in an organisational setting. Semantic technologies have indeed simplified the development and deployment. More evaluations are forthcoming to focus on quantitative usability aspects.

Acknowledgements

This work is supported under the MATURE IP funded by EU Framework 7.

References

1. Baader, F., Calvanese, D., McGuinness, D., Nardi, D., Patel-Schneider, P. (eds.): The Description Logic Handbook: Theory, Implementation and Applications. Cambridge University Press, Cambridge (2003)
2. Bechhofer, B., van Harmelen, F., Hendler, J., Horrocks, I., McGuiness, D.L., Patel-Schneider, P., Stein, L.A.: OWL Web Ontology Language Reference. W3C (February 2004)
3. Cohen, W., Ravikumar, P., Fienberg, S.: A comparison of string distance metrics for name-matching tasks. In: IIWeb, pp. 73–78 (2003)
4. Kalfoglou, Y., Hu, B.: CROSI Mapping System (CMS) Results of the 2005 Ontology Alignment Contest. In: Proceedings of the 3rd International Conference on Knowledge Capture (KCap 2005) workshop on Integrating Ontologies, Banff, Canada (October 2005)
5. Krötzsch, M., Vrandečić, D., Völkel, M.: Semantic mediawiki. In: Cruz, I., Decker, S., Allemang, D., Preist, C., Schwabe, D., Mika, P., Uschold, M., Aroyo, L.M. (eds.) ISWC 2006. LNCS, vol. 4273, pp. 935–942. Springer, Heidelberg (2006)
6. Lovàsz, L., Plummer, M.: Matching Theory. North-Holland, Amsterdam (1986)
7. Rahm, E., Bernstein, P.A.: A survey of approaches to automatic schema matching. The VLDB Journal 10, 334–350 (2001)
8. Riss, U.V., Cress, U., Kimmerle, J., Martin, S.: Knowledge transfer by sharing task templates: two approaches and their psychological requirements. Knowledge management Research & Practice (4), 287–296 (2007)
9. Riss, U.V., Rickayzen, A., Maus, H., van der Aalst, W.M.P.: Challenges for business process and task management. Journal of Universal Knowledge Management, Special Issue on Knowledge Infrastructures for the Support of Knowledge Intensive Business Processes, 77–100 (2005)
10. Sauermann, L., van Elst, L., Dengel, A.: Pimo - a framework for representing personal information models. In: Pellegrini, T., Schaffert, S. (eds.) Proceedings of I-Semantics 2007, pp. 270–277. JUCS (2007)
11. Wang, J., Zhang, K., Jeong, K., Shasha, D.: A system for approximate tree matching. IEEE Transactions on Knowledge Data Engineering 6(4), 559–571 (1994)
12. Williams, K.: What is the digital divide? Technical report

Reasoning about Partially Ordered Web Service Activities in PSL

Michael Gruninger and Xing Tan

Semantic Technologies Laboratory,
Department of Mechanical and Industrial Engineering,
University of Toronto

Abstract. Many tasks within semantic web service discovery can be formalized as reasoning problems related to the partial ordering of sub-activity occurrences in a complex activity. We show how the first-order ontology of the Process Specification Language (PSL) can be used to represent both the queries and the process descriptions that constitute the underlying theory for the reasoning problems. We also identify extensions of the PSL Ontology for which these problems are NP-complete and then explicitly axiomatize classes of activities for which the various reasoning problems are tractable.

1   Introduction

Ontologies that represent complex activities (such as composite web services and manufacturing process plans) are required for many applications of automated reasoning. We typically need to specify the occurrence and ordering constraints over the different subactivities; such constraints may either be explicitly represented or they may be entailed by other properties, such as preconditions and effects.

Within domains of semantic web service discovery, resources are associated with composite web service plans, which are partially ordered sequences of processes. In general, such process plans may also be nondeterministic (that is, involve different choices of sequences of processes). At any point in a web service plan, there are multiple activities that can possibly occur next. Furthermore, different web service plans may have processes in common so that an object may participate in an activity that is part of multiple plans. This scenario motivates four queries that are relevant for the discovery and verification of semantic web service plans, and which we will formalize later in this paper:

1. Is it possible for one activity in a web service process to occur before some other activity?
2. Is one activity in a web service process required to occur before some other activity?
3. Given the occurrence of some activity in a web service process, what activities are required to occur later?
4. Given the occurrence of some activity in a web service process, what activities can possibly occur next?

A. Gómez-Pérez, Y. Yu, and Y. Ding (Eds.): ASWC 2009, LNCS 5926, pp. 231–245, 2009.

In this paper, we will focus on the formalization of these four queries as first-order entailment problems which are related to occurrences of complex activities and the ordering constraints on those occurrences. To achieve this objective, we use a first-order ontology in which complex activities and their occurrences are elements of the domain, so that web service discovery can be expressed as entailment problems that are solved by inference techniques that are sound and complete with respect to models of the ontology. Inference is done using the axioms of the ontology alone, without resorting to extralogical assumptions or special algorithms used by interpreters.

In particular, we will use the PSL Ontology to axiomatize the constraints in the antecedents, as well as the queries in the consequents, of the entailment problems.[1] Furthermore, we use the model theory of the PSL Ontology to provide correctness theorems for the specification of the queries and the process descriptions for certain classes of activities. Finally, we will define extensions of the PSL Ontology in which two of the entailment problems are NP-complete, and additional extensions in which these entailment problems are tractable. In this way, tractable classes of the problems are explicitly axiomatized within the ontology itself, and the relationships between different assumptions can themselves be determined by first-order theorem proving.

2   PSL Ontology

As a modular set of theories in the language of first-order logic, the Process Specification Language (PSL) [6,7] has been designed to facilitate correct and complete exchange of process information;[2] whereas it has been adopted by the Semantic Web Services Language (SWSL) Committee of Semantic Web Services Initiative (SWSI)[3] to specify the model-theoretic semantics of Semantic Web Services Ontology (SWSO) ([3], [8]), one of the two major components within Semantic Web Services Framework (SWSF) [2].

Within the PSL Ontology, all core theories are consistent extensions of a theory referred to as PSL-Core, which introduces the basic ontological commitment to a domain of activities, activity occurrences, timepoints, and objects that participate in activities. Additional core theories capture the basic intuitions for the composition of activities, and the relationship between the occurrence of a complex activity and occurrences of its subactivities.

In order to formally specify a broad variety of properties and constraints on complex activities, we need to explicitly describe and quantify over complex activities and their occurrences. Within the PSL Ontology, complex activities and occurrences of activities are elements of the domain and the *occurrence_of*

---

[1] Given the entailment problem $T \models \phi$, we say that T is the antecedent and ϕ is the consequent.

[2] PSL has been accepted as an International Standard (ISO 18629) within the International Organisation of Standardisation. The full set of axioms in the Common Logic Interchange Format is available at http://www.mel.nist.gov/psl/ontology.html

[3] http://www.swsi.org

relation is used to capture the relationship between different occurrences of the same activity.

A second requirement for formalizing the queries is to specify composition of activities and occurrences. The PSL Ontology uses the *subactivity* relation to capture the basic intuitions for the composition of activities. Complex activities are composed of sets of *atomic* activities, which in turn are either *primitive* (i.e. they have no proper subactivities) or they are concurrent combinations of primitive activities.

Corresponding to the composition relation over activities, *subactivity_occurrence* is the composition relation over activity occurrences. Given an occurrence of a complex activity, subactivity occurrences are occurrences of subactivities of the complex activity.

Within the PSL Ontology, concurrency is represented by the occurrence of concurrent activities rather than concurrent activity occurrences. We use the following relation to generalize the notion of occurrence to include any atomic activity that is a subactivity of the activity that occurs:

$$(\forall s, a)\, atocc(s, a) \equiv (\exists a_1)\, atomic(a_1) \wedge occurrence\_of(s, a_1) \wedge subactivity(a, a_1)$$

Finally, we need some way to specify ordering constraints over the subactivity occurrences of a complex activity. The PSL Ontology uses the *soo_precedes*(s_1, s_2, a) relation to denote that subactivity occurrence s_1 precedes the subactivity occurrence s_2 in occurrences of the complex activity a.

The models of the axioms of the PSL Ontology have been characterized up to isomorphism [7]. A fundamental structure within these models is the occurrence tree, whose branches are equivalent to all discrete sequences of occurrences of atomic activities in the domain. Elements of the occurrence tree are referred to as *arboreal* occurrences.

Although occurrence trees characterize all sequences of activity occurrences, not all of these sequences will intuitively be physically possible within a given domain. We therefore consider the subtree of the occurrence tree that consists only of possible sequences of activity occurrences, which we refer to as the legal occurrence tree. The *legal*(o) relation specifies that the atomic activity occurrence o is an element of the legal occurrence tree.

The basic structure that characterizes occurrences of complex activities within models of the ontology is the activity tree, which is a subtree of the legal occurrence tree that consists of all possible sequences of atomic subactivity occurrences; the relation *root*(s, a) denotes that the subactivity occurrence s is the root of an activity tree for a. Elements of the tree are ordered by the *soo_precedes* relation; each branch of an activity tree is a linearly ordered set of occurrences of subactivities of the complex activity. In addition, there is a one-to-one correspondence between occurrences of complex activities and branches of the associated activity trees.

In a sense, an activity tree is a microcosm of the occurrence tree, in which we consider all of the ways in which the world unfolds in the context of an occurrence of the complex activity. Different subactivities may occur on different branches of the activity tree – different occurrences of an activity may have different subactivity occurrences or different orderings on the same subactivity occurrences (see the examples[4] in Figures 1 through 5). This distinction plays a key role in the specification of the entailment problems in this paper.

3   Formalization of the Entailment Problems

We can use the PSL Ontology to specify the queries (informally posed in the introduction) as the consequents of first-order entailment problems. The antecedent of the entailment problems will consist of the following sets of sentences:

- T_{psl} : the axioms of the PSL Ontology, together with the following three sentences:
    - Activity closure (primitive and complex)
      $(\forall a) \, primitive(a) \equiv (a = A_1) \vee ... \vee (a = A_n))$
      $(\forall a) \, \neg atomic(a) \equiv (a = P_1) \vee ... \vee (a = P_m))$
      where $A_1, ..., A_n, P_1, ..., P_n$ are constants denoting activities.
    - Legal Occurrence Assumption[5]
      $(\forall o, a) \, occurrence\_of(o, a) \wedge atomic(a) \supset legal(o)$
- $\Sigma_{pd}(P_i)$: the process description for the complex activity P_i, which specifies the relationship between occurrences of the activity and its subactivities.

In this section, we first focus on the queries in the consequents of the problems, and then define the classes of activities and process descriptions that constitute the antecedents of the problems.

3.1   Queries

The query in the consequent of one of our entailment problems is a first-order sentence that is satisfied by properties of the activity trees within the models of T_{psl} and process descriptions. We can apply the model theory of the PSL Ontology to provide characterizations of the activity trees for each query that we consider, demonstrating the correctness of the sentence with respect to the intended properties of the activity trees. In the motivating scenarios from the introduction, process plans and composite web services are represented in the PSL Ontology as complex activities. The four particular queries that we formalize in this paper focus on the relationship between occurrences of complex activities and their subactivities.

To formalize the first query, we want a sentence that determines whether the subactivity A_1 can possibly occur before the subactivity A_2, whenever the

---

[4] In these examples, we adopt the convention that o_i^a denotes an occurrence of the activity a. In all of the examples, we use the primitive activities *register*, *hotel*, *airplane*, *payment*, and *train*.

[5] We use this assumption in the complexity analysis to focus on the intractability that arises solely from occurrence and ordering constraints, independently of preconditions and effects.

complex activity P occurs; such a sentence is characterized by the following result:

Lemma 1. *Suppose $\mathcal{M}$ is a model of $T_{psl} \cup \Sigma_{pd}(P)$.*

$\mathcal{M} \models (\forall o)\, root(o, P) \supset (\exists o_1, o_2) occurrence\_of(o_1, A_1) \wedge soo\_precedes(o, o_1, P) \wedge$
$occurrence\_of(o_2, A_2) \wedge soo\_precedes(o, o_2, P) \wedge soo\_precedes(o_1, o_2, P)$

iff any activity tree for the complex activity P contains a branch in which a subactivity occurrence o_1 of A_1 precedes a subactivity occurrence o_2 of A_2.

The activity tree in Figure 1 contains a branch in which the subactivity *hotel* occurs before *airplane* and also contains a branch in which the subactivity *airplane* occurs before *hotel*. In the same activity tree, there does not exist a branch in which the *payment* subactivity occurs before the *register* subactivity. In Figure 2, there does not exist any branch containing occurrences of both subactivities *airplane* and *train*. The following lemma characterizes the sentence that is satisfied when the subactivity A_1 is required to occur before the subactivity A_2 in occurrences of the complex activity P:

Lemma 2. *Suppose $\mathcal{M}$ is a model of $T_{psl} \cup \Sigma_{pd}(P)$.*

$\mathcal{M} \models (\forall o, o_1, o_2)\, root(o, P) \wedge occurrence\_of(o_1, A_1) \wedge occurrence\_of(o_2, A_2) \wedge$
$soo\_precedes(o, o_1, P) \wedge soo\_precedes(o, o_2, P) \supset \neg soo\_precedes(o_2, o_1, P)$

iff for any branch $\mathbb{B}$ in any activity tree for the complex activity P, either

1. *every occurrence of the subactivity A_1 in $\mathbb{B}$ precedes every occurrence of the subactivity A_2 in $\mathbb{B}$, or*
2. *$\mathbb{B}$ does not contain occurrences of both A_1 and A_2.*

For example, in Figure 4, the subactivity *hotel* occurs before *payment* on every branch of the activity tree. On the other hand, there is no branch in Figure 2, that contains occurrences of both *train* and *airplane*.

The third query from the introduction determines whether occurrences of the subactivity A_1 are followed by later occurrences of the subactivity A_2 in occurrences of the complex activity P. This sentence is characterized by the next result:

Lemma 3. *Suppose $\mathcal{M}$ is a model of $T_{psl} \cup \Sigma_{pd}(P)$.*

$\mathcal{M} \models (\forall o, o_1)\, root(o, P) \wedge occurrence\_of(o_1, A_1) \wedge soo\_precedes(o, o_1, P) \supset$
$(\exists o_2)\, occurrence\_of(o_2, A_2) \wedge soo\_precedes(o_1, o_2, P)$

iff every occurrence of the subactivity A_1 is the initial element of a subtree of an activity tree for P that contains an occurrence of the subactivity A_2.

Figure 2 illustrates this query, where every occurrence of the subactivity *train* is followed by an occurrence of the subactivity *payment*.

The final query determines which activities can possibly occur next, given the occurrence of some activity A_1 in a process plan P. The following lemma characterizes the sentence that defines this query:

Lemma 4. *Suppose $\mathcal{M}$ is a model of $T_{psl} \cup \Sigma_{pd}(P)$.*

$\mathcal{M} \models (\forall o, o_1)\ root(o, P) \wedge occurrence\_of(o_1, A_1) \wedge soo\_precedes(o, o_1, P) \supset$
$(\exists a, o_2)\ occurrence\_of(o_2, a) \wedge soo\_precedes(o_1, o_2, P) \wedge$
$\neg((\exists o_3)\ soo\_precedes(o_1, o_3, P) \wedge soo\_precedes(o_3, o_2, P))$

iff no occurrence o_1 of the subactivity A_1 is a leaf of an activity tree for P.

In any proof of the above sentence with answer extraction, the variable a binds to one of the successors of the occurrence of A_1 in an activity tree for P. In Figure 2, the next subactivity to occur after o_2^{hotel} is either *airplane* or *train*.

3.2   Classification of Activities

One set of sentences within the antecedent of the entailment problems is the extension of the ontology with restricted classes of activities. Within the PSL Ontology, complex activities are classified with respect to symmetries of their activity trees. Concretely, these are axiomatized by mappings between the different branches of an activity tree or between different activity trees. In this section we introduce the model-theoretic definitions for the classes of activity trees that play a prominent role in this paper; their first-order axiomatization can be found in the PSL Ontology.

Definition 1. *An activity tree τ in a model of T_{psl} is permuted iff for every two branches $\mathbb{B}_1, \mathbb{B}_2 \subseteq \tau$, there exists a bijection $\varphi : \mathbb{B}_1 \to \mathbb{B}_2$ such that for any activity occurrence $\mathbf{o} \in \mathbb{B}_1$ and activity $\mathbf{a}$,*

$$\langle \mathbf{o}, \mathbf{a} \rangle \in \mathbf{occurrence\_of} \Leftrightarrow \langle \varphi(\mathbf{o}), \mathbf{a} \rangle \in \mathbf{occurrence\_of}$$

Figure 1 shows an example of a permuted activity tree; there is a bijection that maps the subactivity occurrences o_2^{hotel}, $o_3^{airplane}$, $o_6^{payment}$ in the branch $\mathbb{B}_1$ to the subactivity occurrences o_5^{hotel}, $o_7^{airplane}$, $o_4^{payment}$, respectively, in the branch $\mathbb{B}_2$. Since the occurrence $o_1^{register}$ is an element of every branch, it is mapped to itself. Intuitively, each branch of a permuted activity tree is a different permutation of the same set of subactivity occurrences; in the example, the same activities (*register*, *hotel*, *airplane*, and *payment*) occur on each branch, although they occur in a different order. On the other hand, the activity tree in Figure 2 is not permuted, since there is no mapping between the branch containing $o_3^{airplane}$ and o_8^{train}.

Definition 2. *An activity tree τ in a model of T_{psl} is folded iff there exists a branch $\mathbb{B}_1 \subseteq \tau$ such that for any branch $\mathbb{B}_2 \subseteq \tau$ there exists a surjection $\varphi : \mathbb{B}_2 \to \mathbb{B}_1$ such that for any activity occurrence $\mathbf{o} \in \mathbb{B}_1$ and activity $\mathbf{a}$,*

$$\langle \mathbf{o}, \mathbf{a} \rangle \in \mathbf{atocc} \Rightarrow \langle \varphi(\mathbf{o}), \mathbf{a} \rangle \in \mathbf{atocc}$$

With folded activity trees, the mappings between branches of the activity tree allow occurrences of atomic subactivities to be mapped to occurrences of concurrent subactivities. Figure 3 shows an example of a folded activity tree; the two

Fig. 1. Example of a permuted activity tree   **Fig. 2.** Example of a nonpermuted activity tree

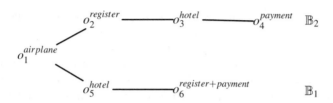

Fig. 3. Example of a folded activity tree

subactivity occurrences $o_2^{register}$ and $o_4^{payment}$ on the branch $\mathbb{B}_2$ are mapped to the subactivity occurrence $o_6^{(register+payment)}$, which is an occurrence of the atomic activity whose primitive subactivities *payment* and *register* are concurrent.

Each activity tree can be associated with a partial ordering that is preserved by the mappings between branches, so that activity trees can be classified with respect to the relationship between this ordering and branches of the trees. This leads to two subclasses of permuted and folded activity trees that are particularly relevant to the specification of manufacturing process plans and semantic web services.

Definition 3. *Within a model of T_{psl}, a permuted activity tree is a strong poset activity tree iff there exists a partial linear such that there is a there is a one-to-one correspondence between its linear extensions and branches of the tree.*

Figure 4 is an example of a strong poset activity tree. The subactivities *hotel* and *airplane* are incomparable in the partial ordering (since the ordering of the occurrences of these two activities is not preserved on each branch), so there are two branches in the activity tree, corresponding to the two possible linear extensions.

Definition 4. *Within a model of T_{psl}, a folded activity tree is a concurrent poset activity tree iff there is a one-to-one correspondence between branches of the tree and the weak orderings of some set.*

Figure 5 is an example of a concurrent poset activity tree. The subactivities *hotel* and *register* are incomparable in the partial ordering, so there are two branches in the activity tree, corresponding to the two possible linear extensions; there

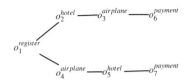
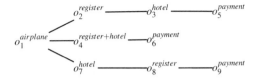

Fig. 4. Example of a strong poset activity tree

Fig. 5. Example of a concurrent poset activity tree

is also a branch in the activity tree containing the occurrence of the activity in which *hotel* and *register* are concurrent. Note that any concurrent poset tree contains a subtree that is a strong poset tree.

Strong poset and concurrent poset activity trees capture the intended semantics of constructs that are present in a wide variety of approaches to process modelling, including OWL-S [11], UML activity diagrams [4], and IDEF3 [10]. In particular, strong posets are equivalent to the AnyOrder control construct in OWL-S and the AND-junctions of IDEF3, while concurrent posets are equivalent to the Split control construct in OWL-S and to forks in UML diagrams. Nevertheless, none of these constructs can be axiomatized in their respective formalisms, either because the underlying language lacks the expressiveness (as with OWL-S) or the formalism lacks a language with a formal model-theoretic semantics. For the same reasons, the relevant queries are not definable in these formalisms.

On the other hand, the class of strong posets has a rather elegant first-order axiomatization in the PSL Ontology that is based on the following property – within a strong poset, there exists a one-to-one mapping between siblings and children in the activity tree that preserves the *occurrence_of* relation, for any two elements that are incomparable in the partial ordering. Thus, the activity tree in Figure 1 is not a strong poset activity tree since the next subactivity occurrence after o_2^{hotel} is not an occurrence of the subactivity *payment*.

Concurrent posets have a similar axiomatization, with the additional condition that for any two siblings in the activity tree there exists another sibling that is an occurrence of the concurrent activity that is composed the activities associated with the siblings.

3.3   Process Descriptions

A process description is an axiomatization of the set of activity trees for an activity within models of the PSL Ontology. The syntactic form of the process description is tightly constrained by the classes of activities in the ontology.

Theorem 1. *A complex activity P has a set of finite permuted activity trees iff its process description $\Sigma_{pd}(P)$ is logically equivalent to a sentence of the form*

$(\forall o)\ occurrence\_of(o, P) \supset$
$[(\exists o_1, ..., o_n)\ occurrence\_of(o_1, A_1) \wedge ... \wedge occurrence\_of(o_n, A_m) \wedge$
$subactivity(A_1, P) \wedge ... \wedge subactivity(A_m, P) \wedge \mathcal{O}(o_1, ..., o_n, P) \wedge$
$((\forall s)\ arboreal(s) \supset subactivity\_occurrence(s, o) \equiv ((s = o_1) \vee ... \vee (s = o_n))))]$

where $\mathcal{O}(o_1, ..., o_n, P)$ is a boolean combination of soo_precedes literals whose only variables are $o_1, ..., o_n$.

Proof. $\Rightarrow$:
Suppose that the complex activity P has finite permuted activity trees.

Since the activity trees are finite, all of their branches are finite. Furthermore, since there is a bijection between the branches in the tree, all branches have the same cardinality n. Thus, each branch consists of n occurrences of subactivities of P, so that we have

$T_{psl} \cup \Sigma_{pd}(P) \models (\forall o)\ root(o, P) \supset [(\exists o_1, ..., o_n)\ occurrence\_of(o_1, A_1) \wedge ... \wedge$
$occurrence\_of(o_n, A_m) \wedge subactivity(A_1, P) \wedge ... \wedge subactivity(A_m, P) \wedge$
$\mathcal{O}(o_1, ..., o_n, P)$

Each of these is an arboreal subactivity occurrence of an occurrence of P, and all arboreal subactivity occurrence of an occurrence of P are elements of a branch of an activity tree. We therefore have

$T_{psl} \cup \Sigma_{pd}(P) \models (\forall o)\ root(o, P) \supset [(\exists o_1, ..., o_n)\ occurrence\_of(o_1, A_1) \wedge ... \wedge$
$occurrence\_of(o_n, A_m) \wedge ((\forall s)\ arboreal(s) \supset subactivity\_occurrence(s, o) \equiv$
$((s = o_1) \vee ... \vee (s = o_n))))]$
$\Leftarrow$:
Suppose that P has the process description Σ_{pd}.

This process description entails that every occurrence of P has exactly n occurrences of atomic subactivities of P. Therefore, each branch of an activity tree for P has occurrences of the same set of atomic subactivities, so that we can define a bijection between any two branches of the activity tree. Hence, the activity tree is permuted.

For example, the process description for the activity P_1 in Figure 1 is

$(\forall o)\ occurrence\_of(o, P_1) \supset [(\exists o_i, o_j, o_k, o_l)\ occurrence\_of(o_i, register)$
$\wedge occurrence\_of(o_j, hotel) \wedge occurrence\_of(o_k, airplane) \wedge$
$occurrence\_of(o_l, payment) \wedge subactivity(register, P_1) \wedge subactivity(hotel, P_1) \wedge$
$subactivity(airplane, P_1) \wedge subactivity(payment, P_1) \wedge soo\_precedes(o_i, o_j, P_1) \wedge$
$soo\_precedes(o_j, o_k, P_1) \wedge (soo\_precedes(o_l, o_k, P_1) \equiv \neg soo\_precedes(o_k, o_l, P_1))$
$\wedge((\forall s)\ arboreal(s) \supset subactivity\_occurrence(s, o) \equiv ((s = o_i) \vee (s = o_j) \vee (s = o_k) \vee (s = o_l))))]$

For folded activity trees, we have a similar, albeit weaker, result:

Theorem 2. *If the complex activity P has a set of finite folded activity trees, then its process description $\Sigma_{pd}(P)$ entails a sentence of the form*

$(\forall o)$ $occurrence\_of(o, P) \supset [(\exists o_1, ..., o_n)\ atocc(o_1, A_1) \wedge ... \wedge atocc(o_n, A_m) \wedge$
$subactivity(A_1, P) \wedge ... \wedge subactivity(A_m, P) \wedge \mathcal{O}(o_1, ..., o_n, P) \wedge ((\forall s)arboreal(s) \supset$
$subactivity\_occurrence(s, o) \equiv ((s = o_1) \vee ... \vee (s = o_n)))]$

where $\mathcal{O}(o_1, ..., o_n, P)$ is a boolean combination of soo_precedes and equality literals whose only variables are $o_1, ..., o_n$.

Proof. Suppose that the complex activity P has finite folded activity trees.

Since the activity trees are finite, all of their branches are finite. Furthermore, since there is a surjection from the branches in the tree into a unique maximal branch, this branch has the maximum cardinality n. Thus, each branch consists of at most n occurrences of subactivities of P.

By the definition of folded activity trees, for each element **s** of a branch of an activity tree that is an occurrence of the subactivity **a** of P, there exists an element **s'** of the maximal branch such that

$$\langle \mathbf{s'}, \mathbf{a} \rangle \in \mathbf{atocc}$$

We therefore have

$T_{psl} \cup \Sigma_{pd}(P) \models (\forall o)root(o, P) \supset [(\exists o_1, ..., o_n)atocc(o_1, A_1) \wedge ... \wedge atocc(o_n, A_m) \wedge$
$subactivity(A_1, P) \wedge ... \wedge subactivity(A_m, P)$

Each element of a branch is an arboreal subactivity occurrence of an occurrence of P, and all arboreal subactivity occurrence of an occurrence of P are elements of a branch of an activity tree:

$T_{psl} \cup \Sigma_{pd}(P) \models (\forall o)root(o, P) \supset [(\exists o_1, ..., o_n)atocc(o_1, A_1) \wedge ... \wedge atocc(o_n, A_m) \wedge$
$((\forall s)\ arboreal(s) \supset subactivity\_occurrence(s, o) \equiv ((s = o_1) \vee ... \vee (s = o_n)))]$

The additional conditions in the definitions of strong poset activities and concurrent poset activities also impose restrictions on their process descriptions.

Theorem 3. *If a complex activity P has finite strong poset or concurrent poset activity trees, then the ordering formula $\mathcal{O}(o_1, ..., o_n, P)$ in its process description is logically equivalent to a conjunction of soo_precedes literals.*

For example, the process description for the activity P_2 in Figure 4 is

$(\forall o)$ $occurrence\_of(o, P_2) \supset [(\exists o_i, o_j, o_k, o_l)\ occurrence\_of(o_i, register)$
$\wedge occurrence\_of(o_j, hotel) \wedge occurrence\_of(o_k, airplane) \wedge$
$occurrence\_of(o_l, payment) \wedge subactivity(register, P_1) \wedge subactivity(hotel, P_1) \wedge$
$subactivity(airplane, P_1) \wedge subactivity(payment, P_1) \wedge soo\_precedes(o_i, o_j, P_2) \wedge$
$soo\_precedes(o_i, o_k, P_2) \wedge soo\_precedes(o_j, o_l, P_2) \wedge soo\_precedes(o_k, o_l, P_2)$
$\wedge ((\forall s)\ arboreal(s) \supset$
$subactivity\_occurrence(s, o) \equiv ((s = o_i) \vee (s = o_j) \vee (s = o_k) \vee (s = o_l)))]$

4   Complexity of Reasoning Problems

We can now consider the computational complexity of the entailment problems, under the assumptions that the process descriptions axiomatize the activity trees in the classes that we have presented above. In particular, we introduce additional assumptions to specify extensions to the PSL ontology, and then determine the complexity of the entailment problems in these extensions.

Definition 5. *The Permuted or Folded Occurrence Assumption (PFOA) is the sentence*[6]:

$(\forall o, a)\ occurrence\_of(o, a) \wedge \neg atomic(a) \supset (permuted(o) \vee folded(o))$

The Strong or Concurrent Poset Assumption (SCPA) is the sentence:

$(\forall o, a)occurrence\_of(o, a)\wedge\neg atomic(a)\supset(strong\_poset(o)\vee concurrent\_poset(o))$

It can be shown that $T_{psl} \models SCPA \supset PFOA$.

The following results show that these two assumptions are close to the boundary between tractability and intractability for the entailment problems that we have defined.

Theorem 4. *Suppose the complex activity P has only finite activity trees. Determining*

$T_{psl} \cup \Sigma_{pd}(P) \cup PFOA \models (\forall o)\ root(o, P) \supset (\exists o_1, o_2)occurrence\_of(o_1, A_1) \wedge$
$soo\_precedes(o, o_1, P) \wedge occurrence\_of(o_2, A_2) \wedge soo\_precedes(o, o_2, P)\wedge$
$soo\_precedes(o_1, o_2, P)$

is NP-complete.

Proof. By Theorem 1 and 2, there are n existentially quantified activity occurrence variables in $\Sigma_{pd}(P)$. By $PFOA$, any branch of an activity tree contains at most n atomic activity occurrences, and the maximum number of branches in any activity tree is equal to the number of weak orderings on a set of n points. Thus, the problem is in NP, since by Lemma 1 we need to check whether the branch contains a subactivity occurrence of A_1 that precedes a subactivity occurrence of A_2.

For folded activity trees, the proof can be found in [14], which first provides a straightforward reduction from an instance I of $Isat$ problem [1] in Interval Algebra (represented as a set of precedence and/or concurrency restrictions between endpoints of intervals of the instance) into $f(I)$, an instance of the problem of determining the existence of a complex activity P (composed of subactivity occurrences with corresponding *soo_precedes* and/or *conc* constraints) occurrences. A new subactivity occurrence o_i that precedes any other occurrences o_j

---

[6] *permuted(o)*, *folded(o)*, *strong_poset(o)*, and *concurrent_poset(o)* are the relations defined within the PSL Ontology to axiomatize the corresponding classes of activity trees.

is then added to construct a new complex activity P'. It is obvious I is satisfiable iff $soo\_precedes(o_i, o_j, P')$.

For permuted activity trees, since all of the occurrence variables denote distinct subactivity occurrences and the ordering formulae in the process description is a boolean combination of $soo\_precedes$ literals, NP-completeness follows from a straightforward reduction from 3SAT.

Thus, this entailment problem (whose query was characterized in Lemma 1) is intractable even when we restrict the activities to have permuted or folded activity trees.

If we strengthen the assumption so that we consider only strong poset or concurrent poset activity trees, then we obtain an extension of the theory in which the entailment problem is tractable.

Theorem 5. *Suppose the complex activity P has only finite activity trees. There exists an $O(n^2)$ algorithm to determine*

$$T_{psl} \cup \Sigma_{pd}(P) \cup SCPA \models (\forall o)\ root(o, P) \supset (\exists o_1, o_2)occurrence\_of(o_1, A_1) \land$$
$$soo\_precedes(o, o_1, P) \land occurrence\_of(o_2, A_2) \land soo\_precedes(o, o_2, P) \land$$
$$soo\_precedes(o_1, o_2, P)$$

where n is the number of existentially quantified activity occurrence variables in $\Sigma_{pd}(P)$.

Proof. Suppose that $\Sigma_{pd}(P)$ contains n activity occurrence variables and m $soo\_precedes$ literals. By Theorem 3, we can construct a directed graph $G = \langle V, E \rangle$, where V is the set of subactivity occurrence variables occ_i, and $(occ_i, occ_j) \in E$ iff the literal $soo\_precedes(occ_i, occ_j, P)$ is in $\Sigma_{pd}(P)$. $\Sigma_{pd}(P)$ is consistent (and hence there exists an occurrence of P) iff there exists a linear ordering on the vertices in V. This can be found using a topological sort algorithm, whose complexity is $O(n+m)$, where the upper bound for m is $n(n-1)/2$ (for a complete graph). Now, let Φ be the existential conjunction that is the consequent of the query. We can define another process description

$$\Sigma_{pd}(P') = \Sigma_{pd}(P) \land \Phi$$

Similarly, we know that checking the consistency of $\Sigma_{pd}(P')$ (which is equivalent to the existence of P') can also be solved in $O(n^2)$ time (as n stays unchanged). And it is straightforward to see that the existence of occurrence of P' implies that it is possible that there exists a subactivity occurrence of A_1 before a subactivity occurrence of A_2 in an activity tree for P.

Note that the algorithm requires a process description with a fixed set of subactivity occurrence variables and ordering constraints that are equivalent to a conjunction of $soo\_precedes$ literals. Although the first condition is satisfied by all permuted and folded activities, only strong poset and concurrent poset activities have process descriptions that satisfy the second condition.

We can also consider the query that we characterized in Lemma 2:

Theorem 6. *Suppose the complex activity P has only finite activity trees. Determining*

$$T_{psl} \cup \Sigma_{pd}(P) \cup PFOA \models (\forall o, o_1, o_2) \, root(o, P) \wedge occurrence\_of(o_1, A_1) \wedge$$
$$occurrence\_of(o_2, A_2) \wedge soo\_precedes(o, o_1, P) \wedge soo\_precedes(o, o_2, P) \supset$$
$$\neg soo\_precedes(o_2, o_1, P)$$

is NP-complete.

Proof. By Lemma 2, the sentence is logically equivalent to

$(\forall o) \, root(o, P) \supset \neg(\exists o_1, o_2) occurrence\_of(o_1, A_1) \wedge soo\_precedes(o, o_1, P) \wedge$
$occurrence\_of(o_2, A_2) \wedge soo\_precedes(o, o_2, P) \wedge soo\_precedes(o_1, o_2, P)$
Since the activity trees for P are either permuted or folded, we know that the same set of activities occur on every branch, so that the above sentence becomes

$(\forall o) \, root(o, P) \supset (\exists o_1, o_2) occurrence\_of(o_1, A_1) \wedge soo\_precedes(o, o_1, P) \wedge$
$occurrence\_of(o_2, A_2) \wedge soo\_precedes(o, o_2, P) \wedge soo\_precedes(o_1, o_2, P)$
which is equivalent to the sentence in Theorem 4.

Once again, if we use the Strong or Concurrent Poset Assumption, then we have an extension of the PSL Ontology in which the entailment problem is tractable.

Theorem 7. *Suppose the complex activity P has only finite activity trees. There exists an $O(n^2)$ algorithm to determine*

$$T_{psl} \cup \Sigma_{pd}(P) \cup SCPA \models (\forall o, o_1, o_2) \, root(o, P) \wedge occurrence\_of(o_1, A_1) \wedge$$
$$occurrence\_of(o_2, A_2) \wedge soo\_precedes(o, o_1, P) \wedge soo\_precedes(o, o_2, P) \supset$$
$$\neg soo\_precedes(o_2, o_1, P)$$

where n is the number of existentially quantified activity occurrence variables in $\Sigma_{pd}(P)$.

Proof. We can use the algorithm from the proof of Theorem 5 to determine whether there exists branch containing a subactivity occurrence of A_1 before a subactivity occurrence of A_2 in an activity tree for P and a branch containing a subactivity occurrence of A_2 before a subactivity occurrence of A_1 in an activity tree for P. If one of these branches does not exist, then the ordering satisfied by the other branch is satisfied on all branches.

The complexity of the entailment problems characterized in Lemmas 3 and 4 is still open.

5   Summary

We have shown how the PSL Ontology can be used to define the antecedents and consequents for first-order entailment problems related to the partial ordering of subactivity occurrences in occurrences of complex activities. The model theory of the PSL Ontology also allows us to prove the correctness of the definitions of the queries, as well as the correctness of the process descriptions for the classes of activities used within this paper.

It is difficult to define these entailment problems using other process modelling ontologies. Approaches such as the Business Process Modelling Notation (BPMN) lack an ontological foundation. Ontologies such as [13] lack a model theory. The ontologies in [5] and [11] lack axiomatizations in their respective languages, so that we cannot formalize the queries as entailment problems. Ontologies such as [12] and [9] provide axiomatizations, but they lack an explicit and complete characterization of the models of the axiomatizations. These approaches also fail to make the distinction between the axioms in ontology and the classes of sentences in the process descriptions. As a result, it is difficult to define classes of activities such as *permuted* and *strong_poset*, and we are unable to prove the correctness of the process descriptions. Finally, approaches such as [9] are unable to quantify over complex activities and their occurrences, which is required by the entailment problems that we considered.

In addition to providing a model-theoretic characterization of the sentences in the antecedents and consequents of the entailment problems, we have also defined extensions of the PSL Ontology in which the associated entailment problems are NP-complete and stronger extensions in which the problems are tractable. This demonstrates that the PSL Ontology can not only be used to axiomatize the assumptions that guarantee tractability, but it can also be used to reason about the logical relationships among these assumptions.

There are several avenues for future work. First, we want to provide a sharper characterization of the boundary between tractable and intractable extensions of the PSL Ontology by finding the maximal classes of ordered activity trees that contain the strong poset and concurrent poset activity trees and for which the entailment problems are still tractable.

Second, there are many other classes of activity trees and activities in the PSL Ontology which are independent of the folded and permuted activity trees; no work has yet been done to characterize the complexity of the entailment problems with these extensions of the PSL Ontology. This includes the entailment of ordering constraints from precondition and effect axioms.

Finally, we can apply the methodology of defining tractable extensions of the PSL Ontology to reasoning problems including temporal projection, plan verification, and plan recognition.

References

1. Allen, J.F.: Maintaining knowledge about temporal intervals. Commun. ACM 26(11), 832–843 (1983)
2. Battle, S., Bernstein, A., Boley, H., Grosof, B., Gruninger, M., Hull, R., Kifer, M., Martin, D., McIlraith, S., McGuinness, D., Su, J., Tabet, S.: Semantic Web Services Framework (SWSF) Overview. Version 1.0 (2005), http://www.daml.org/services/swsf/1.0/overview/

3. Battle, S., Bernstein, A., Boley, H., Grosof, B., Gruninger, M., Hull, R., Kifer, M., Martin, D., McIlraith, S., McGuinness, D., Su, J., Tabet, S.: Semantic Web Services Ontology (SWSO). Version 1.0 (2005),
   http://www.daml.org/services/swsf/swso/
4. Bock, C., Gruninger, M.: PSL: A semantic domain for flow models. Software and Systems Modeling 4, 209–231 (2005)
5. Ghallab, M., McDermott, D.: PDDL: The planning domain definition language v.2. Technical Report Technical Report CVC TR-98-003, Yale Center for Computational Vision and Control (1998)
6. Gruninger, M., Menzel, C.: The process specification language theory and applications. AI Mag. 24(3), 63–74 (2003)
7. Gruninger, M.: Ontology of the Process Specification Language. In: Staab, S., Studer, R. (eds.) Handbook of Ontologies in Information Systems. Springer, Heidelberg (2004)
8. Gruninger, M., Hull, R., McIlraith, S.: A First-Order Ontology for Semantic Web Services. In: Proceedings of W3C Workshop on Frameworks for Semantic in Web Services, Innsbruck, Austria, June 9-10 (2005)
9. Levesque, H., Reiter, R., Lesperance, Y., Lin, F., Scherl, R.: Golog: A logic programming language for dynamic domains. Journal of Logic Programming 31, 92–128 (1997)
10. Mayer, R., Menzel, C., Painter, M., de Witte, P., Blinn, T., Perekath, B.: Information integration for concurrent engineering: IDEF3 process description capture method report. Technical Report Technical Report AL-TR-1995, Knowledge Based Systems Incorporated (1995)
11. McIlraith, S., Son, T., Zeng, H.: Semantic web services. Intelligent Systems 16(2), 46–53 (2001)
12. Pease, A., Niles, I.: IEEE standard upper ontology: A progress report. The Knowledge Engineering Review 17, 65–70 (1998)
13. Roman, D., Keller, U., Lausen, H., de Bruijn, J., Lara, R., Stollberg, M., Polleres, A., Feier, C., Bussler, C., Fensel, D.: Web Service Modeling Ontology. Applied Ontology 1(1), 77–106 (2005)
14. Tan, X.: Computational properties of PSL problems. Technical report, Semantic Technologies Laboratory, Department of Mechanical and Industrial Engineering, University of Toronto (2008)

Two-Fold Service Matchmaking – Applying Ontology Mapping for Semantic Web Service Discovery

Stefan Dietze[1], Neil Benn[1], John Domingue[1], Alex Conconi[2], and Fabio Cattaneo[2]

[1] Knowledge Media Institute, The Open University, MK7 6AA, Milton Keynes, UK
{s.dietze,n.j.l.benn,j.b.domingue}@open.ac.uk
[2] TXT eSolutions, Via Frigia 27, 20126 Milano, Italy
{alex.conconi,fabio.cattaneo}@txt.it

Abstract. Semantic Web Services (SWS) aim at the automated discovery and orchestration of Web services on the basis of comprehensive, machine-interpretable semantic descriptions. Since SWS annotations usually are created by distinct SWS providers, semantic-level mediation, i.e. mediation between concurrent semantic representations, is a key requirement for SWS discovery. Since semantic-level mediation aims at enabling interoperability across heterogeneous semantic representations, it can be perceived as a particular instantiation of the ontology mapping problem. While recent SWS matchmakers usually rely on manual alignments or subscription to a common ontology, we propose a two-fold SWS matchmaking approach, consisting of (a) a general-purpose semantic-level mediator and (b) comparison and matchmaking of SWS capabilities. Our semantic-level mediation approach enables the implicit representation of similarities across distinct SWS by grounding service descriptions in so-called Mediation Spaces (MS). Given a set of SWS and their respective grounding, a SWS matchmaker automatically computes instance similarities across distinct SWS ontologies and matches the request to the most suitable SWS. A prototypical application illustrates our approach.

Keywords: Semantic Web Services, Matchmaking, Mediation, Vector Spaces.

1 Introduction

The increasing availability of a broad variety of Web services raises the need to automatically discover and orchestrate appropriate services for a given need. *Semantic Web Services (SWS)* [11] aim at addressing this challenge on the basis of comprehensive, machine-interpretable semantic descriptions. However, since Web services usually are provided by distinct and independent parties, the actual Web service interfaces as well as their semantic representations are highly heterogeneous. This strongly limits interoperability and raises the need of mediating between SWS descriptions as well as the actual Web services. However, despite the importance of mediation for widespread dissemination of SWS technologies, approaches to mediation are still limited and widely ignored by current SWS matchmakers [23].

In this paper, we propose a two-fold SWS matchmaking approach which implicitly tackles *semantic-level mediation* during SWS discovery. Semantic-level mediation

A. Gómez-Pérez, Y. Yu, and Y. Ding (Eds.): ASWC 2009, LNCS 5926, pp. 246–260, 2009.

refers to the resolution of heterogeneities between semantic representations of services – the actual SWS descriptions – as opposed to data-level mediation, i.e. mediation related to the structure, values or formats of input and output (I/O).

In our vision, semantic-level mediation can be perceived as a particular instantiation of the *ontology mapping* problem. In that, we argue that semantic-level mediation strongly relies on identifying *semantic similarities* between entities across different SWS ontologies [21][31]. However, semantic similarity is not an implicit notion within existing SWS representations (e.g. based on WSMO [30] and OWL-S [22]). Moreover, automatic similarity-detection as demanded by semantic mediation requires semantic meaningfulness. But the symbolic approach – i.e. describing symbols by using other symbols without a grounding in the real world - of established SWS representations does not fully entail semantic meaningfulness, since meaning requires both the definition of a terminology in terms of a logical structure (using symbols) and grounding of symbols [14]. Current approaches to mediation usually foresee the manual development of rather ad-hoc one-to-one mappings or the application of semi-automatic ontology mapping methodologies, mostly based on identifying (a) linguistic commonalities and/or (b) structural similarities between entities [20][5]. Since manually or semi-automatically defining similarity relationships is costly, current approaches are thus not capable to support SWS discovery on a web scale.

In our work, we investigate a mediation mechanism that is based on fuzzy similarity computations between instances as part of SWS ontologies in order to overcome the need for manual or semi-automatic mappings between distinct SWS representations. In this respect, we propose a general purpose matchmaking approach which implicitly addresses semantic-level mediation through (a) a representational approach allowing to implicitly represent similarities and (b) a general-purpose mediator exploiting similarities as represented through (a). In particular, we introduce the concept of *Mediation Spaces (MS)* to enable the implicit representation of semantic similarities across heterogeneous SWS representations through a grounding of SWS descriptions into vector spaces. We will demonstrate that refining heterogeneous SWS descriptions in multiple shared MS supports similarity-based mediation at the semantic level and implicitly facilitates SWS discovery. The provided general-purpose mediator – implemented as a dedicated mediation Web service – supports SWS discovery and is deployable for any semantic-level mediation scenario together with our proposed representational approach.

The remainder of the paper is organized as follows: Section 2 introduces the SWS matchmaking problem, while our two-fold matchmaking approach is proposed in Section 3. In Section 4, we a vector-based approach for semantic-level mediation and the implementation of a generic mediator is being presented in Section 5. Its deployment in a proof-of-concept application is proposed in Section 6 while we discuss and conclude our work in Section 7.

2   Semantic Web Services Mediation

Before formally introducing the SWS mediation problem, we report below the abstract definition of SWS as used throughout the remainder of the paper and a description of the SWS mediation problem, together with background information on current mediation approaches.

Semantic Web Services: A SWS description (either the description of the Web service or the description of the service request) is formally represented within a particular ontology that complies with a certain SWS reference model such as OWL-S [22] or WSMO [30]. Following the formalisation of [9][9], we define a populated *service ontology O* – as utilised by a particular SWS representation – as a tuple:

$$O = \{C, I, P, R, A\} \subset SWS$$

With C being a set of n *concepts* in O where each concept C_i is described through $l(i)$ *concept properties pc*. I represents all m *instances* where each instance I_{ij} represents a particular instance of a concept C_j and consists of $l(i)$ *instantiated properties pi* instantiating the concept properties of C_j. The properties P of an ontology O represent the union of all concept properties PC and instantiated properties PI of O.

Given these definitions, we would like to point out that properties here exclusively refer to so-called data type properties. Hence, we define properties as being distinctive to relations R. The latter describe relations between concepts and instances. In addition, A represents a set of *axioms* which define constraints on the other introduced notions. Since certain parts of a SWS ontology describe certain aspects of the Web service, such as its capability Cap, interface If or non-functional properties Nfp [6], a SWS ontology can be perceived as a conjunction of ontological subsets:

$$Cap \cup If \cup Nfp = O \subset SWS$$

The capability description, as central element of a SWS description, consists of further subsets, describing the assumptions As, effects Ef, preconditions Pre and postconditions $Post$. However, for simplification reasons we prefer the exclusive consideration of assumptions/effects:

$$As \cup Ef = Cap \subset O \subset SWS$$

The SWS mediation problem: Mediation aims at addressing heterogeneities among distinct SWS to support all stages that occur at SWS runtime, namely *discovery*, *orchestration* and *invocation*. In contrast to [23][6], we classify the mediation problem into (i) *semantic-level* and (ii) *data-level mediation*. Figure 1 illustrates the chronological order of different mediation tasks at SWS runtime. Whereas (i) refers to the resolution of heterogeneities between concurrent semantic representations of services – e.g. by aligning distinct SWS representations – (ii) refers to the mediation between mismatches related to the Web service implementations themselves, i.e. related to the structure, value or format of I/O messages. Hence, semantic-level mediation primarily supports the discovery stage, whereas data-level mediation occurs during orchestration and invocation. Please note that, for the sake of simplification, Figure 1 just depicts mediation between a SWS request and multiple SWS, while leaving aside mediation between different SWS or between different requests.

Several approaches, such as [1][2][3][19][25][28][31], aim at addressing the mediation issue partially by dealing either with (i) or (ii). For instance, [2] proposes a semantic mediation framework for scientific workflows relying on the notion of semantic type and structural type, defined in a shared ontology. The semantic type gives a meaning to data, and the structural type is the data schema. As in [28] their work adapts data with a common semantic type but different structural types. In contrast, [31] provides an attempt to support similarity detection for mediation within SWS composition by exploiting syntactic similarities between SWS representations. However, it can be stated that all the above mentioned approaches rely on the definition of a priori mappings, the agreement of a shared ontology or the exploitation of semi-automatic ontology mapping approaches. Hence, providing a generic solution to mediation between heterogeneous SWS remains a central challenge to be solved by SWS matchmaking approaches.

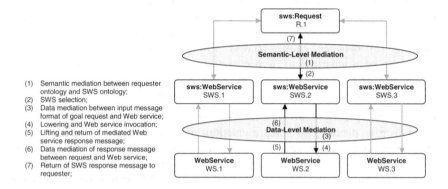

Fig. 1. Semantic-level and data-level mediation as part of SWS discovery, orchestration and invocation

3   SWS Matchmaking as a Two-Fold Process

In order to better understand the needs of semantic-level mediation, it is necessary to understand the requirements of the SWS discovery task to which semantic-level mediation is supposed to contribute. In order to identify whether a particular SWS S_1 is potentially relevant for a given request S_2, a SWS broker has to compare the capabilities of S_1 and S_2, i.e. it has to identify whether the following holds true:

$$As_2 \subset As_1 \cup Ef_2 \subset Ef_1$$

However, in order to compare distinct capabilities of available SWS which each utilise a distinct vocabulary, these vocabularies have to be mapped. For instance, to compare whether an assumption expression $As_1 \equiv \neg I_1 \cup I_2$ of one particular SWS_1 is the same as $As_2 \equiv I_3 \cup \neg I_4$ of another SWS_2, where I_i represents a particular instance, matchmaking engines have to perform two steps:

S1. Semantic-level mediation: alignment of concepts/instances involved in distinct SWS representations;
S2. Matchmaking: evaluation whether the semantics of the SWS expressions match each other.

Whereas current SWS execution environments exclusively focus on *S1*, SWS matchmaking also requires mediation between different SWS ontologies, as in *S1*.

3.1 Semantic-Level Mediation as an Ontology Mapping Problem

Semantic-level mediation can be perceived as a particular instantiation of the *ontology mapping* problem [31]. With respect to [5] and [24], we define *ontology mapping* as the creation of structure-preserving relations between multiple ontologies. I.e. the goal is, to establish formal relations between a set of knowledge entities E_1 from an ontology O_1 – used to represent a particular SWS S_1 - with entities E_2 which represent the same or a similar semantic meaning in a distinct ontology O_2 [9] which is used to represent an additional SWS S_2. The term *set of entities* here refers to the union of all concepts C, instances I, relations R and axioms A defined in a particular SWS ontology. In that, semantic mediation strongly relies on identifying *semantic similarities* [1] between entities across different SWS ontologies. Hence, the identification of similarities is a necessary requirement to solve the mediation problem for multiple heterogeneous SWS representations [21][31]. However, in this respect, the following issues have to be taken into account:

Symbolic SWS representations lack meaningfulness and are ambiguous: similarity-detection across distinct SWS representations requires semantic expressions rich enough to inherently represent semantic similarity between represented entities. However, the symbolic approach, i.e. describing symbols by using other symbols, without a grounding in the real world, of established SWS representation standards, leads to ambiguity issues and does not fully entail semantic meaningfulness, since meaning requires both the definition of a terminology in terms of a logical structure (using symbols) and grounding of symbols to a conceptual level [14].

Lack of automated similarity-detection methodologies: Describing the complex notion of specific SWS capabilities in all their facets is a costly task and may never reach semantic completeness due to the issue described above. While capability representations across distinct SWS representations – even those representing the same real-world entities – hardly equal another, semantic similarity is not an implicit notion within SWS representations. But manually or semi-automatically defining similarity relationships is costly. Moreover, such relationships are hard to maintain in the longer term.

Given the lack of inherent similarity representation, current approaches to ontology mapping could be applied to facilitate SWS mediation. These approaches aim at semi-automatic similarity detection across ontologies mostly based on identifying linguistic commonalities and/or structural similarities between entities of distinct ontologies [20][5]. Work following a combination of such approaches in the field of ontology mapping is reported in [17][10][13][16][20][7]. However, it can be stated, that such approaches require manual intervention, are error-prone, and hence, similarity-computation remains as central challenge. In our vision, instead of semi-automatically formalising individual mappings, methodologies to automatically compute or implicitly represent similarities across distinct SWS representations are better suited to facilitate SWS mediation.

3.2 Alternative Approaches to Similarity-Computation

Distinct streams of research approach the automated computation of similarities through spatially oriented knowledge representations. *Conceptual Spaces (CS)* [12] follow a theory of describing entities in terms of their quality characteristics similar to natural human cognition in order to bridge between the neural and the symbolic world. [12] proposes the representation of concepts as multidimensional geometrical *Vector Spaces* which are defined through sets of quality dimensions. Instances are represented as vectors, i.e. particular points in a CS. For instance, a particular color may be defined as a point described by vectors measuring the quality dimensions hue, saturation, and brightness. Describing instances as points within vector spaces where each vector follows a specific metric enables the automatic calculation of their semantic similarity by means of distance metrics such as the Euclidean, Taxicab or Manhattan distance [16] or the Minkowsky Metric [28]. Hence, in contrast to the costly formalisation of such knowledge through symbolic representations, semantic similarity is implicit information carried within a CS representation. This is perceived as the major contribution of the CS theory. *Soft Ontologies (SO)* [15] follow a similar approach by representing a knowledge domain D through a multi-dimensional *ontospace A*, which is described by its so-called *ontodimensions*. An item I, i.e. an instance, is represented by scaling each dimension to express its impact, presence or probability in the case of I. In that, a SO can be perceived as a CS where dimensions are measured exclusively on a ratio-scale.

However, although CS and SO aim at solving SW(S)-related issues, several issues still have to be taken into account. For instance, similarity computation within CS requires the description of concepts through quantifiable metrics even in case of rather qualitative characteristics. Moreover, CS as well as SO do not provide any notion to represent any arbitrary relations [27], such as *part-of* relations which usually are represented within first-order logic (FOL) knowledge models. In this regard, it is even more obstructive that the scope of a dimension is not definable, i.e. a dimension always applies to the entire CS/SO [27].

4 A Vector-Based Approach to Semantic-Level Mediation

To overcome the issues introduced in Section 3.1, we propose a mediation approach which utilises a novel representation mechanism that extends the expressiveness of SWS representations with implicit similarity information.

In particular, we claim that basing service models on either SWS or CS is not sufficient and propose a representational approach which grounds a SWS representation into so-called *Mediation Spaces (MS)*. MS are inspired by CS and enable the implicit representation of semantic similarities across heterogeneous SWS representations provided by distinct agents. MS propose the representation of concepts which are used as part of SWS descriptions as CS defined through sets of quality dimensions. Instances as part of SWS descriptions are represented as vectors (members) in a MS where similarity between two vectors is indicated by their spatial distance. Hence, refining heterogeneous SWS descriptions into multiple shared MS supports similarity based mediation at the semantic-level and consequently facilitates SWS selection.

Whereas CS allow the representation of semantic similarity as a notion implicit to a constructed knowledge model, it can be argued, that representing an entire SWS through a coherent MS might not be feasible, particularly when attempting to maintain the meaningfulness of the spatial distance as a similarity measure. Therefore, we claim that MS are a particularly promising model when being applied to individual concepts – as part of SWS descriptions – instead of representing an entire SWS ontology in a single MS. In that, we would like to highlight that we consider the representation of a set of n concepts C of a SWS ontology O through a set of n MS. Hence, instances of concepts are represented as members (i.e. vectors) in the respective MS. While still taking advantage of implicit similarity information within a MS, our hybrid approach – combining ontology-based SWS descriptions with multiple vector-based MS – allows to overcome CS-related issues, such as the lack of expressivity for arbitrary relations, by maintaining the advantages of ontology-based SWS representations. Please note that our approach relies on the agreement on a common set of MS for a given set of distinct SWS ontologies, instead of a common agreement on the used ontologies/vocabularies themselves. Thus, whereas in the latter case two agents have to agree on a common ontology at the concept and instance level, our approach requires just agreement at the schema level, since instance similarity becomes an implicit notion. Moreover, we assume that the agreement on ontologies at the schema level becomes an increasingly widespread case, due, on the one hand, to increasing use of upper-level ontologies such as DOLCE[1], SUMO[2] or OpenCyc[3] which support a certain degree of commonality between distinct ontologies, and on the other hand, to SWS ontologies often being provided within closed environments where a common agreement to a certain extent is ensured. In such cases, the derivation of a set of common MS is particularly applicable and straightforward.

In order to refine and represent SWS descriptions within a set of MS, we formalised the MS model into an ontology, currently being represented through OCML [18]. The ontology enables the instantiation of a set of MS to represent a given set of concepts as part of SWS descriptions. Referring to [26], we formalise a MS as a vector space defined through quality dimensions d_i of MS. Each dimension is associated with a certain metric scale, e.g. ratio, interval or ordinal scale. To reflect the impact of a specific quality dimension on the entire MS, we consider a prominence value p for each dimension [26]. Therefore, a MS is defined by

$$MS^n = \{(p_1 d_1, p_2 d_2, ..., p_n d_n) | d_i \in MS, p_i \in \Re\}.$$

Please note that we enable dimensions to be detailed further in terms of subspaces. Hence, a dimension within one MS may be defined through another MS by using further dimensions. In such a case, the particular quality dimension d_j is described by a set of further quality dimensions. In this way, a MS may be composed of several subspaces and consequently, the description granularity can be refined gradually. Furthermore, dimensions may be correlated. Information about correlation is expressed through axioms related to a specific quality dimension instance.

---

[1] http://www.loa-cnr.it/DOLCE.html
[2] http://www.ontologyportal.org/
[3] http://www.opencyc.org/

A member M – representing a particular instance – of the MS is described through through a vector defined by the set of valued dimensions v_i:

$$M^n = \{(v_1, v_2, ..., v_n) | v_i \in M\}$$

With respect to [7], we define the semantic similarity between two members of a space as a function of the Euclidean distance between the points representing each of the members. However, we would like to point out that different distance metrics could be considered, dependent on the nature and purpose of the MS. Given a MS definition MS and two members v and u, defined by vectors v_0, v_1, ...,v_n and u_1, u_2,...,u_n within MS, the distance between v and u can be calculated as:

$$dist(u, v) = \sqrt{\sum_{i=1}^{n} p_i ((\frac{u_i - \bar{u}}{s_u}) - (\frac{v_i - \bar{v}}{s_v}))^2}$$

where $\bar{u}$ is the mean of all values of data set U and s_u is the standard deviation of U.

The formula above already considers the so-called Z-transformation or standardization which facilitates the standardization of distinct measurement scales utilised by different quality dimensions in order to enable the calculation of distances in a multi-dimensional and multi-metric space. Please refer to [8], for a detailed description on how distinct MS can be derived for arbitrary SWS, i.e. a methodology to represent SWS through MS.

5   Implementing Two-Fold SWS Matchmaking Based on WSMO and IRS-III

The representational model described above had been implemented by and aligned to established SWS technologies based on WSMO [30] and the Internet Reasoning Service IRS-III [4]. However, please note that in principle the representational approach described above could be applied to any SWS reference model and is particularly well-suited to support rather light-weight approaches such as SAWSDL or WSMO Lite [29].

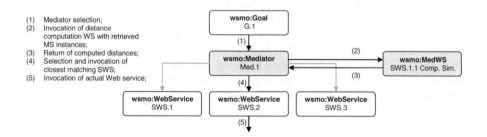

Fig. 2. WSMO SWS matchmaking utilizing a similarity-based Mediator for semantic-level Mediation

To facilitate our MS-based approach, we provided a general-purpose matchmaking approach (Fig. 2) utilising a semantic-level mediator which implemented as a particular mediation service. Given the ontological refinement of SWS descriptions into MS as introduced above, the mediation service is reusable and can be deployed to solve all sorts of semantic-level mediation scenarios. Please note that our current Mediator assumes logical SWS capability expressions to be defined through simple conjunctions of instances. Arbitrary logical expressions will be considered within a revised implementation.

When attempting to achieve match a SWS request (*wsmo:Goal* in Figure 2), our mediator is provided with the actual SWS request SWS_i, named *base,* and the SWS descriptions of all x available services that are potentially relevant for the base – i.e. linked through a dedicated mediator:

$$SWS_i \cup \{SWS_1, SWS_2, ..., SWS_x\}$$

Each SWS contains a set of concepts $C=\{c_1..c_m\}$ and instances $I=\{i_1..i_n\}$. We first identify all members $M(SWS_i)$ – in the form of valued vectors $\{v_1..v_n\}$ refining the instance i_l of the base as proposed in Section 4. In addition, for each concept c within the base the corresponding conceptual space representations $MS=\{MS_1..MS_m\}$ are retrieved. Similarly, for each SWS_j related to the base, members $M(SWS_j)$ – which refine capabilities of SWS_j and are represented in one of the CS $CS_1..CS_m$ – are retrieved:

$$CS \cup M(SWS_i) \cup \{M(SWS_1), M(SWS_2), ..., M(SWS_x)\}$$

Based on the above ontological descriptions, for each member v_l within $M(SWS_i)$, the Euclidean distances to any member of all $M(SWS_j)$ which is represented in the same space MS_j as v_l are computed. In case one set of members $M(SWS_j)$ contains several members in the same MS – e.g. SWS_j targets several instances of the same kind – the algorithm just considers the closest distance since the closest match determines the appropriateness for a given goal. For example, if one SWS supports several different locations, just the one which is closest to the one required by SWS_i determines the appropriateness.

Consequently, a set of x sets of distances is computed as follows $Dist(SWS_i)=\{Dist(SWS_i,SWS_1), Dist(SWS_i,SWS_2) .. Dist(SWS_i,SWS_x)\}$ where each $Dist(SWS_i,SWS_j)$ contains a set of distances $\{dist_1..dist_n\}$ and any $dist_i$ represents the distance between one particular member v_i of SWS_i and one member refining one instance of the capabilities of SWS_j. Hence, the overall similarity between the base SWS_i and any SWS_j could be defined as being reciprocal to the mean value of the individual distances between all instances of their respective capability descriptions and hence, is calculated as follows:

$$Sim(SWS_i, SWS_j) = \left(\overline{Dist(SWS_i, SWS_j)}\right)^{-1} = \left(\frac{\sum_{k=1}^{n}(dist_k)}{n}\right)^{-1}$$

Finally, a set of x similarity values – computed as described above – which each indicates the similarity between the base SWS_i and one of the x target SWS is computed:

$$\{Sim(SWS_i, SWS_1), Sim(SWS_i, SWS_2), ..., Sim(SWS_i, SWS_x)\}$$

As a result, the most similar SWS_j, i.e. the closest associated SWS, can be selected and invoked. In order to ensure a certain degree of overlap between the actual request and the invoked functionality, we also defined a *threshold similarity value T* which determines the similarity threshold for any potential invocation.

Within our current implementation, we provided a new matchmaking function within IRS-III which automatically performs the similarity computation described above as part of the matchmaking procedure and hence, realizes our two-fold matchmaking approach.

6 Application – Similarity-Based Selection of Video Retrieval Services

We provided a prototypical implementation which aims at similarity-based retrieval of public multimedia (MM) content exposed via Web services. Our prototypical application utilizes our approach to annotate (Web) services which operate on top of distributed MM metadata repositories. These services had been created in the context of the EC-funded project NoTube[4] and make use of the Youtube-API[5] as well as data feeds provided by BBC- Backstage[6] and Open Video[7]. The available services were annotated following the representational approach proposed in Section 4. We make use of standard SWS technology based on WSMO and IRS-III which had been extended with our two-fold matchmaking mechanism to tackle the semantic-level mediation problem.

6.1 Representing Video Retrieval Services through Multiple MS

In fact, five different Web services had been provided, each able to retrieve content from distinct repositories through keyword-based searches. WS_1 is able to retrieve content from the Youtube channel of The Open University[8], while WS_2 provides Youtube content associated with the *entertainment* category following the Youtube vocabulary. WS_3 performs keyword-based searches on top of the Open Video repository, while WS_4 operates on top of the news metadata feeds provided by BBC Backstage. In addition, WS_4 provides Youtube content suitable for mobiles.

Based on the SWS reference model WSMO, we provided service annotations following the approach described above. Each service has distinct constraints, and thus distinct SWS metadata. In particular, we annotated the Web services in terms of the *purpose* they serve MM content for and the technical *environment* supported by

---

[4] http://projects.kmi.open.ac.uk/notube/
[5] http://code.google.com/intl/en/apis/youtube/
[6] http://backstage.bbc.co.uk/
[7] http://www.open-video.org/
[8] http://www.youtube.com/ou

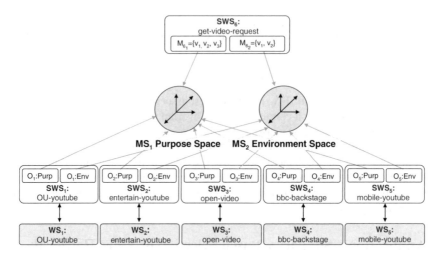

Fig. 3. MM service metadata refined in two distinct CS

the delivered content. In that, a simplified space (*MS₁: Purpose Space* in Figure 3) was defined to refine the notion of purpose by using three dimensions indicating the intended purpose of a particular piece of MM content: $\{((p_1*information),$ $(p_2*education), (p_3*leisure))\} = MS_1$. The dimensions of MS_1 are measured on a ratio scale ranging from 0 to 100. For instance, a member P_1 in MS_1 described by vector $\{(0, 100, 0)\}$ would indicate a rather educational purpose. In addition, a second space (*MS₁: Environment Space* in Figure 3) was provided to represent technical environments in terms of dimensions measuring the available resolution and bandwidth $\{((p_1*resolution), (p_2*bandwidth))\} = MS_2$. For simplification, also the dimensions of MS_2 were ranked on a ratio scale. However, it is intended to refine the resolution dimension to apply an interval scale to both dimensions to be able to represent actual resolution and bandwidth measurements. Each dimension was ranked equally with a prominence of 1 in all cases.

By applying the representational approach proposed here, each concept of the involved heterogeneous SWS representations of the underlying services was refined as shared MS, while instances – used to define SWS and SWS requests – were defined as members, i.e. vectors. No explicit relations were formalised across SWS representations. Instead, similarities are computed by means of distance calculation following the algorithm proposed in Section 5. In that, assumptions (*Ass*) of available MM services had been described independently in terms of simple conjunctions of instances which were individually refined as vectors in shared MS as shown in Table 1. Each MM service was associated with a set of members (vectors) in MS_1 and MS_2 to represent its purpose and the targeted environment. For instance, SWS_3 which provides resources from the Open Video repository, which in fact are of rather educational or information nature, was associated with a corresponding purpose vector $\{(50, 50, 0)\}$. While SWS_5 represents a Web service dedicated to video content suitable for mobiles, a vector $\{(10,10)\}$ indicating low resolution and bandwidth values was associated with SWS_5.

Table 1. Assumptions of involved SWS (requests) described as vectors in MS_1 and MS_2

	Assumption $Ass_{SWSi} = (P_{1SWSi} \cup P_{2SWSi} \cup .. \cup P_{nSWSi}) \cup (E_{1SWSi} \cup E_{2SWSi} \cup .. \cup E_{mSWSi})$	
	Members P_i in MS$_1$ (purpose)	Members E_j in MS$_2$ (environment)
SWS$_1$	$P_{1(SWS1)}=\{(0, 100, 0)\}$	$E_{1(SWS1)}=\{(100, 100)\}$
SWS$_2$	$P_{1(SWS2)}=\{(0, 0, 100)\}$	$E_{1(SWS2)}=\{(100, 100)\}$
SWS$_3$	$P_{1(SWS3)}=\{(50, 50, 0)\}$	$E_{1(SWS3)}=\{(100, 100)\}$
SWS$_4$	$P_{1(SWS4)}=\{(100, 0, 0)\}$	$E_{1(SWS4)}=\{(100, 100)\}$
SWS$_5$	$P_{1(SWS5)}=\{(100, 0, 0)\}$ $P_{2(SWS5)}=\{(0, 100, 0)\}$	$E_{1(SWS5)}=\{(10, 10)\}$

6.2 Similarity-Based Matchmaking

An AJAX-based user interface (Fig. 4) was provided which allows users to define requests by providing measurements describing their context, i.e. the purpose and environment, and WS input parameters, i.e. a set of keywords. Fig. 4 depicts a screenshot of the Web interface after our mediator computed a ranking of most suitable SWS based on distances in MS.

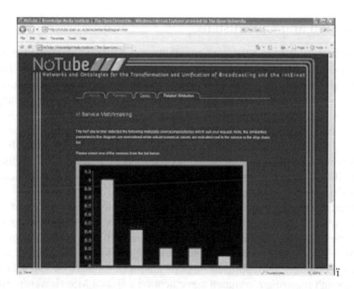

Fig. 4. Screenshot of AJAX interface depicting a suitability ranking of available services to match a given request

For instance, a user provides a request R with the input parameter keyword "Aerospace" together with context measurements which correspond to the following vectors: $P_1(R)=\{(60, 55, 5)\}$ in MS_1 and $P_2(R)=(95, 90)\}$ in MS_2. These vectors indicate the need for content which serves the need for education or information and which supports a rather high resolution environment. Though no SWS matches these criteria exactly, at runtime similarities are calculated between R and the related SWS (SWS_1-SWS_5) through the similarity computation service described in Section 5.

This led to the calculation of the similarity values shown in Table 2. Given these similarities, our reasoning environment automatically selects the most similar MM service (SWS_3) and triggers its invocation.

Table 2. Automatically computed similarities between request R and available SWS

	Similarities
SWS_1	0.023162405
SWS_2	0.014675636
SWS_3	0.08536871
SWS_4	0.02519804
SWS_5	0.01085659

Eventually, the most similar service is invoked and retrieves MM metadata records from the Open Video repository which match the requested search term "Aerospace". As illustrated above, our application utilises our two-fold matchmaking mechanism to support matchmaking of distributed SWS while tackling the semantic-level mediation problem.

7 Discussion and Conclusions

In order to further facilitate SWS interoperability we proposed a two-fold matchmaking approach which implicitly tackles the semantic-level mediation problem. Note, while our approach utilises a general-purpose mediation service which utilises SWS refinements in MS, different SWS alignment methodologies could be applied and combined to further optimise SWS alignment, i.e. semantic-level mediation. The introduced two-fold matchmaking approach supports implicit representation of similarities between instances across heterogeneous ontologies through dedicated representations in MS, and consequently, provides a means to facilitate SWS interoperability. To evaluate our approach, we deployed a prototypical application based on WSMO in a video metadata retrieval scenario.

The proposed approach has the potential to significantly reduce the effort required to mediate between distinct heterogeneous SWS ontologies and the extent to which two distinct parties have to share their conceptualisations. Whereas traditional matchmaking methodologies rely on either manual formalisation of one-to-one mappings or subscription to a common ontology, our approach supports automatic similarity-computation between instances though requiring a common agreement on a shared MS. However, even for the case of heterogeneous MS, traditional semi-automatic mapping methodologies could be applied to initially align distinct spaces. In addition, incomplete similarities are computable between partially overlapping MS. Given the nature of our approach - aiming at mediating between sets of concepts/instances which are used to annotate particular SWS - we argue that our solution is particularly applicable to SWS frameworks which are based on rather light-weight service semantics such as WSMO-Lite [29] or OWL-S [22]. Moreover, by representing SWS through vectors which are independent from the underlying representation language, we believe that our approach also has the potential to bridge between SWS across concurrent SWS reference models and modeling languages.

However, the authors are aware that our approach requires a considerable amount of additional effort to establish MS-based representations. Future work has to investigate on this effort in order to further evaluate the potential contribution of the proposed approach. Moreover, whereas defining instances, i.e. vectors, within a given MS appears to be a straightforward process of assigning specific quantitative values to quality dimensions, the definition of the MS itself is not trivial and dependent on individual perspectives and subjective appraisals. Furthermore, whereas the size and resolution of a MS is indefinite, defining a reasonable MS may become a challenging task. Nevertheless, distance calculation relies on the fact that resources are described in equivalent geometrical spaces. However, particularly with respect to the latter, traditional ontology and schema matching methods could be applied to align heterogeneous spaces. In addition, we would like to point out that the increasing usage of upper level ontologies, such as DOLCE or SUMO, and the progressive reuse of ontologies, particularly in loosely coupled organisational environments, leads to an increased sharing of ontologies at the concept level what also applies to SWS representations. As a result, our proposed hybrid representational model and mediation approach becomes increasingly applicable by further enabling similarity-computation at the instance-level towards the vision of interoperable ontologies.

References

[1] Bicer, V., Kilic, O., Dogac, A., Laleci, G.B.: Archetype-based semantic interoperability of web service messages in the health care domain. International Journal of Semantic Web and Information Systems (IJSWIS) 1(4), 1–23 (2005)

[2] Bowers, S., Ludäscher, B.: An ontology-driven framework for data transformation in scientific workflows. In: Rahm, E. (ed.) DILS 2004. LNCS (LNBI), vol. 2994, pp. 1–16. Springer, Heidelberg (2004)

[3] Cabral, L., Domingue, J.: Mediation of semantic web services in IRS-III. In: First International Workshop on Mediation in Semantic Web Services (MEDIATE 2005) held in conjunction with the 3rd International Conference on Service Oriented Computing (ICSOC 2005), Amsterdam, The Netherlands, December 12 (2005)

[4] Cabral, L., Domingue, J., Galizia, S., Gugliotta, A., Norton, B., Tanasescu, V., Pedrinaci, C.: IRS-III: A Broker for Semantic Web Services Based Applications. In: Cruz, I., Decker, S., Allemang, D., Preist, C., Schwabe, D., Mika, P., Uschold, M., Aroyo, L.M. (eds.) ISWC 2006. LNCS, vol. 4273, pp. 201–214. Springer, Heidelberg (2006)

[5] Choi, N., Song, I., Han, H.: A survey on ontology mapping. SIGMOD Rec. 35(3), 34–41 (2006)

[6] Cimpian, E., Mocan, A., Stollberg, M.: Mediation Enabled Semantic Web Services Usage. In: Mizoguchi, R., Shi, Z.-Z., Giunchiglia, F. (eds.) ASWC 2006. LNCS, vol. 4185, pp. 459–473. Springer, Heidelberg (2006)

[7] Dietze, S., Gugliotta, A., Domingue, J.: Conceptual Situation Spaces for Semantic Situation-Driven Processes. In: Bechhofer, S., Hauswirth, M., Hoffmann, J., Koubarakis, M. (eds.) ESWC 2008. LNCS, vol. 5021, pp. 599–613. Springer, Heidelberg (2008)

[8] Dietze, S., Gugliotta, A., Domingue, J.: Exploiting Metrics for Similarity-based Semantic Web Service Discovery. In: IEEE 7th International Conference on Web Services (ICWS 2009), Los Angeles, CA, USA (2009)

[9] Ehrig, M., Sure, Y.: Ontology Mapping - An Integrated Approach. In: Bussler, C.J., Davies, J., Fensel, D., Studer, R. (eds.) ESWS 2004. LNCS, vol. 3053, pp. 76–91. Springer, Heidelberg (2004)

[10] Euzenat, J., Guegan, P., Valtchev, P.: OLA in the OAEI 2005 Alignment Contest. In: K-Cap 2005 Workshop on Integrating Ontologies, pp. 97–102 (2005)

[11] Fensel, D., Lausen, H., Polleres, A., de Bruijn, J., Stollberg, M., Roman, D., Domingue, J.: Enabling Semantic Web Services – The Web service Modelling Ontology. Springer, Heidelberg (2006)

[12] Gärdenfors, P.: Conceptual Spaces - The Geometry of Thought. MIT Press, Cambridge (2000)

[13] Giunchiglia, F., Shvaiko, P., Yatskevich, M.: S-Match: An Algorithm and an Implementation of Semantic Matching. In: Bussler, C.J., Davies, J., Fensel, D., Studer, R. (eds.) ESWS 2004. LNCS, vol. 3053, pp. 61–75. Springer, Heidelberg (2004)

[14] Harnad, S.: The Symbol Grounding Problem. CoRR cs.AI/9906002 (1999)

[15] Kaipainen, M., Normak, P., Niglas, K., Kippar, J., Laanpere, M.: Soft Ontologies, spatial Representations and multi-perspective Explorability. Expert Systems 25(5) (November 2008)

[16] Krause, E.F.: Taxicab Geometry. Dover (1987)

[17] Mitra, P., Noy, F.N., Jaiswals, A.: OMEN: A Probabilistic Ontology Mapping Tool. In: Gil, Y., Motta, E., Benjamins, V.R., Musen, M.A. (eds.) ISWC 2005. LNCS, vol. 3729, pp. 537–547. Springer, Heidelberg (2005)

[18] Motta, E.: An Overview of the OCML Modelling Language. In: The 8th Workshop on Methods and Languages (1998)

[19] Mrissa, M., Ghedira, C., Benslimane, D., Maamar, Z., Rosenberg, F., Dustdar, S.: A context-based mediation approach to compose semantic web services. ACM Trans. Internet Techn. 8(1) (2007)

[20] Noy, N.F., Musen, M.A.: The PROMPT Suite: Interactive Tools for Ontology Merging and Mapping. International Journal of Human-Computer Studies 59, 983–1024 (2003)

[21] Qu, Y., Hu, W., Cheng, G.: Constructing Virtual Documents for Ontology Matching. In: WWW 2006, Edinburgh, Scotland, May 23-26. ACM, New York (2006); 1595933239/06/0005

[22] OWL-S 1.0 Release, http://www.daml.org/services/owl-s/1.0/

[23] Paolucci, M., Srinivasan, N., Sycara, K.: Expressing WSMO Mediators in OWL-S. In: Proceedings of 3rd International Semantic Web Conference (ISWC 2004) (2004)

[24] Pease, A., Niles, I., Li, J.: The suggested upper merged ontology: A large ontology for the semanticweb and its applications. In: AAAI 2002 Workshop on Ontologies and the Semantic Web. Working Notes (2002)

[25] Radetzki, U., Cremers, A.B.: Iris: A framework for mediator-based composition of service-oriented software. In: ICWS, pp. 752–755. IEEE Computer Society, Los Alamitos (2004)

[26] Raubal, M.: Formalizing Conceptual Spaces. In: Varzi, A., Vieu, L. (eds.) Formal Ontology in Information Systems, Proceedings of the Third International Conference (FOIS 2004). Frontiers in Artificial Intelligence and Applications, vol. 114, pp. 153–164. IOS Press, Amsterdam (2004)

[27] Schwering, A.: Hybrid Model for Semantic Similarity Measurement. In: Meersman, R., Tari, Z. (eds.) OTM 2005. LNCS, vol. 3761, pp. 1449–1465. Springer, Heidelberg (2005)

[28] Spencer, B., Liu, S.: Inferring data transformation rules to integrate semantic web services. In: McIlraith, S.A., Plexousakis, D., van Harmelen, F. (eds.) ISWC 2004. LNCS, vol. 3298, pp. 456–470. Springer, Heidelberg (2004)

[29] Vitvar, T., Kopecký, J., Viskova, J., Fensel, D.: WSMO-Lite Annotations for Web Services. In: Bechhofer, S., Hauswirth, M., Hoffmann, J., Koubarakis, M. (eds.) ESWC 2008. LNCS, vol. 5021, pp. 674–689. Springer, Heidelberg (2008)

[30] WSMO Working Group, D2v1.0: Web service Modeling Ontology (WSMO). WSMO Working Draft (2004), http://www.wsmo.org/2004/d2/v1.0/

[31] Wu, Z., Ranabahu, A., Gomadam, K., Sheth, A.P., Miller, J.A.: Automatic Composition of Semantic Web Services using Process Mediation. In: Proceedings of the 9th International Conference on Enterprise Information Systems (ICEIS 2007), Funchal, Portugal, June 2007, pp. 453–461 (2007)

An Approach to Analyzing Dynamic Trustworthy Service Composition[*]

Guisheng Fan[1,2], Huiqun Yu[1,2], Liqiong Chen[3], and Dongmei Liu[1]

[1] Department of Computer Science and Engineering,
East China University of Science and Technology, Shanghai 200237, China
{gsfan,yhq,dmliu}@ecust.edu.cn
[2] Shanghai Key Laboratory of Computer Software,
Evaluating and Testing, Shanghai 201112, China
[3] Department of Computer Science and Information Engineering,
Shanghai Institute of Technology, Shanghai 200235, China
lqchen@sit.edu.cn

Abstract. Service composition and related technologies have provided favorable means for building complex Web software systems. It may span multiple organizational units requires particular considerations on trustworthy issues. However, the distributive and heterogeneous characteristics of services make it hard to guarantee trustworthiness of service composition. This paper presents a method for analyzing dynamic trustworthy service composition according to the characteristics and requirements of service composition. Petri nets are used to precisely describe the composition process in order to describe the logic relation between different components. Based on this, the concept of trust matrix is given to represent the relationships between states. A trustworthy service composition strategy and its enforcement method are proposed. A case study of Travel Service demonstrates the feasibility of proposed method.

1 Introduction

Service-Oriented Architecture (SOA) is a loosely coupled architecture designed to meet business needs of an organization. Web service is a well known and widely used technology for implementing the SOA[1]. Using Web services, it is possible to send any type of information in any form. For example, in e-business, tourism and other service areas, more and more services have been published in the form of Web service. As a single Web service can provide limited function, in order to increase Web service sharing, it is necessary to compose Web service to provide a more powerful service[2, 3].

In recent year, the number of Web service is exponentially growing. Service consumers can enjoy the convenience of service composition, but they also faces some difficulties: (1)How to guarantee that component service behaves correctly

[*] This work was partially supported by the NSF of China under grants No. 60773094 and 60473055, Shanghai Shuguang Program under grant No. 07SG32, Fund of Key Laboratory of Shanghai Science and Technology under grant No. 09DZ2272600.

A. Gómez-Pérez, Y. Yu, and Y. Ding (Eds.): ASWC 2009, LNCS 5926, pp. 261–275, 2009.

when they are composed together and the result of composition can meet the requirements of service consumers; (2)Involved services in composition may come from different service providers and run on different platform, the choice of any service may affect the quality of service, which make get high-quality service composition that meet user's requirements very difficult. While many network applications such as financial services, online transaction or e-commerce are running in an unpredictable environment, but they requires a higher level of trustworthy. Failure response to these requests will cause the loss of customers and economic, which make the application with a specific trustworthy has become more and more important. Therefore, how to dynamically construct a trustworthy service composition has become increasingly important.

In order to effectively address these problems, we propose a Trust Service Composition Net (TSCN) model based on Petri nets, and use it to simulate the process of service composition. According to the characteristics of service composition and available service, TSCN is used to model for available service, component, the relationships between component, the operation mechanism of service composition. Based on this, the concept of Trust matrix is given to represent the relationships between state, and dynamic trustworthy service composition strategy and its enforcement are also proposed, the semantic and state space of Petri nets help prove the effectiveness of strategy. A case study of Travel Service demonstrates the approach can not only describe Web service resource on the Internet, but also can guarantee the highest trustworthy of service composition.

The remainder of this paper is organized as follows: Section 2 gives the definition and semantics of TSCN model; In section 3, we construct the TSCN model of service composition. Section 4 is the analysis technologies. In Section 5, we explain the feasibility and practicability of our methods by a specific example. Section 6 presents some related works while section 7 is conclusion.

2 Computation Model

Petri nets are formal languages for describing the concurrency systems. Some recent researches indicate that Petri nets are powerful and expressive enough to describe the behavior features of service composition[4, 5]. At the same time, the behaviors of Web service are the collection of orderly operations, which can directly map into Petri nets, the basic concepts of Petri nets can refer to [6].

2.1 Definition of TSCN

In this section, we will give formal definition of TSCN model based on the characteristics of trustworthy service composition.

Definition 1: A four-tuple $TN = (PN, I, Pr, M_0)$ that meets the following conditions is called Trustworthy Service Net(TSN) model, such that:

(1) $PN = (P, T, F, W)$ is a basic Petri net;
(2) $I \subset P$ is a special place, which is called the interface of Σ and denoted by dotted circle;

(3) $Pr : T \rightarrow N \times (0, 1]$ is the attribution function of transition, N is a natural number, $Pr(t_i) = (\gamma_i, \beta_i)$, γ_i, β_i describe priority and trustworthy of transition t_i and its default value is $(1, 1)$.

(4) M_0 is the initial marking of TN.

TSN model is mainly used to model for available service, where place describes the position of service's operation, while transition describes the possible operation of service; the interface describes the outside input and output of service. The attribution function describes the priority and credibility of transition, the default value is 0 and 1.

Definition 2: A seven tuple $\Omega = \{TN, \Sigma, \Gamma, TI, TA, PI, PA\}$ is called a Trustworthy Service Composition net($TSCN$), such that:

(1) TN is the TSN model;

(2) $\Sigma = \{\Sigma_i | i \in N\}$ is a finite set of $TSCN$ and TSN model, each element is called a page of Ω;

(3) $\Gamma : \Sigma \rightarrow T^*$ is the operation set of each page, which is called the operation of page.

(4) $TI \subset T$ is the set of substituted operation and denoted by double rectangle;

(5) $TA : TI \rightarrow \Gamma$ is the operation allocation function, that is, allocating the corresponding page to the substituted operation, and the relationship between page and substituted operation is one by one;

(6) $PI \subset P$ is the set of interface node, which describes the input and output of substituted node, and denoted by double circle;

(7) PA is the mapping function of interface, whose function is to map the interface node into the input and output of the corresponding page's operation;

$TSCN$ is mainly used to model for service composition and its function module(component). Each page represents a component or an available service, and substituted operation represents a sub-page of $TSCN$ model, which can operate only mapping into the specific component or service.

The distribution of token in each place is called the marking of $TSCN$ model, denoted by M. The marking $M(p)$ denotes the number of tokens in place p. $\forall x \in (P \cup T)$, we denote the pre-set of x as $^{\bullet}x = \{y | y \in (P \cup T) \wedge (y, x) \in F\}$ and the post-set of x as $x^{\bullet} = \{y | y \in (P \cup T) \wedge (x, y) \in F\}$.

2.2 Semantics of *TSCN* Model

In order to describe the trustworthy characterizes of $TSCN$ model clearly, we introduce credibility to describe the state of $TSCN$ model. A tuple $S = (M, TV)$ is called a state of $TSCN$ model, where M is a marking and TV is the credibility of reaching the state, which is called reach credibility of state S. Initial state $S_0 = (M_0, TV_0)$ and $TV_0 = 1$. We will give the semantics of $TSCN$ model based on the definition of state.

Definition 3: Let Ω be a $TSCN$ model, $S = (M, TV)$ is a state of Ω. For transition $t_i \in T$, if t_i meets the following conditions: $\forall p_j \in P : p_j \in^{\bullet} t_i \rightarrow M(p_j) \geq W(p_j, t_i)$. Then transition t_i has the right to fire under state S, denoted by $S[t_i >$. All transitions that have the right to fire under state S are denoted by $ET(S) = \{t_i | t_i \in T \wedge S[t_i >\}$.

Definition 4: Let Ω be a *TSCN* model, $S = (M, TV)$ is a state of Ω. The firing of transition t_i is effective if it meets the following condition:

$\gamma_i \le min(\gamma_j)$, where $t_j \in ET(S)$

The firing of transition t_i under state S is effective means that the enabled transition has the highest priority. All effective firing transitions under state S are denoted by set $FT(S)$.

Definition 5: Let Ω be a *TSCN* model. $S = (M, TV)$ is a state of Ω. The model Ω will reach a new state $S' = (M', TV')$ by effectively fire the transition in FT, denoted by $S[t_i > S'$, S' is called the reachable state of S. The computation of M', TV' are based on the following rules:

(1) Computing marking:
$\forall p_j \in^\bullet t_i \cup t_i^\bullet$: $M'(P_j) = M(P_j) - W(P_j, t_i) + W(t_i, P_j)$
(2) Computing credibility TV': $TV' = TV * \beta_i$

If there exist firing sequences $\delta = t_1, t_2, \ldots, t_k$ and state sequences $S_1, S_2, \ldots, S_k$, which make $S[t_1 > S_1[t_2 > S_2 \ldots S_{k-1}[t_k > S_k$, then S_k is reachable from state S, denoted by $S[\delta > S_k$. $TS(\delta) = TS(t_i)$ represents the credibility of firing sequence δ. All the possibly reachable states of S are denoted by $R(S)$ and $S \in R(S)$.

3 Modeling *TSCN*

Service composition must bind its functional modules to specific available services before operating, thus generating the composition schemas. In this section, we will firstly analyze the requirements of trustworthy service composition, and use *TSCN* model to describe it.

3.1 Requirements of Trustworthy Service Composition

Based on reference [7, 8] and taking into account the practicality and scalability, we will mainly consider the credibility of service in this paper.

The credibility of service $WS_{i,j}$ refers to the probability of successful completion service $WS_{i,j}$, denoted by $\tau_{i,j}$. Successful completion means that the results of service composition can meet *QoS* requirements of service consumers, such as availability, reliability, credibility, running time, etc. This paper does not address the calculation of credibility, which can refer to [9].

Because the function of service composition can be realized by a number of independent sub-function, which is called component in this paper, and each component will have a number of available services.

Definition 6: The requirement model of trustworthy service composition is a five-tuple $\Xi = \{C, WS, RL, TW, RT\}$, where:

(1) C is the finite component set;
(2) WS is the available service set;

(3) $RL : C \times C \rightarrow \{>, +, \|, n\}$ is the relationship between components;

(4) $TW : C \rightarrow WS^*$ is the function of component's available service, $TW(C_i) = WS_i = \{WS_{i,1}, WS_{i,2}, \ldots, WS_{i,m}\}$ represents the available service set of C_i, where service $WS_{i,j}$ represent the jth available service of component C_i;

(5) $RT : WS \rightarrow (0, 1]$ is the attribution function of available services, $RT(WS_{i,j}) = \tau_{i,j}$ describes the credibility of service $WS_{i,j}$.

3.2   Modeling Trustworthy Service Composition

In this section, we will use $TSCN$ to model for service, component, the relationship between component based on the requirements of trustworthy service composition. In order to distinguish the input and output interface, which are marked by using superscript I and O respectively.

Modeling available services. The $TSCN$ model of available service is shown in Fig.1. The operation process of service $WS_{i,j}$ is: when the service is in the initial position $p_{s,i,j}$, if the place $p_{u,i,j}$ has tokens, then calling initial operation $(t_{e,i,j})$ to make it be in the running operation $p_{a,i,j}$; if service has operated successfully $t_{e,i,j}$, then the system will be in the termination position $(p_{e,i,j})$ and output the execution results$(p_{o,i,j})$. Let the credibility of transition $t_{e,i,j}$ be equal to the credibility of service $WS_{i,j}$, that is, $\beta_{e,i,j} = \tau_{i,j}$, the rest is 1.

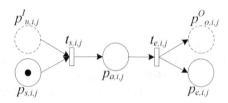

Fig. 1. $TSCN$ model of service $WS_{i,j}$

Modeling components. The $TSCN$ model of component C_i is shown in Fig.2, where the substituted transition $\{WS_{i,1}, WS_{i,2}, \ldots WS_{i,m}\}$ corresponds to the page of all services in the available service set WS_i, the interface node $P_{u,i,j}$ and $P_{o,i,j}$ describe the input and output of service $WS_{i,j}$, the specific modeling steps are:

(1) Input interface $p_{s,i}^I$ and $p_{ps,j}^I$ represent the operation condition and the forward component's credibility of C_i;

(2) Transition $t_{u,i,1}, t_{u,i,2}, \ldots, t_{u,i,m}$ represent the invoked operations of the corresponding available services, the priority of these transitions may be different according to the actual requirement. For every transition $t_{u,i,j}$, we can set $^\bullet t_{u,i,j} = p_{s,i}^I, t_{u,i,j}^\bullet = \{P_{u,i,j}, p_{f,i,j}^I\}$. According to the definition of effective firing, we can get the available service that has the highest priority will be invoked.

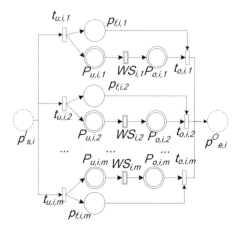

Fig. 2. *TSCN* model of component C_i

(3) Introducing place $p_{f,i,j}$ to represent the invocation of service $WS_{i,j}$, $WS_{i,j}$ is called the execution service. If service $WS_{i,j}$ has finished operating, then invoking transition $t_{r,i,j}$ to transfer the results $(P_{u,i,j})$ of service to the output interface $p^O_{e,i}$. We can set $^{\bullet}t_{o,i,j} = \{p_{u,i,j}, p_{f,i,j}\}$, $t^{\bullet}_{o,i,j} = \{p^O_{e,i}\}$.

In the specific application, we can regard service composition as a function module (component). Then reusing the model of existing service composition to realize more complex service.

Modeling basic relationships. In this section, we will model for sequence, parallel, choice and loop relationships between components. Let components C_f, C_k meet $C_f = Fw(C_i) \cap Fw(C_j)$, $C_k = Bck(C_i) \cap Bck(C_j)$.

The *TSCN* model of sequence relationship is shown in Fig.3(a), we introduce transition t_{ou} to transfer the output parameters of the forward component C_i to the input interface of component C_j: $^{\bullet}t_{ou} = P_{e,i}$, $t^{\bullet}_{ou} = P_{s,j}$.

The *TSCN* model of parallel relationship $C_i \parallel C_j$ is shown in Fig.3(b). We use transition $t_{f,ij}$ to transfer the output results $(P_{e,f})$ of the forward component C_f to the input interface $P_{s,i}$, $P_{s,j}$ of C_i and C_j, while transition $t_{ij,k}$ is to transfer the output results of C_i and C_j to the forward component C_k.

The *TSCN* model of choice relationship $C_i + C_j$ is shown in Fig.3(c), we use transition $t_{f,i}$, $t_{f,j}$ to compete the output results $(P_{e,f})$ of the forward component C_f, while transition $t_{i,k}$ and $t_{j,k}$ is to transfer the output results $P_{e,i}$, $P_{e,j}$ of C_i and C_j to the afterward component C_k.

The *TSCN* model of loop relationship nC_i is shown in Fig.3(d), we introduce transition $t_{s,ni}$, $t_{e,ni}$ to describe the beginning and termination operation of nC_i, while transition $t_{s,i}$ and $t_{e,i}$ present the beginning and termination operation of C_i. Place p_c is used to store the uncompleted times, while place p_{ch} is used to store the completed times of C_i.

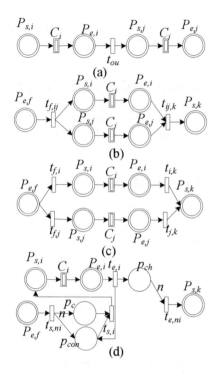

Fig. 3. Modeling basic relationship

3.3 Modeling Service Composition

According to the characteristics of service composition, the corresponding *TSCN* model is shown in Fig.4, the specific modeling steps are:

(1) Introducing the initial place p_s and beginning transition t_s, which makes ${}^\bullet t_s = \{p_s\}$, $t_s^\bullet = \{P_{s,i} | Fw(C_i) = null\}$, ${}^\bullet p_s = null$, $p_s^\bullet = \{t_s\}$, $M_0(p_s) = 1$;

(2) Constructing the *TSCN* model of available service according to the attributes of service, and modeling for each component of service composition;

(3) Composing the *TSCN* model of each component based on the relation function RT between component;

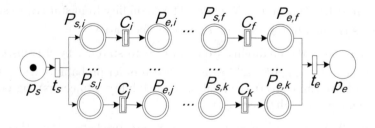

Fig. 4. Modeling service composition

(4) Introducing the termination place p_e and transition t_e, which makes ${}^{\bullet}t_e = \{p_{e,i}|Bck(C_i) = null\}$, $t_e^{\bullet} = \{p_e\}$, ${}^{\bullet}p_e = \{t_e\}$, $p_e^{\bullet} = null$.

4   Analysis Technologies of TSCN Model

In this section, we will analyze the relationships between state based on the *TSCN* model and proposed trust matrix. Based on this, the dynamic composition strategy is proposed and its effectiveness is proved.

4.1   Trust Matrix of TSCN Model

Because the state space of large scale service composition may be complicated, and analysis by direct computation may be hard, it is necessary to further abstract state graph. According to the operation mechanism of *TSCN* model, its state space will have some special states. Among them, S_{end} is the normal termination state of *TSCN*, which makes $\forall S_i \in S^F(\Omega)$, the reach credibility from state S_i to S_{end} is equal to 1. Where $S^F(\Omega) = \{S|\forall t_i \in T, \neg S[t_i >\}$. Let S be a state of Ω, then the reach credibility from state S to S_{end} is denoted by $TVE(S)$, which is called the credibility of state S. We will propose transfer matrix of *TSCN* based on the relationships between state.

Definition 7: Let Ω be a *TSCN* model. The credibility a_{ij} from state S_i to S_j meets the following conditions:

$$a_{ij} = \begin{cases} 1: & S_i \in S^F(\Omega), S_j = S_{end} \\ \beta_{ij}: & if\, t_{ij} \in T, S_i[t_{ij} > S_j \\ 0: & otherwise \end{cases}$$

Let the number of reachable state in *TSCN* model Ω be L, the L-order square matrix A is called transfer matrix of Ω if it meets the following conditions: $A = [a_{ij}]_{L \times L}$, where a_{ij} is the transformation probability from state S_i to state S_j. Denoted $A^{(n)}$ as n-power of matrix A, while $a_{ij}^{(n)}$ is the element in the *ith* row and *jth* column of matrix $A^{(n)}$.

According to the definition of transfer matrix and operation mechanism of *TSCN* model, there has the following conclusions, the specific proof we can refer to our previous work [10].

Theorem 1: Let Ω be a *TSCN* model. The credibility from state S_i to state S_j by n steps is equal to the value of $a_{ij}^{(n)}$ in transfer matrix $A^{(n)}$.

Theorem 1 shows the credibility from state S_i to state S_j by n steps is equal to the value of $a_{ij}^{(n)}$ in $A^{(n)}$. $a_{ij}^{(n)}$ is also called n-order probability from state S_i to state S_j. We can convert the analysis of credibility into computing power of transfer matrix through Theorem 1.

Theorem 2: $\forall i, j < L'$, if state S_i and S_j are reachable, then there exists $K_{ij} \in N$ which makes $\forall E \in N$, $a_{ij}^{(K_{ij}+E)} = 0 \wedge a_{ij}^{(K_{ij})} \neq 0$.

Theorem 2 illustrates that we can analyze the credibility of service composition by computing stop conditions of transfer matrix's power. K_{ij} and $a_{ij}^{(K_{ij})}$ are called stable order and stable credibility based on matrix A from state S_i to S_j. The max stable order between state is called max stable order.

For a complicated service composition, the corresponding transfer matrix of *TSCN* model is relative complexity. Therefore it is necessary to further simplify transfer matrix while maintaining credibility unchanged. If the reach credibility from state S_i to state S_j is 1, and only S_i can reach state S_j, mapping into transfer matrix is: $a_{IJ} = 1$ and $\sum_{k=1}^{L-1} a_{ik} = \sum_{k=1}^{L-1} a_{kj} = 1$, then we can eliminate the *ith* row and *jth* column from matrix A. That is, two states S_i and S_j in composition meet following condition: S_I can reach state S_J and state S_J is only reached by S_I; the credibility of transition t_{IJ} is $\beta_{I,J} = 1$. For any two states S_p and S_q, there has two cases in the path from state S_I to S_J: (1)S_I is included in δ, because states S_I and S_J have met above conditions, then $\delta = \{S_i, t_1, S_1, \ldots, t_k, S_I, t_{IJ}, S_J, \ldots, t_n, S_j\}$, that is, the firing probability of δ is $P = \lambda_1 * \lambda_2 \ldots * \lambda_k * \lambda_{IJ} \ldots * \lambda_n = \lambda_1 * \lambda_2 \ldots * \lambda_k \ldots * \lambda_n$. Therefore, eliminating the *Ith* row and *Jth* column from matrix A does not interfere the computation of P; (2) S_I is not included in firing sequence δ, then eliminating the *Ith* row and *Jth* column from matrix A does not interfere the computation of P.

Definition 8: Let Ω be a *TSCN* model. A is the transfer matrix of Ω, K is the max stable order of A, matrix B: $B = \sum_{r=1}^{K} A^r$. Then B is call trust matrix of Ω.

Each element in the corresponding column $R_{end,B}$ of trust matrix B represents the termination credibility $TVE(S)$ of state S.

4.2   Dynamic Trustworthy Service Composition Strategy

In the complicated service composition, the function of component can be completed by a number of available services and these firings are feasible, because each firing my cause service composition have different credibility, therefore, it is necessary to choose the service which has the highest credibility.

Definition 9: Let S be a state of Ω, the service set $AWF(C_i, S) \subseteq WS_i$, if: $\forall WS_{i,j} \in AWF(C_i, S)$, there has $t_{u,i,j} \in FT(S)$. Then $AWF(C_i, S)$ is the feasible service set of component C_i under state S.

Definition 10: Let S be a state of Ω, $AWF(C_i, S)$ is the feasible service set of component C_i under state S, if S can reach S' by firing service $WS_{i,j}$, then

$$Ter(S, WS_{i,j}) = \frac{\tau_{i,j} * TVE(S')}{|AWF(TK_i, S)|}$$ is the credibility after firing $WS_{i,j}$.

Credibility refers to the reach credibility of whole application after firing service $WS_{i,j}$ under state S. In the same state, the choice of service with different credibility will cause the credibility of whole application difference. We will give dynamic composition strategy based on the definition of credibility.

Definition 11: Let S be a state of Ω, $AWF(C_i, S)$ is the feasible service set of C_i under state S, dynamic trustworthy service composition strategy is:

(1) Setting priority to the invoked available service:

If $\forall C_j \in C$: $RT(C_j, C_i) \neq +$, then sorting the credibility of available service descending, and allocating priority to the available service according to their position: If $\tau_{i,j} > \tau_{i,k}$, then $\alpha_{i,j} > \alpha_{i,k}$

(2) $\exists C_j \in C$: $RT(C_j, C_i) = +$, transition $t_{c,i}$ and $t_{c,j}$ correspond to the selected operation of component C_i and C_j, then for every service in the available service set $AWF(C_i, S) \cup AWF(C_i, S)$, we will compute its credibility under state S: If $max\{Ter(S, WS_{i,k})\} > max\{Ter(S, WS_{j,f})\}$, then $\alpha_{c,i} > \alpha_{c,j}$

Dynamic composition strategy is allocating the highest priority to the service which has the highest credibility after mapping into Ω. According to the definition of the termination credibility and credibility, we can draw that dynamic composition strategy is selecting the service which has the highest credibility from $AWF(C_i, S)$ to realize the function of C_i.

Theorem 3: Using dynamic trustworthy service composition strategies, the composted service will have the highest credibility.

Proof (Reduction to absurdity): Assuming the implementation sequence of service composition be δ by using dynamic service composition strategy, and there exists another implementation sequence δ' for the same composition process, which make $TS(\delta) > TS(\delta')$. Therefore, there exists C_i which meets: implementation sequence δ and δ' is the invoked available services of C_i respectively, we can assume they are $WS_{i,j}$ and $WS_{i,k}$. Because $TS(\delta) > TS(\delta')$, there has $Ter(S, WS_{i,k}) > Ter(S, WS_{i,j})$ according to the definition of credibility. From the service's selection rules of dynamic trustworthy service composition, we can draw that the model will choose service $WS_{i,k}$ to complete the function, which is contradicted with the assumption, so the assumption does not hold.

4.3   Enforcement of Dynamic Trustworthy Service Composition

From the definition of dynamic service composition strategy, each component must choose the service which has highest credibility to perform. The specific steps of constructing service composition are as follows:

(1) Constructing *TSCN* model based on the requirements of service composition, and computing its trust matrix;

(2) Cutting trust matrix based on the operation characteristics of *TSCN* model;

(3) Computing trust matrix based on the cut transfer matrix, thus getting the credibility of each service, and allocating different priority to service. The enforcement algorithm of dynamic service composition strategy is shown in Table 1. The algorithm establishes the available service set WS that user can access based on the current Web service, then establishing the trust matrix of *TSCN* model based on the actual available service.

Table 1. Enforcement algorithm of dynamic service composition

```
1: Comp Any(Ω,A)
2: { A=0;
3: For i=0, i<|R(S₀) |,i++ do
4:    For j=1, j≤| R(S)|,j++ do
5: {If S∈S^F(Ω) and S=S_end do
6: a[i,j]=1;
7: If ∃t_k∈T , S[t>S` do
8:a[i,j]=βk;}}
9: Analysis(A)
10:{ Cut (A);
11: B=A;K=2;
12:while(A^K!=0) do
13:{B=+A^k; K++;}
14: if a[S₀,S_end]=0 do
15: Output "This web service is not composition";
16: else return B;}
17: Computering Pr(Ω,B)
18: { C=0;
19: For i=0, i<|B |,i++ do
20: For j=1, j≤|C|,i++ do
21: {TWS_i,j=Computering_WS(S_i,C_j);}
22: ∀WS_j,k∈ TWS_i,j do
23: C[j,k]=Computering_Ter(S_i,WS_j,k);}
24: Return C;}
25: Ass  pr(Ω,B)
26: { For i=0, i<|B |,i++ do
27: For j=1, j≤C, j++ do
28: ∀WS_j,k∈ TWS_i,j do
29: if C[j,k]>C[j]∧ γ_j,k≤γ_j,k
30: γ_j,k=γ_j,k+1;}
```

5 Experiments

This section shows the analysis process through a simplified Travel Service. The specific service composition process is: looking up information and choosing destination (C_1), train tickets reservations (C_2) responses for handling customer's train tickets, airline tickets reservation (C_3) is to purchase a suitable destination flights, passage booking(C_4) is to order the appropriate passage in accordance with the requirements of consumers, tourism planning (C_5) responses for specific travel arrangement, car reservation (C_6) arranges the custom to arrived at the railway station or airport, hotel reservation (C_7) arranges for the local living, finally, the tourism service(C_8) responses for the customer's local tourism-related matters. The composition process can be represented by expression $C_1 > (C_2 + C_3 + C_4) > C_5 > (C_6||C_7) > C_8$. The specific service and its attributes is shown in Table 2.

According to the process of Travel Service and the available services, we construct the corresponding $TSCN$ model, which is shown in Fig.5. Using algorithm 1 to compute the trust matrix of model Ω, and computing the credibility of the available service of component C_2, C_3, C_4, which is shown in Table 3. Based on the credibility of service, we can allocate priority to $t_{1,2}$, $t_{1,3}$, $t_{1,4}$, which is 1, 0, 0. The implementation service of Travel Service is $(WS_{1,1}, WS_{2,1}, WS_{3,3}, WS_{4,2}, WS_{5,1}, WS_{6,2}, WS_{7,4}, WS_{8,2})$ by using dynamic composition strategy, and the credibility is 88.65%. We can get the schema has the highest credibility by analyzing the remaining composition schemas.

Table 2. Service composition process and success probability

WS	τ	WS	τ	WS	τ	WS	τ
$WS_{1,1}$	0.9756	$WS_{2,1}$	0.9644	$WS_{3,1}$	0.8963	$WS_{4,1}$	0.9247
$WS_{1,2}$	0.9274	$WS_{2,2}$	0.9395	$WS_{3,2}$	0.9751	$WS_{4,2}$	1
$WS_{1,3}$	0.8644	$WS_{2,3}$	0.8718	$WS_{3,3}$	1	$WS_{4,3}$	0.8381
$WS_{1,4}$	0.7756	$WS_{2,4}$	0.7336	$WS_{3,4}$	0.7579	$WS_{4,4}$	0.7529
$WS_{5,1}$	0.9759	$WS_{6,1}$	0.8732	$WS_{7,1}$	0.9689	$WS_{8,1}$	0.8381
$WS_{5,2}$	0.7892	$WS_{6,2}$	1	$WS_{7,2}$	0.7338	$WS_{8,2}$	0.9529
$WS_{5,3}$	0.9513	$WS_{6,3}$	0.9424	$WS_{7,3}$	0.8915	$WS_{8,3}$	0.9268
$WS_{5,4}$	0.8716	$WS_{6,4}$	0.7473	$WS_{7,4}$	0.9772	$WS_{8,4}$	0.8877

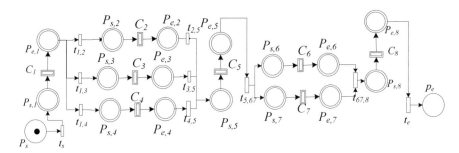

Fig. 5. TSCN model of Travel Service

Table 3. The credibility of feasible services

WS	Ter	WS	Ter	WS	Ter
$WS_{2,1}$	0.219095	$WS_{3,1}$	0.203624	$WS_{4,1}$	0.210076
$WS_{2,2}$	0.213439	$WS_{3,2}$	0.221526	$WS_{4,2}$	0.227183
$WS_{2,3}$	0.198058	$WS_{3,3}$	0.227183	$WS_{4,3}$	0.190402
$WS_{2,4}$	0.166662	$WS_{3,4}$	0.172182	$WS_{4,4}$	0.171046

In order to effectively estimate the approach proposed in the paper, we conduct an experiment to analyze the performance of model. We randomly generated 44000 services as service resource. Each service has the basic information such as service name, credibility. Based on the constructed service resource, we will do simulation for the Travel Service, the specific steps are:

(1) Dividing service resource into four grades: each component has 10, 20, 50 and 100 available service, and doing step 2 for each component;

(2) Diving each grade into ten groups and computing the credibility of Travel Service.

The experimental results are shown in Fig.6. From this figure, we can get: (1) When the available service set of each component are relatively large (≥ 50), then the credibility of service composition are more higher (88.33% of composition

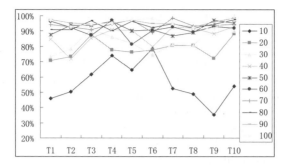

Fig. 6. The result of experiment

results is higher than 90%); (2)The credibility of service composition will not increase with the available services increasing, however, the credibility of service composition will increase when high credibility of service resource increasing; (3) For the same size of service resource, the difference of service composition's credibility may be larger when service resource is inadequate. For example, when the available service of each component is equal to 10, then the highest credibility is 78.76%, while the lowest credibility is 35.48%. However, when service resource is abundant($\geq$ 50), the difference between the experimental results of the same group is little, which is less than 10%.

6   Related Works

In the existing works, formal analysis of service composition by using process algebra are given in [11,12]. Reference [11] claims that process algebras can provide a very complete and satisfactory assistance to the whole process of Web service development. A similar approach is given in [12], which presents a framework for the design and the verification of Web service using process algebras and their tools, thus automatically obtaining the corresponding *BPEL4WS* code. However, they do not take trustworthy properties of composition behaviors into account, and the priority between service was not considered too.

Another formal analysis method of service composition is developed in [13, 14], these works use Finite State Machine (*FSM*) as their analysis tools. The authors in [13] address the issues of service composition with characteristic of transaction, a formal model of transaction service composition ground on *FSM* is proposed. *FSM* is also used in [14] to provide a precise and well defined semantic framework for establishing the key language attributes. These works ignore dynamic characteristics of service, and lack of analyzing QoS properties.

Similarly, *Petri nets* are used in [15-17] for modeling and analyzing service composition to support Web service management. The authors in [15] transform *BPEL* into service workflow net which is a kind of colored Petri net and analyzing the compatibility of two services, then propose an approach to check whether there exists any message mediation so that their mediation-aided composition will not violate the constraints imposed by either side. A checking tool for translating *BPEL* specifications into the input language of the Petri net model called LoLA has been proposed in [16], which demonstrates that the semantics is well suited for computer aided verification purposes. CP-nets are used in [17] to analyze and verify effectively the net to investigate several behavioral properties.

To some extent, our work has been influenced by the above research results. Below are some of the key differences when comparing the above approaches with the one presented in this paper: (1) Comparing with finite state machines, Petri nets provide a much broad basis for computer aided verification, so we think it is more natural to model different input and output information of a composition process by means of Petri nets. (2) The work described in [15-17] is also based on Petri nets, but the importance of these works isn't involve *QoS* attributes of service composition and no guarantee the composition has the highest credibility. Hence our proposed approach is suitable to analyze and verify the composition not only at design time but also at execution time.

7   Conclusions

In this paper, we have made research on dynamic trustworthy service composition, and propose Petri nets based method to model requirements of trustworthy service composition. The dynamic trustworthy service composition strategy and its enforcement are also advanced. The advantages of work are as follows: (1) Using formal methods to describe the process, where Petri nets and its related theory can accurately describe different states of service, and can clearly express the logic of service composition, the use of related tools can simulate the composition process, which makes the method easily promote; (2) Proposing the concept of trust matrix based on the state space, which can convert the analysis of credibility into computing trust matrix, thereby reducing the complexity of analysis; (3) Analyzing the constructed model by mathematical reduction, which make analyze and verify the established model easily.

This paper has made progress in modeling and analyzing dynamic trustworthy service composition. However, we do not consider resource scheduling of composition process. In addition, the reasoning mechanisms and tools are also not covered. We will make research on these areas in the future work.

References

1. Yu, Q., Liu, X.M., Athman, B., Brahim, M.: Deploying and managing Web services: issues, solutions, and directions. The Int. J. Very Large Data Bases 17(3), 537–572 (2008)
2. Beek, M., Bucchiarone, A., Gnesi, S.: Web Service Composition Approaches: From Industrial Standards to Formal Methods. In: Proc. Int. Conf. Internet and Web Applications and Services, pp. 15–20 (2007)
3. Bultan, T., Su, J., Fu, X.: Analyzing conversations of web services. IEEE Internet Computing 10(1), 18–25 (2006)

4. van der Aalst, W.M.P., Dumas, M., Ouyang, C.: A conformance checking of service behavior. Tran. on Internet Technology 8(3), 1–30 (2008)
5. Xiong, P.C., Zhou, M.H., Pu, C.: A Petri Net Siphon Based Solution to Protocol-level Service Composition Mismatches. In: Proc. IEEE Int. Conf. Web Services, pp. 952–958 (2009)
6. Girault, C., Valk, R.: Petri Nets for System Engineering: A Guide to Modeling, Verification, and Applications. Springer, Heidelberg (2003)
7. Ahamed, S.I., Sharmin, M.: A trust-based secure service discovery (TSSD) model for pervasive computing. Computer Communications 31(18), 4281–4293 (2008)
8. Avizienis, A., Laprie, J.C., Randell, B.: Basic concepts and taxonomy of dependable and secure computing. IEEE Tran. Dependable and Secure Computing 1(1), 11–33 (2004)
9. Ran, S.: A model for Web services discovery with QoS. ACM SIGecom Exchanges 4(1), 1–10 (2003)
10. Fan, G.S., Yu, H.Q., Chen, L.Q., Liu, D.M.: Analyzing Reliability of Time Constrained Service Composition. In: Proc. IEEE/ACIS Int. Conf. Computer and Information Science, pp. 1155–1160 (2009)
11. Salaun, G., Bordeaux, L., Schaerf, M.: Describing and reasoning on Web services using process algebra. In: Proc. IEEE Int. Conf. Web Services, pp. 43–50 (2004)
12. Ferrara, A.: Web services: a process algebra approach. In: Proc. Int. conf. Service Oriented Computing, pp. 242–251 (2004)
13. Hu, J.J., Zhao, X., Cao, Y.D., Zhou, R.T.: A Service Composition Model with Characteristic of Transaction based on Finite State Machine. In: Proc. Int. Conf. Computer and Electrical Engineering, pp. 450–454 (2008)
14. Farahbod, R., Glasser, U., Vajihollahi, M.: An abstract machine architecture for web service based business process management. Int. J. Business Process Integration and Management 1(4), 279–291 (2006)
15. Tan, W., Rao, F., Fan, Y., Zhu, J.: Compatibility analysis and mediation-aided composition for BPEL services. In: Kotagiri, R., Radha Krishna, P., Mohania, M., Nantajeewarawat, E. (eds.) DASFAA 2007. LNCS, vol. 4443, pp. 1062–1065. Springer, Heidelberg (2007)
16. Hinz, S., Schmidt, K., Stahl, C.: Transforming BPEL to petri nets. In: van der Aalst, W.M.P., Benatallah, B., Casati, F., Curbera, F. (eds.) BPM 2005. LNCS, vol. 3649, pp. 220–235. Springer, Heidelberg (2005)
17. Yang, Y.P., Tan, Q.P., Xiao, Y., Liu, F., Yu, J.: Transform BPEL Workflow into Hierarchical CP-Nets to Make Tool Support for Verification. In: Zhou, X., Li, J., Shen, H.T., Kitsuregawa, M., Zhang, Y. (eds.) APWeb 2006. LNCS, vol. 3841, pp. 275–284. Springer, Heidelberg (2006)

Improving Folksonomies Using Formal Knowledge: A Case Study on Search

Sofia Angeletou, Marta Sabou, and Enrico Motta

Knowledge Media Institute (KMi),
The Open University, Milton Keynes, United Kingdom
{S.Angeletou,R.M.Sabou,E.Motta}@open.ac.uk

Abstract. Search in folksonomies is impeded by lack of machine understandable descriptions for the meaning of tags and their relations. One approach to addressing this problem is the use of formal knowledge resources (KS) to assign meaning to the tags, most notably WordNet and (online) ontologies. However, there is no insight of how the different characteristics of such KS can contribute to improving search in folksonomies. In this work we compare the two KS in the context of folksonomy search, first by evaluating the enriched structures and then by performing a user study on searching the folksonomy content through these structures. We also compare them to cluster-based folksonomy search. We show that the diversity of ontologies leads to more satisfactory results compared to WordNet although the latter provides richer structures. We also conclude that the idiosyncrasies of folksonomies can not be addressed by only using formal KS.

1 Introduction

Folksonomies are a convenient medium to publish, annotate and share content on the web. Their basic entities are the **users**, who annotate (i.e., tag) **resources** with **tags** (text labels). Fig.1a shows a snippet of an example folksonomy where resources (R_1, R_2 and R_3) are tagged with a number of tags. Due to the lack of tagging restrictions the following phenomena have been observed [3], which hamper the process of search in folksonomies:

Tag synonymy arises when lexically different tags express the same concept, e.g., *cake* and *dessert*. Synonymy may cause exclusion of results if these are tagged with synonym(s) of the search keyword, e.g., in Fig.1a, searching for *cake* will return only R_2 and not R_1.

Basic level variation. Tags with different levels of specificity are used to describe resources that relate to the same concept. For example, *apple* and *fruit* can both describe resources about apples. The **lack of structure** in tagspaces, does not allow for explicit declaration of the fact *"apple is a fruit"*. This limits the potential of querying for resources tagged with related tags. For example, for the folksonomy of Fig.1a, querying for *fruit* only returns R_1, although R_2 is also tagged with a fruit. Moreover, the lack of structure does not allow for result

A. Gómez-Pérez, Y. Yu, and Y. Ding (Eds.): ASWC 2009, LNCS 5926, pp. 276–290, 2009.

diversity. Result diversity [10] is the grouping of similar results in distinctive sets (e.g., querying for fruit returns one set with apples, one for oranges e.t.c.).

Tag polysemy occurs among lexically identical tags that denote different meanings. For example, *apple* may refer to *fruit* or a *company*. Tag polysemy causes the retrieval of unwanted results when the tag is used with a different meaning than the search keyword, e.g., searching for *apple* in the folksonomy of Fig.1a will return both R_2 and R_3, although, depending on the meaning of *apple* in the context of the query, only one of these resources is relevant.

To overcome these issues and improve search, the research community has focused its efforts on two research lines. The first line relies on statistical methods applied to the dynamics of folksonomies and the distribution of tags, resources and users in order to address the above phenomena. Adapted ranking algorithms for improving folksonomy information retrieval [4], vector space models for reducing data sparsity and improving search [1] and relevance based models that allow better result diversification [10] are just some of the efforts in this direction.

The second line aims to address the common underlying cause of the above phenomena, which is **the lack of machine understandable descriptions for the meaning of the tags and their relations**. Ideally, in order to overcome the above impediments, each folksonomy should be associated with a semantic structure that provides explicit meaning for its tags and their relations. In such a structure (Fig.1b), the meanings of tags are made explicit by being linked to appropriate concepts (see dotted lines). The relations among these concepts are also explicitly stated. Such a structure contains the required information to solve polysemy (e.g., *fruit != company*) and synonymy (e.g., *cake = dessert*) issues, and to allow for query expansion based on the concept relations. It also offers the possibility to present the results in an intuitive way allowing for result diversity.

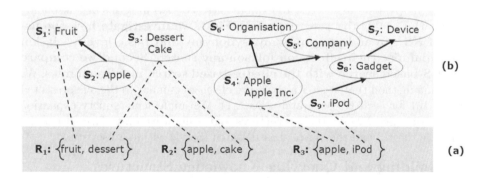

Fig. 1. (a) example folksonomy (b) semantically enriched folksonomy

The efforts reported in this research line, deal with the above issues by performing explicit alignment of tags to concepts from formal **knowledge sources (KS)**, most notably WordNet and ontologies.

WordNet is a long term, continuously maturing project used for information retrieval, text classification and sense disambiguation and spans over several knowledge domains. As an initiative of a closed research team, WordNet's evolution depends on this team and may be slow. However, due to the same reason it is of high quality and contains limited errors.

One of the first works exploiting WordNet to resolve tag ambiguity is [7]. They map the user queries to relevant WordNet senses, rely on the user to confirm the intended sense of the query and retrieve the resources tagged with the synonyms of this sense. [6] use WordNet to apply structure to clusters of related tags and facilitate hierarchical browsing. They create a concept tree for each cluster by mapping tags to senses, extracting the WordNet paths of the senses and finally merging them into one tree.

An alternative paradigm utilises **(online) ontologies** to improve folksonomies. Different types of ontologies exist, varying in scope and context as they are built to serve the purposes of specific tasks, applications or describe certain domains. Because of that, they cover different knowledge domains in variable levels of detail. Ontologies are more recent than WordNet and contain new terms due to the fact that they are created and updated by many knowledge experts. As a result of this, the knowledge defined in ontologies is diverse and at times redundant. Furthermore, due to the lack of quality control, ontologies suffer of modelling errors (e.g. *China subClassOf Asia*) as described in our previous work [9].

[8] use ontologies from an ontology repository to perform query expansion and limit the effect of tag ambiguity on search. One effort combining WordNet and ontologies was presented in [2]. WordNet was used to disambiguate the meaning of tags and to expand them with additional lexical information. The expanded tags were enriched with entities from online ontologies. However, we observed that this sequential combination was suboptimal, as the WordNet based step ruled out a high number of tags that existed in online ontologies.

This observation and the lack of comparative studies on WordNet and ontologies motivated our interest to explore these two KS in terms of folksonomy search. Our goal in this paper is twofold. **First, we investigate how the usage of KS can address polysemy, synonymy and basic level variation and minimise their effects on folksonomy search. Second we compare the KS-based search with the cluster-based search in folksonomies.** We present a method that creates a structure of senses (similar to the one presented in Fig.1b) for each KS separately (Sec.2.1). We implement a query expansion mechanism (Sec.2.2) for these structures and using a web interface (Sec.3) we compare the KS and the cluster-based search (Sec.4) within a user study.

2   Building and Querying Knowledge Structures

The goal of this work is to compare the characteristics of WordNet and ontologies that contribute to the creation of semantic structures (e.g., Fig.1b) and to the resolution of the search impediments caused by polysemy, synonymy and basic level variation. Although, context investigation and disambiguation mechanisms are required in order to define the precise meaning and relations of tags, the analysis of disambiguation techniques is out of the scope of this paper. In this work we focus on studying which characteristics of the KS can deal with the above phenomena. We start by defining the basic elements of our approach.

Definition 1. A folksonomy consists of a set of resources $R = \{r_1, ..r_{|R|}\}$ and a set of tags $T = \{t_1, ..t_{|T|}\}$. For a resource $r \in R$, $tags(r)$ represents its set of tags. For a tag $t \in T$, $res(t) = \bigcup_{i=0}^{|R|} r_i \ \forall r_i : t \in tags(r_i)$ is the set of resources tagged with tag t. For example, in Fig.1a, $tags(R1) = \{fruit, dessert\}$ and $res(apple) = \{R_2, R_3\}$.

Definition 2. A Knowledge Source KS consists of a set of senses $S = \{s_1, ..s_{|S|}\}$. Each sense s has a number of more specific, $sub(s)$, and more generic, $sup(s)$, senses. For example, in Fig.1b $sub(S_1) = \{S_2\}$. In addition, a sense s has a set of words that define its meaning, its set of synonyms, and it is denoted by $syn(s)$. For example, in Fig.1b $syn(S_3) = \{cake, dessert\}$.

Definition 3. We define the relation between a tag $t \in T$ and a sense $s \in S$ as $Dfn(t, s)$, $\forall t$, $\forall s : t \in syn(s)$. This means that t is potentially defined by s if it belongs to its set of synonyms. In addition, $senses(t) = \bigcup_{i=0}^{|S|} s_i : Dfn(t, s_i)$ is the set of senses assigned by the KS to the tags representing its possible meanings. For example, in Fig.1b $senses(apple) = \{S_2, S_4\}$.

As previously mentioned, in Fig.1a, searching for *cake* excludes R_1 due to **tag synonymy**. *Query Results* $= res(cake) = R2$. The existence of a semantic structure allows for the inclusion of results tagged with synonym tags to *cake*, i.e., *Query Results* $= \bigcup_{i=0}^{T} res(t_i) : t_i \in syn(s_j) \forall s_j : Dfn(cake, s_j) = \{R_1, R_2\}$. Intuitively, the richer a sense s in terms of synonyms, the higher the probability of retrieving resources tagged with a tag that is equivalent to a synonym. Therefore, to compare the two KS in terms of dealing with tag synonymy we measure the average number of synonyms per sense assigned to tags: (I) $\overline{|syn(S)|} = \frac{\sum_{i=0}^{|S|} |syn(s_i)|}{|S|}$. For example, in Fig.1b $\overline{|syn(S)|} = \frac{1+1+2+2+1}{5} = 1.4$.

In the same scenario, searching for *apple* retrieves both R_2 and R_3 due to **tag polysemy**. To allow for querying mechanisms to deal with polysemy the structure should have enough senses representing the meanings of polysemous tags. The more senses a KS provides for a tag the more likely to match all the possible senses of the tag. Therefore, we compare the two KS in terms of dealing with tag polysemy by measuring the average number of senses per tag: (II) $\overline{|senses(T)|} = \frac{\sum_{i=0}^{|T|} |sense(t_i)|}{|T|}$. E.g., in Fig.1 $\overline{|senses(T)|} = \frac{1+1+2+1+1}{5} = 1.2$

To deal with problems caused by lack of structure and **basic level variation** we investigate how tags can be mapped onto the hierarchical structure of senses in each KS. First, we measure the mean number of direct sub/super-senses. The higher the number of sub/super-senses the higher the probability to map more specific and more generic tags of a tag into the structure. The mean numbers of subsenses and supersenses are : (III) $\overline{|sub(s)|} = \frac{\sum_{i=0}^{|S|} |sub(s_i)|}{|S|}$ and (IV) $\overline{|sup(s)|} = \frac{\sum_{i=0}^{|S|} |sup(s_i)|}{|S|}$. In Fig.1b $\overline{|sup(S)|} = \frac{0+1+0+2+1}{5} = 0.8$.

Formulas (I to IV) calculate the richness of the KS, however, to decide which KS performs better during search one more measure needs to be defined. This represents how well the expansion of t using each KS is mapped to the tagspace and is reflected by the ratio of the resources retrieved using only t to the resources retrieved using the expansion of t. The expansion of t is defined as $exp(t) = \{syn(s) + syn(sub(s)) + syn(sup(s))\}, \forall s : Dfn(t, s)$. The increase ratio for tag t is defined as: $inc(t) = \frac{|res(exp(t)) - res(t)|}{|res(exp(t)) + res(t)|}$. E.g., $inc(cake) = \frac{|\{R1\} - \{R2\}|}{|\{R1\} + \{R2\}|} = \frac{1}{2} = 0.5$. We measure the mean increase for T which is: (V) $\overline{|inc(T)|} = \frac{\sum_{i=0}^{|T|} |inc(t_i)|}{|T|}$.

Finally, to obtain an estimation of how well a KS can solve the above phenomena, the set of tags covered by this KS, $T_{KS} \subseteq T$ should be measured. Intuitively, the higher the number of tags defined by the KS the better the KS performs in this task. (VI) $|T_{KS}| = |\bigcup_{i=0}^{|T|} t_i : \exists\ s \in\ S : Dfn(t_i, s)$.

To evaluate measures (I to VI), first, we use each KS to assign senses to the tags and to link the senses with their hierarchical information extracted from each KS. This **semantic enrichment** process yields one semantic structure per KS as described in Sec.2.1. Second, we implement a **query mechanism** for these semantic structures as described in Sec. 2.2.

2.1   Step 1: Semantic Enrichment

We use two different strategies to enrich tagspaces with semantic structure. Strategy A uses WordNet and Strategy B uses online ontologies. The intended output for both strategies is a structure similar to the one depicted in Fig.1b. This structure is built in two stages, common to both strategies.

First, the **potential meanings of a tag** are made explicit by aligning it to appropriate senses. While in previous work [2] we use disambiguation algorithms to precisely identify the meaning of a tag in a certain context, for the purposes of this comparative study we assign all possible senses to a tag. As mentioned above, we are interested in the richness and coverage of the KS over a tagspace and want to rule out any bias introduced by disambiguation methods. As a result, we assign $senses(apple) = \{S_2, S_5\}$ (Fig.1). Strategy A relies on WordNet's synsets to find such senses, while in the case of Strategy B we developed a clustering mechanism which identifies a possible set of senses for a tag by combining information from multiple online ontologies. To estimate $\overline{|syn(S)|}$ (I) for each KS, we assign as synonyms $(syn(s))$ to each sense s all the available lexical information from the respective KS that can possibly describe the sense (e.g., synonyms from WordNet, ids and labels from ontologies).

Second, we include **structural information among the senses** by reusing knowledge from the KS. First, we select all possible ancestors for each sense, i.e., S_4:*Apple* is defined as a S_5:*Company* and an S_6:*Organisation*. The reason for selecting all possible ancestors is the heterogeneity of the structures of the knowledge sources. In order to achieve high connectivity among the senses we import the subsumption path to the highest possible ancestor e.g., S_9:*iPod* is a subsense of S_8:*Gadget* which is a subsense of S_7:*Device*. We currently restrict the method to subsumption relations, as these are present in both KS.

Strategy A: WordNet-Based Enrichment. WordNet is a hierarchy of synsets each describing a sense. Most synsets are subsumed by at least one hypernym synset; they subsume a set of hyponym synsets and contain a set of words describing the same sense (synonyms).

For **sense selection**, we consider all the synsets that contain a given tag in their list of synonyms. Note that we consider only noun synsets as these have richer hierarchical information than other parts of speech. For each sense, we import in the structure the corresponding synonyms of the sense. WordNet's matching mechanism automatically caters for lexical variations and plurals. To create a **structure of senses**, we import each sense's ancestor path till the root of the WordNet hierarchy and their first level of hyponyms.

Strategy B: Online-Ontology Based Enrichment. In order to enrich the tagspace, we explore online ontologies through the Watson[1] Semantic Web gateway. The sense selection is more difficult in this case, because, unlike WordNet, the Semantic Web does not contain an established set of senses. To overcome this limitation, we build a clustering algorithm, which groups together entities that are sufficiently similar and therefore might denote the same sense.

First, all ontological concepts containing the tag in their localname (id) or label(s) are selected. We focus on concepts as they have a richer hierarchy than properties and individuals and also to maintain comparability with the structure generated from WordNet. We use Watson's API and we strictly match the tags against the id or label(s) of ontological concepts, e.g., *berry* is not matched against *berry_fruit* neither is *water* against *water_container*. This is done to restrict additional noise. By using multiple ontologies, the same concept may be defined more than once thus leading to different types of redundancies such as:

1. Redundancy of the same entity. Several ontologies declare the same URI.
2. One entity with the same id is declared in two different versions of the same ontology, e.g., *O1.daml:plant* and *O1.owl:plant*.
3. The same concept is declared in different ontologies in the same manner, namely it is subsumed by the same concept(s) and has the same ontological neighbourhood (relations, literals and so on) but different URI.
4. The same concept is defined in different ontologies by two different entities with different neighbourhood, e.g., $O1{:}Banana \overset{subClassOf}{\longrightarrow} \{O1{:}GroceryProduce,$ $O1{:}TropicalFruit\}$ and $O2{:}Banana \overset{subClassOf}{\longrightarrow} O2{:}Tree\text{-}Fruit$

The clustering algorithm minimises these redundancies by grouping sufficiently similar semantic descriptions of entities together and merging them into a new description, a cluster of entities, which we consider one sense. The algorithm is repeated until all obtained senses are sufficiently different from each other. To compute the similarity between two entities we compare their semantic neighbourhoods (superclasses, subclasses, disjoint and equivalent classes and named relations) as well as their lexical information (localnames, labels). The similarity $Sim(e_1, e_2)$ for two entities e_1 and e_2 is computed as:

---

[1] http://watson.kmi.open.ac.uk. The ontologies indexed in Watson during the experiment (May-June 2009) were approximately 9.000 and contained a total number of 460.000 classes (including redundancies).

$$Sim(e_1, e_2) = W_L \times Sim_L(e_1, e_2) + W_G \times Sim_G(e_1, e_2)$$

$Sim_L(e_1, e_2)$ is the similarity of the lexical information of the two entities computed using the Levenshtein metric. $Sim_G(e_1, e_2)$ is the similarity of the entities' neighbourhood graphs. For example, the superclasses of e_1 are compared against the superclasses of e_2. This is repeated for all the neighbour entities of e_1 and e_2. The similarity among the neighbour entities is computed based on string similarity too. For this experiment we set a low similarity threshold of 0.3 in order to achieve a maximum clustering result. In addition, we set $W_G = W_L = 0.5$ in order to cater for the heterogeneity of online ontologies in terms of the richness of their lexical and structural information. For example, for *banana* we obtained a single cluster, because, according to our clustering algorithm there is only one sense of banana in all online ontologies. This cluster of entities contributes to the sense of *banana* with synonyms derived from the localnames and labels $\{L_1:$ *"banana"*, $L_2:$*"an elongated yellowish fruit which grows on palm trees"*$\}$ and leads to the structural information $Banana \xrightarrow{subClassOf} \{Fruit, Tropical\ Fruit,$ *GroceryProduce, Tree Fruit.*

L_2 was the label of one of the clustered entities. Different ontologists have different representation styles and may include a comment as a label. In addition, unlike in the case of WordNet, mapping of inflections is not covered by the Watson API's search mechanism and therefore they will denote two different senses if they are not clustered by our algorithm. Issues such as lexical matching and entity redundancy need to be dealt with in Strategy B. All these are effects of the heterogeneity of ontologies.

Once the entity clustering is complete, for all the direct superclasses of the cluster's entities we iteratively get their superclasses till the root of an ontology. For example, we obtain *Tropical Fruit* $\xrightarrow{subClassOf}$ *Fruit*. We notice that by adding this knowledge there is then one direct and one indirect relation between *Fruit* and *Banana*. We maintain as many subsumption relations as possible regardless if they are implicitly redundant in order to support query expansion.

2.2   Step 2: Query Mechanism

The query mechanism allows the exploration of the structures created by using each KS. Algorithm 1 (Alg.1) describes a querying mechanism which a) maps query keywords to appropriate senses; b) retrieves the resources tagged with tags associated to these senses, and c) groups the returned resources into meaningful groups, which are used as a basis for the presentation of the results.

Our algorithm is based on the following intuition: users are primarily interested in resources tagged with the exact keyword, as well as with tags denoting more specific concepts. However, if only a few resources are returned for these cases, the user might also be interested in exploring resources tagged with more generic tags. For example, when searching for *fruit*, a user is likely to also be looking for resources annotated with the various types of *fruit*, such as *apple* or *tropical fruit*. Alternatively, if few results are returned, he may be interested in broader notions e.g., *plant*. Accordingly, for a query keyword k, Alg.1 retrieves all relevant senses and all their subsenses (Alg.1, 3-5). For all these senses, it retrieves the resources that are tagged with tags mapped to these senses (Alg.1, 6-11). The algorithm also categorises all the retrieved resources into groups according to the overlap of the resources' tags with the synonyms of the subsenses. For example, in a query for *animal*, items tagged with *animal* and *zebra* and items tagged with *zebra* are grouped together into a group described with *zebra* (Alg.1, 12-18). If the number of subsenses is less than four, then the same process is repeated with the supersenses (Alg.1, 20 -27). The threshold of four is selected because we further compare the KS-based querying with the cluster-based querying (Sec.4) and the mean number of clusters per tag is 3.4.

Algorithm 1. KS-based Querying

1: **for all** $k \in QueryKeywords$ **do**
2:     create group g_k ▷ This is to place all the uncategorised results
3:     retrieve $S : \forall s \in S, Dfn(k,s)$ ▷ S=senses(k)
4:     **for all** $s \in S$ **do**
5:         retrieve $sub(s)$
6:         **for all** $\acute{s} \in sub(s)$ **do**
7:             retrieve $\acute{R} = \bigcup_{\acute{t} \in syn(\acute{s})} res(\acute{t})$,
8:             create group $\acute{g}$
9:             $\forall \acute{r} \in \acute{R}$, put $\acute{r}$ in $\acute{g}$
10:        **end for**
11:        retrieve $R = \bigcup_{t \in syn(s)} res(t)$,
12:        **for all** $r \in R$ **do**
13:            $Overlap(r,\acute{s}) := |tags(r) \cap syn(\acute{s})| / |tags(r) \cup syn(\acute{s})|,\ \acute{s} \in sub(s)$
14:            put r in $\acute{g} : Overlap(r,\acute{s}) = \max Overlap(r,\acute{s}_j),\ \acute{s}_j \in sub(s)$
15:            **if** $Overlap(r,\acute{s}) = 0 \ \forall \ \acute{s} \in sub(s)$ **then**
16:                put r in g_k
17:            **end if**
18:        **end for**
19:    **end for**
20:    **if** $|\bigcup_{i=0}^{|S|} sub(s_i)| < 4$ **then**
21:        retrieve $sup(s)$
22:        **for all** $\grave{s} \in sup(s)$ **do**
23:            retrieve $\grave{R} = \bigcup_{\grave{t} \in syn(\grave{s})} res(\grave{t})$
24:            create group $\grave{g}$
25:            $\forall \grave{r} \in \grave{R}$, put $\grave{r} \in \grave{g}$
26:        **end for**
27:    **end if**
28: **end for**

---

3   System Implementation

We implemented Alg.1 into two web interfaces to perform KS-based search on the structures acquired from WordNet (**System 2, S2**) and ontologies (**System 3, S3**). We also developed a web interface to simulate the cluster-based presentation of results as provided currently by folksonomies, **System 1 (S1)**[2]. To do that, the clusters of the query keyword are retrieved using the folksonomy API. All the resources tagged with the keyword are divided into groups according to the number of tags they share with each cluster.

Fig. 2. Result screenshot for the query *sport* in system S1

All systems display the results grouped in meaningfully named groups. Fig.2 shows a screenshot with results from S1 for the query *sport*. For each group, there is a descriptive header which contains the title and the number of results per group. For S1 the title consists of the three most popular tags of the cluster (in accordance to the folksonomy clustering paradigm). For S2 and S3 the titles consist of the synonyms of the sense under which the results are clustered. For example, for this query, S2 returned groups described as *track and field*, *skiing* or *judo*, while S3 had groups named *golf*, *hunting* and *horseback riding*. The "see all" link allows the user to view all the results of the group when there are more than five results per group.

---
[2] http://www.flickr.com/photos/tags/TAG/clusters/

4   Experiments

As a basis for our experiments, we used the MIRFLICKR-25000 [5] dataset proposed for ImageCLEF 2009. This contains 25000 images from Flickr with 69099 distinct tags. Although this is a dataset proposed for image analysis and 9% of the images are not tagged, the rest are tagged with a number of tags ranging from one to 75, spanning various domains.

We conducted three experiments. First, we enriched the dataset with strategies A and B (Sec.2.1) and evaluated the enrichment in terms of quantitative (Sec.2: (I), (II), (III), (IV), (VI)) and qualitative measures (Sec.4.1). Second, we performed a user evaluation on searching with the three systems built in Sec.3 (Sec.4.2). Finally, we used the user queries to measure $|inc(T)|$ (V) (Sec.4.3).

4.1   Step 1. Enrichment Evaluation

The values obtained for metrics (I) to (IV) and (VI), defined for the evaluation of the enrichment, are shown in Tbl.1. In terms of the tagset coverage of the two knowledge sources, we observe that WordNet covers more tags than online ontologies (26.3% of all distinct tags, vs. 16%). We also note that a larger amount of tags were exclusively enriched by WordNet (12%) than ontologies (2.3%). Indeed, WordNet was better than ontologies in enriching concrete instances (*Hannover*) which was a limitation of Strategy B, as well as tags in a plural form. In a complementary fashion, ontologies outperformed WordNet by enriching non-noun tags (*alpine*), which was a limitation of Strategy A, or non-English tags (*collezione*).

One of the reasons for the low coverage by both sources is that, 71.4% of the tags were not mapped to any of the KS. This was due to phenomena such as compound tag concatenation (*rowingboats*), misspellings (*rasberry*), non English tags (*chaminá*), idiosyncratic tags (*:D*), tags that are not defined in either source (*augor*) and phrases (*daughters of the american revolution*).

Additionally, the major difference in coverage between WordNet and ontologies can be (at least partially) explained by the difficulty of matching tags to the concepts of these sources. Indeed, while the matching mechanisms of WordNet support the mapping of plurals and lexical variations to correct senses, matching is more difficult in ontologies. We found, for example, that ontologies use different **modelling styles** to express the names of entities, using one or more of the following mechanisms: the local name, rdf:label, rdf:comment or even locally specified properties (e.g., *O2::name =fiat*). In addition, the delimitation of compound labels (e.g., *O1#zantedeschia__genus_zantedeschia, O2#FloweringPlant*) is inconsistent across ontologies .

In terms of the richness of the created structures, WordNet provides, on average, more senses per tag (2.9) than ontologies (1.8). The amount of synonyms per senses is comparable in both sources, but important differences can be observed in the average number of more generic and more specific senses created in the two structures. Indeed, the structure created with WordNet, has a higher number of subsenses (2.7) on average than ontologies (1.5). Inversely, ontologies

Table 1. Quantitative results of the enrichment evaluation

	Measure	WordNet	Ontologies
(I)	$|syn(S)|$	2.3	2.2
(II)	$|senses(T)|$	2.9	1.8
(III)	$|sub(S)|$	2.7	1.5
(IV)	$|sup(S)|$	1.0	1.5
(V)	$|inc(T)|$	38%	39%
(VI)	$|T_{KS}|$	26.3%	16%

lead to more supersenses (1.5) than WordNet (1.0). One explanation for this is that online ontologies often express different points of views, or cover more domains than WordNet does and therefore lead to more supersenses. For example, mosquito has three more generic senses in ontologies and only one in WordNet:

ontologies: $Mosquito \overset{subSense}{\longrightarrow} \{BloodFeedingArthropod,\ Insect,\ Vermin\}$
WordNet: $Mosquito \overset{subSense}{\longrightarrow} Dipterous\ Insect \overset{subSense}{\longrightarrow} Insect$

We also observe **variable hierarchical granularity** between ontologies and WordNet. Indeed, as shown for *mosquito*, the WordNet definitions of terms tend to be more fine-grained than in ontologies. Additionally, differences in the granularity of the definitions can also be observed within WordNet itself. For example:

WordNet: $Orange \overset{subSense}{\longrightarrow} Citrus \overset{subSense}{\longrightarrow} Edible\ Fruit$
WordNet: $Apple \overset{subSense}{\longrightarrow} Edible\ Fruit$

4.2   Step 2. User-Based Search Evaluation

In the second experiment we performed a user study. The user group consisted of 25 expert and non expert users with basic knowledge of image search on the web. Their task was to post at least three single keyword queries to systems S1, S2 and S3 without domain or any other restrictions. We limited the search to single keywords because we were interested in comparing the richness of the structures created in Step 1 per keyword. In addition we maintained the same terms to compare with S1, which simulates the cluster-based search which is only available for single keywords. We obtained 88 distinct queries and the evaluators had to report on their comparative experience on using S1, S2 and S3. More

Table 2. User Questions and Responses

Question	S1	S2	S3
Q1: Did you find what you were looking for?	90%	85%	84%
Q2: How helpful was the presentation of the results and why?	2,9	2,8	2,8
Q3: Rate the number of correct versus incorrect results	3,3	2,8	3,1
Q4: Which is the best performing system?	35%	32%	33%

specifically, they had to report on the questions of Tbl. 2. In Q2 and Q3 they had to select from a scale of 1 to 4; 1 being very unhelpful/all incorrect and 4 very helpful/all correct. They also had to report which results were most (ir)relevant/(in)correct and why for each query and system.

Overall, S1 performs better than S3 which performs better than S2 as seen in Tbl.2. Considering that in S2 and S3 none of the results of S1 are excluded (Alg.1), a possible explanation for this result is the reported decrease in precision (Tbl.2, Q3). The users stated that **S1 performed better because there were less groups** and it was easier to navigate through the results. It should be stressed that all the results returned from S1 were tagged with the query keyword (see Sec.3). S2 and S3 included results tagged with tags related to the keyword, thus increasing the number of groups. In some cases the users reported that the results of some of these additional groups were irrelevant.

The effect of irrelevant results was maximised by an additional factor. In some cases the users reported that the **photos were tagged incorrectly**. This can be further justified from the result of Q3:S1 = 3.3/4 (Tbl.2). An example of this is the query *tiger*. Among the groups containing photos of tigers, S1 returned a group headed with {*butterfly, shallowtail*} containing one image of a tiger butterfly. This was reported as incorrect because the user was unaware of this sense of the word *tiger* and no further explanation was given from the system. This is a common phenomenon arising from categorising photos based on clusters of tags derived from co-occurrence. The **relations among the tags are not clear** (e.g., *tiger is a type of butterfly*) and it is not possible to give a justification for the retrieval of results and their categorisation in a particular group. This, however, would be possible if the knowledge *Tiger Butterfly* $\overset{subSense}{\longrightarrow}$ *Butterfly* was provided by a KS.

In some cases the users reported that the presentation of S2 and S3 was more helpful even when the results returned by S1 were almost the same. According to them, the **images were presented under a meaningful category**. For example, for the query *horse*, S2 and S3 returned different groups for *colt, palomino* as opposed to the groups returned by S1 *italy, cavallo, england*. They found this distinction of results helpful for understanding the kind of horse depicted.

In cases that the query keyword did not return meaningful results in S1, the users reported that S2 and S3 returned **more and a higher variety of results**. For example, querying for *soap*, most S1's results depicted bubbles but S3 returned results depicting shampoo because *Shampoo* $\overset{subSense}{\longrightarrow}$ *Soap* was found in online ontologies. Equally, for *doggy* S1 retrieved only two images while S2 retrieved all images tagged with *dog* because *doggy* is one of the synonym terms for the sense of *dog* in WordNet. Finally the users were asked to select the system that performed better in all their queries (Q4). 35% of the users selected S1, 33% S3 and 32% selected S2. The responses to the rest of the questions of Tbl.2 justify this too. S1 performed better due to less groups of results. S2 and S3 returned better group descriptions and in addition S3 groups were judged to be more relevant.

4.3   Step 3. Quantitative Search Evaluation

Taking into account the user's queries and comments, we measured the approximate average $inc(T)$ for WordNet and ontologies. In Fig.3 WordNet's increase is represented with dark lines and ontologies' increase with light lines. $inc(T)$ is the ratio of additional correct results returned by the expansion of the query keyword divided by the total results returned as described in Sec.2 (V). Fig.3 demonstrates the $inc(T)$ for the user queries for which S2 and S3 returned additional correct results. The rest of the queries were either not mapped to any KS or their related senses did not map to the tagset.

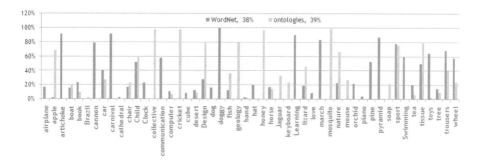

Fig. 3. Increase in results on the user entered keywords

$inc(T)$ is affected by two factors as shown in Sec.2 (V). The first, depends on the tagset and is not relevant to the KS used. This is **how popular is the search keyword k_i in the tagset**, i.e., how many resources are tagged with it (e.g., *soap, doggy*). The second factor, is the number of **correct additional resources** retrieved when expanding k_i with synonyms, subsenses and super-senses. This depends on the number of synonyms, sub/super-senses and how well these are covered in the tagset. A representative case of related senses that were not well covered is the one of *nature*. S2 returned four groups, one for each of the subsenses of *nature* in WordNet, i.e., {*animality, complexion, disposition, sociality*}. The users reported that the results of S2 were not significantly more than S1, they were generic and quite irrelevant to nature. On the other hand, S3 returned five groups for {*sky, fire, mountain, reef, rice*} which are ontological subclasses of *nature*. All the above were well covered by the tagset and, with the exception of *rice*, the users reported that results and their grouping were meaningful and satisfactory. As a result the increase of *nature* from S3 was significant. The phenomenon of irrelevant groups was more frequent in S2 than S3, justifying the lower increase for S2 (Tbl.1) and the lower user satisfaction this particular system (Tbl.2).

In cases of querying for *apple*, system S3 returned one group representing the sense of fruit. S2 returned two groups for apple derived from the two senses of apple in WordNet, i.e., *fruit* and *fruit tree*. However, none of the two senses was relevant to the the sense of *computer company*. This shows that the number of senses is not enough information to decide if a KS can deal with polysemy. To decide this information on the coverage of the senses from the tagspace is required.

Another useful outcome emerged with querying for *may*. While S2 and S3 did not present the results in any meaningful manner nor did they return any additional results, S1 performed quite satisfactorily. Four clusters were returned grouping together images tagged with {*england, london*}, {*spring, flowers*}, {*sky, cloud*} and {*paris, france*}. A plethora of photos shot in and tagged with *may*, can depict flowers, sky, cities and so on but may depict nothing that can symbolise the month May. This is a type of idiosyncratic tagging and no KS can supply formal relations between *may* and these tags since there are no formal relations among them. Nevertheless, for this type of **idiosyncratic tagging the clustering of results based on frequent tag co-occurrence is quite efficient**.

5   Conclusions and Future Work

In this paper we explore how formal knowledge sources, WordNet and online ontologies, can improve folksonomy search and which of them performs better. We evaluate them qualitatively and quantitatively in terms of tagspace enrichment and user satisfaction comparing the KS-based search to cluster-based search.

In terms of tagspace enrichment, WordNet outperformed ontologies in most of the measures. It provided more senses per tag and more synonyms per sense than ontologies which is an indication of better addressing **tag polysemy** and **tag synonymy** respectively. Additionally, WordNet covered a higher percentage of tags than ontologies. WordNet and ontologies returned comparable measures for subsenses and supersenses which served as an indication for deciding which KS can project the **basic level variation** of the tagspace to a semantic structure.

However, in the user evaluation the ontologically created structure performed better in search than the WordNet created structure. The expansion of the query keyword with terms from ontologies returned a higher number of results compared to the expansion provided from WordNet despite the fact that WordNet provided a richer structure. This indicates that the tagset coverage of the created semantic structure is more important in the context of search than the richness of the structure.

In a nutshell, comparing the KS-based search to folksonomy search, indicated that **users prefer the number of groups to be short and concise** similar to cluster-based search but the **explanation of the results to be more intuitive** according to KS-based search. In addition, search problems caused by idiosyncratic tagging can be addressed better by statistical methods rather than formal knowledge sources.

In future work we aim to combine information from folksonomies, ontologies and WordNet to achieve better sense discovery for tags. In particular, we aim to use appropriate disambiguation techniques in order to assign the tags to the most relevant senses using both knowledge sources. In addition, we plan to extend strategies A and B in a way that they exploit more entities and relations from each KS. To reduce the number of irrelevant results, we aim to allow for multiple keyword queries and ranking of results based on the popularity of tags and KS-retrieved senses.

Acknowledgements

This work was funded by the NeOn project sponsored under EC grant number IST-FF6-027595. The authors would like to thank Mr. Evan Karapanos for his invaluable help and the evaluators for dedicating their valuable time.

References

1. Abbasi, R., Staab, S.: Richvsm: enriched vector space models for folksonomies. In: Proc. of the 20th ACM conf. on Hypertext and hypermedia (2009)
2. Angeletou, S., Sabou, M., Motta, E.: Semantically enriching folksonomies with FLOR. In: Bechhofer, S., Hauswirth, M., Hoffmann, J., Koubarakis, M. (eds.) ESWC 2008. LNCS, vol. 5021. Springer, Heidelberg (2008)
3. Golder, S., Huberman, B.: Usage patterns of collaborative tagging systems. Journal of Information Science 32, 198–208 (2006)
4. Hotho, A., Jarschke, R., Schmitz, C., Stumme, G.: Information retrieval in folksonomies: Search and ranking. In: Sure, Y., Domingue, J. (eds.) ESWC 2006. LNCS, vol. 4011, pp. 411–426. Springer, Heidelberg (2006)
5. Huiskes, M., Lew, M.: The mir flickr retrieval evaluation. In: Proc. of the ACM Int. Conf. on MIR (2008)
6. Laniado, D., Eynard, D., Colombetti, M.: Using WordNet to turn a folksonomy into a hierarchy of concepts. In: Proc.of 4th SWAP (2007)
7. Lee, S., Yong, H.: Tagplus: A retrieval system using synonym tag in folksonomy. In: Int. Conf. on Multimedia and Ubiquitous Engineering (2007)
8. Pan, J., Taylor, S., Thomas, E.: Reducing ambiguity in tagging systems with folksonomy search expansion. In: Aroyo, L., Traverso, P., Ciravegna, F., Cimiano, P., Heath, T., Hyvönen, E., Mizoguchi, R., Oren, E., Sabou, M., Simperl, E. (eds.) ESWC 2009. LNCS, vol. 5554, pp. 669–683. Springer, Heidelberg (2009)
9. Sabou, M., Gracia, J., Angeletou, S., d'Aquin, M., Motta, E.: Evaluating the semantic web: A task-based approach. In: Aberer, K., Choi, K.-S., Noy, N., Allemang, D., Lee, K.-I., Nixon, L.J.B., Golbeck, J., Mika, P., Maynard, D., Mizoguchi, R., Schreiber, G., Cudré-Mauroux, P. (eds.) ASWC 2007 and ISWC 2007. LNCS, vol. 4825, pp. 423–437. Springer, Heidelberg (2007)
10. van Zwol, R., Murdock, V., Garcia Pueyo, L., Ramirez, G.: Diversifying image search with user generated content. In: Proc. of the ACM Int. Conf. on MIR (2008)

Querying the Web of Data: A Formal Approach*

Paolo Bouquet[1], Chiara Ghidini[2], and Luciano Serafini[2]

[1] University of Trento, Italy
bouquet@disi.unitn.it
[2] Fondazione Bruno Kessler, Trento, Italy
{serafini,ghidini}@fbk.eu

Abstract. The increasing amount of interlinked RDF data has finally made available the necessary building blocks for the web of data. This in turns makes it possible (and interesting) to query such a collection of graphs as an open and decentralized knowledge base. However, despite the fact that there are already implementations of query answering algorithms for the web of data, there is no formal characterization of what a satisfactory answer is expected to be. In this paper, we propose a preliminary model for such an open collection of graphs which goes beyond the standard single-graph RDF semantics, describes three different ways in which a query can be answered, and characterizes them semantically in terms of three incremental restrictions on the relation between the domain of interpretation of each single component graph.

1 Introduction

One of the most important recent trends in the Semantic Web community is to publish on the Web large collections of interlinked semantic data. The expected result is what is generally referred to as the *web of data*. This is how this vision is very clearly expressed in the home page of the W3C Semantic Web Activity:

> *The Semantic Web is a web of data. There is lots of data we all use every day, and it is not part of the web. I can see my bank statements on the web, and my photographs, and I can see my appointments in a calendar. But can I see my photos in a calendar to see what I was doing when I took them? Can I see bank statement lines in a calendar?*
>
> *Why not? Because we don't have a web of data. Because data is controlled by applications, and each application keeps it to itself.*
>
> *The Semantic Web is about two things. It is about common formats for integration and combination of data drawn from diverse sources, where on the original Web mainly concentrated on the interchange of documents. It is also about language for recording how the data relates to real world objects. That allows a person, or a machine, to start off in one database, and then move through an unending set of databases which are connected not by wires but by being about the same thing.* (http://www.w3.org/2001/sw/)

---

* This work is partially supported by the by the FP7 EU Large-scale Integrating Project OKKAM – Enabling a Web of Entities http://www.okkam.org/ (contract no. 215032).

A. Gómez-Pérez, Y. Yu, and Y. Ding (Eds.): ASWC 2009, LNCS 5926, pp. 291–305, 2009.

The idea is that the web of data is a giant source of (semi-) structured information, which can be used to answer queries that require the integration of bits and pieces coming from different sources. However, this vision is far from becoming true. If, on the one hand, the relevant W3C recommendations (mainly RDF and OWL) have provided the necessary common format for data representation, the way data are published does not always allow for a simple integration, and this for several reasons, including the fact that it is not completely clear what issuing a query to the open web of data really means in practice[1].

In this paper, we propose a preliminary model of the web of data as a graph of graphs, which is based on the framework of *Distributed First Order Logic* or DFOL [5]. The main contribution of the model is to offer a clean semantics of interlinked data on the web. We use this model to characterize three possible modes of querying the web of data, and to formally describe the corresponding (expected) answer; technically, each mode corresponds to a different restriction on the general model proposed.

2 Preliminary Definitions on the Web of Data

In this section we recall the basic definitions of RDF syntax, semantics from [6] and the definition of conjunctive query and query answer over an RDF graph. Then we formally introduce the syntactic notion of graph space as a set of RDF graphs.

2.1 Preliminaries on RDF

RDF distinguishes three sets of syntactic entities. Let Σ denote a set of URIs, B a set of blank nodes, where the single nodes are denoted with x, y, z, and L a set of literals.

Definition 1 (RDF graph). *An RDF graph g is defined to be a subset of the set*

$$\Sigma \cup B \times \Sigma \times \Sigma \cup B \cup L.$$

URI references contained in g are denoted with $i : x$, where i is a URI, called the *prefix*, and is used to identify a dataset; and x is the *local reference* of $i : x$ within the dataset i. In this paper, without any loss of generality, we only consider RDF graphs g which do not contain literals. This is because literals have a standard local interpretation and do not allow to connect different graphs, which is the characteristic we want to concentrate upon. From now on, we use $\Sigma(g)$ and $B(b)$ to denote the set of URIs and blank nodes that occur in g respectively.

A *merge* of a set of RDF graphs $g_1, \ldots g_n$, denoted by $\mathsf{merge}_{i \in \{1, \ldots, n\}}(g_i)$ (we will use the abbreviation $\mathsf{merge}(g_1, \ldots g_n)$), is defined is the the union of the set of triples contained in the graphs $g'_1, \ldots, g'_n$, where each g'_i is obtained by renaming the blank nodes of g_i such that $g'_1, \ldots, g'_n$ don't share any blank node[2].

---

[1] In this paper we focus on the formal aspects of the problem, and we disregard other essential issues like, for example, provenance and trust. Of course a full solution should cover these aspects as well.

[2] For the definition of merging RDF graphs, refer to http://www.w3.org/TR/rdf-mt/

Definition 2 (Interpretation of an RDF graph). *An interpretation of an RDF graph g is a triple* $(\Delta, I, \mathcal{E})$*, where* Δ *is a non empty set,* $I : \Sigma(g) \rightarrow \Delta$*, and* $\mathcal{E} : \Delta \rightarrow 2^{\Delta \times \Delta}$*.*

Given an interpretation, we need to define when an interpretation *satisfies* a statement (a triple). Let a *model* of an RDF graph g be an interpretation that satisfies all statements in g. In the following, we use the symbol "$\equiv$" to denote the URI `owl:SameAs` and the notation a^f to denote the application of any function f to a. With an abuse of notation, we use $(i : x)^{\mathcal{E}}$ to denote $\left((i : x)^I \right)^{\mathcal{E}}$. Since the interpretation of a graph g does not provide any meaning for blank nodes, to define satisfiability we need to provide the interpretation of blank nodes.

Definition 3 (Assignment to blank nodes). *Given an interpretation* $m = \langle \Delta, I, \mathcal{E} \rangle$ *of* g*, an assignment to the blank nodes of g is a function* $\mathbf{a} : B(g) \rightarrow \Delta$*. For each* $n \in \Sigma(g) \cup B(g)$*, we define* $(n)^I_{\mathbf{a}}$ *as follows:*

$$(n)^I_{\mathbf{a}} = \begin{cases} n^I & \text{if } n \text{ is a URI} \\ \mathbf{a}(n) & \text{if } n \text{ is a blank node} \end{cases}$$

Definition 4 (Satisfiability). *Let* m *be an interpretation of a graph* g*,* $\mathbf{a}$ *an assignment to* $B(g)$ *and* $(a.b.c)$ *a triple on the signature* $\Sigma(g) \cup B(g)$*.* m *satisfies* $(a.c.b)$ *under the assignment* $\mathbf{a}$*, in symbols,* $m \models (a.b.c)[\mathbf{a}]$ *if*

$$\left((a)^I_{\mathbf{a}}, (c)^I_{\mathbf{a}} \right) \in (b)^{\mathcal{E}}$$

Given a set of triples $\Gamma = \{\gamma_1, \ldots, \gamma_n\}$*,* $m \models \gamma_1 \wedge \cdots \wedge \gamma_n[\mathbf{a}]$*, if* $m \models \gamma_k[\mathbf{a}]$ *for* $1 \leq k \leq n$*.*

Definition 5 (Model of an RDF graph). *An interpretation* $m = (\Delta, I, \mathcal{E})$ *of* g *is a model of* g*, in symbols* $m \models g$*, if there is an assignment* $\mathbf{a}$ *such that*

1. *for any* $(a.b.c) \in g$*,* $m \models (a.b.c)[\mathbf{a}]$
2. $(\equiv)^{\mathcal{E}}$ *is the identity relation (formally,* $(\equiv)^{\mathcal{E}} = \text{id}(\Delta) = \{(d,d) \mid d \in \Delta\}$*).*

Definition 6 (Logical consequence in an RDF graph). *A triple* $(a.b.c)$ *in* $\Sigma(g)$ *is a logical consequence of* g*, in symbols* $g \models (a.b.c)$ *if, for any interpretation* m*, if* $m \models g$ *then there is an assignment* $\mathbf{a}$ *such that* $m \models (a.b.c)[\mathbf{a}]$*. A graph* g' *is a logical consequence of a graph* g*, in symbols* $g \models g'$ *if for any interpretation* m*, if* $m \models g$ *then there is an assignment* $\mathbf{a}$ *such that* $m \models (a.b.c)[\mathbf{a}]$ *for all* $(a.b.c) \in g'$*.*

Notice that it is possible that $g \models (a.b.c)$ for all $(a.b.c) \in g'$ but $g \not\models g'$. In fact, $g \models g'$ is true only if all the triple of g' is satisfied by the models of g w.r.t., a unique assignment; while $g \models (a.b.c)$ and $g \models (a'.b'.c')$ can be true w.r.t. different assignments. To empha-sise this fact, we use the notation $\bigwedge_{k=1}^{n} (a_k.b_k.c_k)$ to denote the RDF graph composed of the n triples $(a_1.b_1.c_1), \ldots, (a_n.b_n.c_n)$.

Query languages, such as SPARQL, are used to access knowledge contained in an RDF graph. In this paper we consider the simplest RDF query language constituted by the class of *conjunctive queries*[3]. Notationally we use $\mathbf{x}$ for a n-tuple $(x_1, \ldots, x_n)$ of variables (or blank nodes). Similarly, $\mathbf{c}$ is used to denote an n-tuple $(c_1, \ldots c_n)$ of URIs.

---

[3] We are aware that this is a very strong simplification, but it does not affect the general model we propose in this paper. The extension to more expressive queries will be part of our future work.

Definition 7 (Conjunctive Query). *A conjunctive query, or simply a query, $q(\mathbf{x})$ on an RDF graph g is an expression of the form*

$$q(\mathbf{x}) = \{\mathbf{x} \mid \bigwedge_{i=1}^{k} (a_i.r_i.b_i)\}$$

where $\mathbf{x}$ is a subset of the blank nodes occurring in $\bigwedge_{i=1}^{k}(a_i.r_i.b_i)$, and $(a_i.r_i.b_i)$ is a triple in $\Sigma(g) \cup B(g)$.

As an example, the conjunctive query

$$(\langle n,c \rangle \mid (n, \text{ex:attended}, c) \wedge (c, \text{rdf:type}, \text{dbpedia:conference}) \wedge \\ (c, \text{ex:held\_in}, p) \wedge (p, \text{ex:located}, \text{gr:greece})) \tag{1}$$

corresponds to the following SPARQL query

```
SELECT ?name ?conf
WHERE
  {?name ex:attended  ?conf
   ?conf  rdf:type  dbpedia:conference
   ?conf  ex:held_in  ?place
   ?place  ex:located  gr:greece}
```

The result of a query $q(\mathbf{x})$ on a graph g is an n-ary relation (or equivalently a set of n-tuples of URI's) containing the row $(c_1,\ldots,c_n)$ if the RDF graph (i.e. set of triples) obtained by replacing x_i with c_i in all $(a_i.r_i.b_i)$ with $1 \leq i \leq k$ is entailed by g. We recall the notion of entailment among RDF graphs and RDF triples, as introduced in W3C Recommendation on RDF semantics[4].

Let $\mathbf{c}$ be a set of URIs in $\Sigma(g)$. We use $q(\mathbf{c})$ to denote the conjunction of tuples obtained by uniformly replacing x_1 with $c_1, \ldots, x_n$ with c_n, in $\bigwedge_{i=1}^{k}(a_i.r_i.b_i)$.

Definition 8 (Query answer). *The answer of $q(\mathbf{x})$ in an RDF graph g is defined as*

$$ans(q(\mathbf{x}),g) = \{\mathbf{c} \in \Sigma(g)^n \mid g \models q(\mathbf{c})\}$$

2.2   The Graph Space

The web of data can be thought of as a multiplicity of graphs, each of which may contain references to nodes in different graphs. To capture this complex distributed structure of interlinked (addressble) graphs, the syntax of single RDF graph is not enough. Therefore we introduce the notion of *graph space*.

Definition 9 (Graph space). *Given a set of URIs I, a* graph space *on I is a family of RDF graphs $\mathcal{G} = \{g_i\}_{i \in I}$.*

---

[4] http://www.w3.org/TR/rdf-mt/

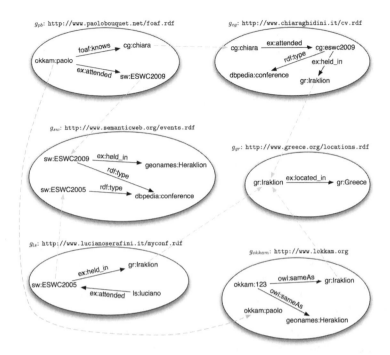

Fig. 1. A simple example

A graph space represents a specific state of the web of data, where I is the set of URIs that can be dereferenced into an RDF graph. The signature of a graph space $\mathcal{G}$, denoted by $\Sigma(\mathcal{G})$, is the union $\bigcup_{i \in I} \Sigma(g_i)$ of the signatures of the graphs in $\mathcal{G}$.

An example of a graph space composed of six graphs is depicted in Figure 1[5]. In this example the graph space $\mathcal{G}_0$ contains the six graphs g_{pb}, g_{cg}, g_{ls}, g_{sw}, g_{okkam}, and g_{gr}, where: g_{pb}, g_{cg} and g_{ls} identify the graphs of Paolo Bouquet, Chiara Ghidini, and Luciano Serafini, respectively, g_{sw} is the graph of semantic web, g_{okkam} identifies a graph containing OKKAM IDs[6] and a few identity statements, and g_{gr} is a graph containing data about Greece. Thus, g_{pb} intuitively states that Paolo Bouquet has attended ESWC09 and that he knows Chiara Ghidini. Similarly for the other graphs. Even if the graphs are distinct, elements of the RDF triples often refer to foreign URIs as illustrated by the dashed lines.

At a first glance, a graph space looks very similar to a set of named graphs. However, there is however a very important difference. In a graph space, the URIs associated to

---

[5] For the sake of readability, in what follows we will use name spaces instead of full prefixes to refer to RDF URIs. So, for example, `pb:me` is an abbreviation for a longer URI, e.g. `http://www.paolobouquet.net/me`. When it is not necessary, we will not even define the full name space, as the examples should be clear anyway. In square brackets we put the abbreviation of the prefix for the corresponding graph.

[6] See `http://www.okkam.org/`

each graph are not interpreted in the graph they contain, as it happens in the standard interpretation of named graphs described in [3]. Instead they are used only to access resources. We further elaborate on the the difference between named graphs and a graph space in the related work section.

As a final remark, we observe that a graph space is an ideal structure, which describes knowledge available on the web of data at a given time. In practice, it is virtually impossible to compute such an ideal structure, to store it in a centralized repository, and to make it available for access, even though semantic web crawlers[7] are able to compute an approximation of it. As a consequence, access to the web of data cannot be modeled through global queries on the web of data as a single global graph space, but only on portions of it.

3   Querying the Web of Data

Imagine we want to query the web of data to know who attended which conference in Greece. The request can be expressed as in the query $q_0(\mathbf{x})$ shown in (1). Suppose that the current status of the web of data is the graph space G_0 depicted in Figure 1. Unlike the typical controlled situations, understanding how $q_0(\mathbf{x})$ can be answered in a web of data requires to address at least the two folowing problems:

1. since the relevant datasets for answering the query may not be explicitly listed, we need a method for collecting the relevant datasets by other means;
2. since each RDF graph has been produced independently, it is not necessarily the case that RDF graphs can be easily integrated, as the same real world object (e.g. the conference or the location in our example) may be denoted by different (disconnected) URIs.

In what follows, we propose three very general methods that one can think of to answer the first question. We call them the *bounded*, the *navigational* and the *direct access* method respectively.

3.1   The Bounded Method

This is our baseline method, as it assumes that the graphs which will be used to process the query are explicitly listed in the query itself. This corresponds to what in the SPARQL specifications is defined as the RDF dataset for a query[8] and is passed to the query processor through the FROM keyword. Nothing else is taken into account outside the given dataset.

More formally, in the bounded method the query $q(\mathbf{x})$ is submitted to the graph g which is obtained by *merging* the graphs g_j with $j \in J$ explicitly listed in the query itself. The resulting dataset is not extended with additional information from graphs with indexes not in J.

---

[7] See e.g. Sindice at http://sindice.com, Falcons at http://iws.seu.edu.cn/services/falcons/ or Swoogle at http://swoogle.umbc.edu/

[8] See definition in http://www.w3.org/TR/rdf-sparql-query/#rdfDataset. Here we are not concerned with the distinction between RDF graphs and named graphs, so we will not make use of this distinction.

Definition 10 (Bounded answer). *The* bounded answer *of the query $q(\mathbf{x})$ submitted at J is*

$$b\_ans_J(q(\mathbf{x}), G) = ans(q(\mathbf{x}), merge_{j \in J}(g_j))$$

Going back to our example, if the query $q_0(\mathbf{x})$ is submitted with the bounded method to the set $J = \{pb, cg, sw, gr\}$ in G_0, then we are able to retrieve that *chiara* attended *eswc2009*, i.e.,

$$b\_ans_{\{pb,cg,sw,gr\}}(q_0(\mathbf{x}), G_0) = \{\langle \mathrm{cg:chiara, cg:eswc2009} \rangle\}$$

by querying the merged graph in Figure 2. The same answer can be obtained by restricting J to contain only $\{cg, gr\}$.

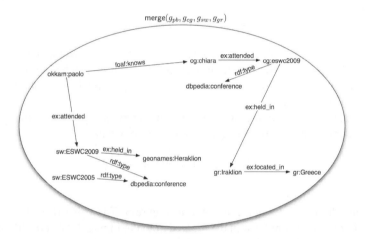

Fig. 2. Querying with the bounded method

3.2 The Navigational Method

In the navigational method, the query does not define the boundaries of the dataset to be used, but only the starting point (graph) from which other (possibly relevant) graphs can be reached. The key idea is that the other graphs are reached by following the links across RDF datasets in a way which is similar to what happens with the navigation on the web. As such, the navigational mode is our attempt of modeling the ideas behind the Linked Data approach[9].

More in detail, this method starts starts from a resource r in a graph g and navigates the web of data by following the links which are found between resources. When a fixpoint is reached, that is, when all the reachable graphs are collected, the query is evaluated against the resulting merged graph.

---

[9] See http://www.w3.org/DesignIssues/LinkedData.html for the description of the approach, and http://www4.wiwiss.fu-berlin.de/bizer/pub/LinkedDataTutorial/ for a tutorial on how to publish Linked Data on the Web).

To formally define how the query is answered in the navigational method we first need to introduce the notion of reachability between graphs. This in turn is based on the notion of foreign reference.

Definition 11 (Local and foreign URI reference). *The occurrence of $i : x$ in the graph g_j is a* local reference *if $i = j$, a* foreign reference *otherwise.*

Definition 12 (Reachable graph). *Given a graph space G, g_j is directly reachable from g_i, denoted by $i \to j$ if i contains a foreign reference to j. g_j is reachable from g_i, in symbols $i \xrightarrow{*} j$ if there is a sequence $i = h_1, h_2, h_3 \ldots, j = h_n$ such that $h_k \to h_{k+1}$ for $1 \le k \le n - 1$. For any $i \in I$, $i^* = \{j | i \xrightarrow{*} j\}$.*

Definition 13 (Navigational answer). *The* navigational answer *of the query $q(\mathbf{x})$ submitted at i is*

$$n\_ans_i(q(\mathbf{x}), G) = ans(q(\mathbf{x}), merge_{j \in i^*}(g_j))$$

Intuitively the definition of navigational answer says that to answer a query on a graph g_i, one needs first to collect all the information that can be reached by following the links originating from g_i (i.e. to compute $merge_{j \in i^*}(g_j)$), and then to submit the query on this extended dataset.

Going back to our example, if we submit the query $q_0(\mathbf{x})$ to the graph G_0 with a navigational method starting from g_{pb} we retrieve the two pairs

$$\langle \texttt{okkam:paolo, sw:ESWC2009} \rangle \tag{2}$$

$$\langle \texttt{cg:chiara, cg:eswc2009} \rangle \tag{3}$$

by querying the merged graph illustrated in Figure 3. The pair (3) is obtained using the same information contained in the merged graph in Figure 2. What is different here is the fact that also the graph g_{okkam} can be used to solve the query as it is reachable starting from g_{pb}. This is due to the fact that the link connecting $okkam : paolo$ from g_{bp} to g_{okkam} is followed to navigate G_0 and to compute the reachable graph from g_{pb}. This allows to obtain also the pair (2). Note that if we submit the same query $q_0(\mathbf{x})$ to the graph G_0 starting from g_{cg} we only obtain an answer containing only the pair (3). This because only the graph g_{gr} can be reached following the directional links from g_{cg}.

3.3 The Direct Access Method

In the thirs method, which we call direct access, the query is processed against the dataset which results from merging all the relevant graphs which can be found on the Web. Of course, there is the problem of defining what relevant means, but we can ignore this orthogonal problem here, as the essence of the method does not change. For example, in the case of our query $q_0(\mathbf{x})$ we can imagine that an oracle (e.g. a semantic search engine) returns all the graphs about Greece and about conferences (or their intersection). The important point with this method is that the relevant graphs are collected not through browsing, but through some kind of global index.

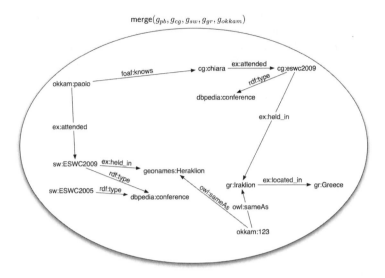

Fig. 3. Querying with the navigational method

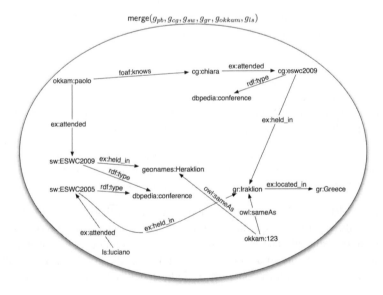

Fig. 4. Querying with the direct access method

Definition 14 (Direct Access answer). *The* direct access answer *at i of the query* $q(\mathbf{x})$ *over the graph space* $\mathcal{G}$ *is defined as follows*

$$d\_ans_i(q(\mathbf{x}), \mathcal{G}) = ans\left(q(\mathbf{x}), merge_{i \in I}(g_i)\right)$$

Intuitively the definition of direct access answer states that answering a query submitted to a graph g_i is the same as submitting the query to the entire (merged) graph space at once. Indeed from the definition 14 we immediately have that $d\_ans_i(q(\mathbf{x}), \mathcal{G}) = d\_ans_j(q(\mathbf{x}), \mathcal{G})$ for every $i, j \in I$; that is, the graph at which the query is submitted is irrelevant in the computation of the answer. Note also that answering a query in a direct access mode is equivalent to answering that in bounded mode over the entire graph space $\mathcal{G}$.

The complete merged graph for $\mathcal{G}_0$ is shown in Figure 4. This time we can obtain an answer that contains also the pair

$$\langle \mathtt{ls{:}luciano}, \mathtt{sw{:}ESWC2005} \rangle \tag{4}$$

in addition to the pairs described by Equation (2) and Equation (3).

In the rest of the paper we provide a precise formalisation of the notion of giant graph and of these three methods of querying it.

4 A Semantics for the Graph Space

To provide a formal semantics for the for the web of data viewed as a graph space, one could follow the idea of considering the graph space as the merge of all the available graphs it "contains", and then to interpret the resulting merged graph using the standard RDF semantics. Let us call this approach the *merged semantics* for the graph space. This semantics is quite simple, intuitive, and it captures the first intuitions behind the idea of the graph space used to model the giant graph of data. However, we follow a different approach. In fact, the merged semantics does not provide an adequate support for the model proposed by the linked data approach as it does not adequately support the notion of *reachability*. To understand the importance of reachability in the linked data let us recall a quote from [2] on how to publish linked data on the web:

> [. . .] after you have published your information as Linked Data, you should ensure that there are external RDF links pointing at URIs from your dataset, so that RDF browser and crawlers can find your data.

Clearly, if an RDF graph is well connected, or easily reachable, form the other RDF graphs then the information contained in it will be "exploited" much more than if the graph is almost isolated. The merged semantics approach does not take into account the degree of connectivity of a single RDF graph as all the graphs of the graph space are uniformly merged in the global graph, independently from the fact that they are reachable or not.

In this section we propose a more structured formal semantics for the graph space where the notion of data resource (identified by a URI) is explicitly modelled. It is based on the semantics of RDF as presented in [6], and it extends it in order to deal with multiple linked graphs. This extension is based on these two facts:

1. we move from a model of a single RDF graph, to a set of *local models* for a set of graphs.
2. we provide a semantics for links between resources belonging to different graphs.

In other words, we see the web of data as a graph space G composed of a family of graphs $g_1, \ldots, g_n$, and we provide the semantics of G in terms of suitable compositions of the semantics of the component graphs. To do this we exploit the framework of local models semantics and Distributed First Order Logic [5].

Definition 15 (Interpretation of a graph space). *An interpretation M for the graph space $G = \{g_i\}_{i \in I}$ is a pair $(\{m_i\}_{i \in I}, \{r_{ij}\}_{i,j \in I})$ where $m_i = (\Delta_i, I_i, \mathcal{E}_i.)$ is an interpretation of $\Sigma(G)$ associated to g_i, and r_{ij}, is a subset of $\Delta_i \times \Delta_j$. r_{ij} is called the domain relation from i to j.*

The interpretation of a graph space G associates to each component graph g_i an interpretation m_i which is defined over the entire set of URIs of G. This is justified by the fact that potentially any URI of the web of data can be reached from any graph. This semantic is consistent with the *open world assumption* usually done in the semantic web. The domain relation represents a form of inter-graph equality. Intuitively the fact that $(d, d') \in r_{ij}$ means that, from the point of view of g_j, d and d' represents the same real world object.

Definition 16 (Model for a graph space). *An interpretation M for G is a model for G, in symbols $M \models G$, if $m_i \models g_i$.*

Definition 17 (Global logical consequence). *Let g be a set of triples in $\Sigma(G)$. Then $G \models i : g$ if for all interpretations M of G, $M \models G$ implies that $m_i \models g$ with m_i the i-th model of M.*

In a graph space queries are submitted to a specific graph g_i and then propagated through semantic links to retrieve a global answer.

Definition 18 (Global answer). *The global answer $g\_ans_i(q(\mathbf{x}), G)$ of a query $q(\mathbf{x})$ submitted at i, is defined as follows:*

$$g\_ans_i(q(\mathbf{x}), G) = \{\mathbf{c} \in (\Sigma(G))^n \mid G \models i : q(\mathbf{c})\}$$

Definition 18 provides a logical definition of what the answer to a query is with respect to the most generic class of models for a graph space. We now move to the illustration and formalization of different ways of using the web of data for answering a query, each of which provides a different result set. Our objective is to characterize each modality of query answering in terms of a restricted class of models for a graph space. More precisely, for each method X which was introduced in Section 3, we will define a restricted class of models for G, called X-models, such that the query answer relative to X is equal to the logical consequence of G w.r.t., the restricted class of the X-models.

5   Formalizing the Query Methods

In this section we provide a formal semantics that completely characterizes the query results in the three different methods defined in section 3. This semantics captures the differences between the three methods by means of different restrictions imposed on the domain relation[10].

Intuitively the domain r_{ij} relation represents the translation of objects in the domain of j into objects of the domain of i: if $(d, d') \in r_{ji}$, it means that the object d' of the interpretation domain of i is the translation of the object d of the interpretation domain of j. Being able to translate object from j allows i to import information about this object which is stored in j. This is formalized by the following general requirement that we impose on models:

Definition 19 (Linked data model for a graph space). *A model M of a graph space $\mathcal{G}$ is a* linked data model *for $\mathcal{G}$ if and only if the following condition holds:*

$$\text{If } (d, d'), (e, e'), (f, f') \in r_{ij} \text{ and } (d, e) \in (f)^{\mathcal{E}_i}, \text{ then } (d', e') \in (f')^{\mathcal{E}_j} \tag{5}$$

Condition (5) in the definition above formalizes the fact that the properties stated in one graph propagate to other graphs through the domain relation. In other words, the domain relation is used to model a weak form of inter-graph identity. It is important to observe that at this stage no general identity condition is imposed on the model for the situation in which the same URI occurs in different graphs. This case will be taken into account in Section 5, where we investigate different ways of using the web of data for building the dataset for answering a query over a graph space.

5.1 The Bounded Method

In the bounded method, the query is submitted to the graph g obtained by *merging* the graphs g_j with $j \in J$ and this dataset is not extended with information from graphs with indexes not in J as formally defined in Definition 10.

From the semantic point of view, the bounded method can be modeled by isolating J's resources from the rest of resources (namely the $I \setminus J$-resources). This is done in the following definition.

Definition 20 (Bounded model). *M is a J-*bounded model *for any set $J \subseteq I$, if it is a linked data model for $\mathcal{G}$ and for all $i, j \in I$*

- *if $i, j \in J$ then $r_{ij}((k:x)^{I_i}) = (k:x)^{I_j}$*
- *if $j \in J$ and $i \notin J$, then $r_{ij} = r_{ji} = \emptyset$*

$g\_ans_J^B(q(\mathbf{x}), \mathcal{G})$ is defined as the global answer $g\_ans_i(q(\mathbf{x}), \mathcal{G})$ restricted to the J-bounded models, for some $i \in J$.

Theorem 1. $g\_ans_J^B(q(\mathbf{x}), \mathcal{G}) = b\_ans_J(q(\mathbf{x}), \mathcal{G})$

Theorem 1 formalises the intuition that the answer of a query submitted on $\mathcal{G}$ is only computed by using the local information available in the graphs in J. This is an immediate consequence of the fact that all the domain relations r_{ij}, r_{ji} between resources inside and outside J are empty, while domain relations between resources inside J are an isomorphism.

---

[10] The proofs of the theorems presented in this section can be found at http://dkm.fbk.eu/index.php/Image:IR-KR-2009-TechRep.zip

5.2   The Navigational Method

The navigational answer introduced in Definition 13 states that the answer of a query on a graph g_i, is computed first by collecting all the information that can be reached by following the links originating from g_i, and then to submit the query on this collected dataset.

Definition 21 (Navigational model). *M is a* navigational model *if it is a linked data model for G and for all $i, j \in I$ with $i \overset{*}{\rightarrow} j$, then*

$$r_{ji}((j : x)^{I_j}) = (j : x)^{I_i}$$

$g\_ans_i^N(q(\mathbf{x}), G)$ *is defined as the global answer $g\_ans_i(q(\mathbf{x}), G)$ restricted to the navigational models.*

Theorem 2. $g\_ans_i^N(q(\mathbf{x}), G) = n\_ans_i(q(\mathbf{x}), G)$

Note that the navigational method is intrinsically directional. In fact the domain relation is defined as an injective function. This means that asking the query at different entry points will potentially lead to different results. Indeed, the collection of graphs which are reachable from a graph g is in general different from the collection of graphs which can be reached from a graph g'. This would not be true only if all links were bi-directional, which in practice is not the case.

5.3   The Direct Access Method

The direct access answer introduced in Definition 14 is based on the idea that we answer $q(\mathbf{x})$ by collecting all the graphs in the graph space at once.

Definition 22 (Direct access model). *M is a* direct access model *if it is a linked data model for G and for all $j : x$ and for all $i \in I$, $r_{ji}((j : x)^{I_i}) = (j : x)^{I_j}$.*

$g\_ans_i^D(q(\mathbf{x}), G)$ is defined as the global answer $g\_ans_i(q(\mathbf{x}), G)$ restricted to the direct access models. The following theorem states this property.

Theorem 3. $g\_ans_i^D(q(\mathbf{x}), G) = d\_ans_i(q(\mathbf{x}), G)$

In this final case the domain relation is defined as an isomorphism for all the foreign references of the graph space G. This allows to answer the query by using the merge of the entire graph space.

6   Related Work

To the best of our knowledge, there are no proposals for a formal semantics for the web of data seen as an interlinked set of graphs. However there are some similarities between the semantics presented here and the semantics of *named graphs* presented in [3]. The basic difference is the fact that the semantics of named graphs is based on a single global interpretation for the merge of all the graphs, whereas this is not the case for our graph spaces.

A second approach comparable with the notion of graph space is the one of *RDF molecules* [4]. While the approach presented in this paper is compositional (that is, we start from local graphs and construct the semantics for the global graph), RDF molecules support a lossless decomposition of a (large) graph in a set of (smaller) graphs. This is done for efficiency reasons. However, the similarity between the two approaches resides in the fact that the result of the decomposition can be seen as a graph space which should be equivalent to the initial large graph. Focusing on the semantics we can see that also RDF molecules use an approach based on the merging semantics, while in our work we propose a range of semantics for composing graphs in three different ways.

The Semantic Web Client Library[11] supports the execution of SPARQL-queries over the giant graph. As claimed in the homepage of the library, to answer queries, the library *dynamically retrieves information* from the Semantic Web: *"(i) by dereferencing HTTP URIs [. . .] or (ii) by querying the Sindice search engine."* Modality (i) corresponds to the navigational method and to the corresponding navigational semantics, while modality (ii) corresponds to (an approximation of) what we call direct access. Furthermore, this library can be configured by setting the parameter maxdepth to a number N so that, when modality (i) is used to evaluate a query, the maximum depth of links dereferenced is N. It is easy to see that the case maxdepth=0 corresponds to our bounded method, while maxdepth=N > 0 corresponds to a modification of the navigational method that can be obtained by replacing $i \xrightarrow{*} j$ with $i \xrightarrow{\leq N} j$. The intuition behind this change is that $i \xrightarrow{\leq N} j$ holds if j can be reached from i in N jumps.

The semantics proposed in this paper has several aspects in common with the semantics of Package-based Description Logics (P-DL) [1], a formalism for distributed ontology integration. P-DL supports the partial reuse of ontologies by enabling an ontology to import some of the symbols defined in another ontology. The similarity stands in the fact that the operation of importing the symbol σ of the ontology O_i into the ontology O_j corresponds, in linked data, to the occurrence of the foreign reference $i : \sigma$ in G_j. Furthermore the semantics of P-DL and *semantic import* is also based on the notion of domain relation with restrictions similar to the one given in (5). The difference between P-DL is the fact that we work on RDF while P-DL is defined on OWL, and the fact that we provide a formalization of query answering, which is not available in P-DL.

7   Conclusions

In this paper we have presented a formal model of the web of data as a graph space, and a formalisation of three query methods over this space: the bounded method, the navigational method, and the direct access method. This semantics gives a precise and clear account of the distributed and interlinked nature of the web of data. This formal model is based on the framework of *Distributed First Order Logic* [5] and formalises

---

[11] http://www4.wiwiss.fu-berlin.de/bizer/ng4j/semwebclient/

the differnt methods for quey answering by means of appropriate restrictions over the domain relation.

An interesting issue, which is orthogonal to what we discussed in this paper, is the following: to get the best results from querying the web of data, should we try to maximize the interlinking across local URIs of different RDF graphs, or to maximize the reuse of the same URI for the same resource in any graphs in which the entity is named? On the positive side, the first practice is completely consistent with the idea of navigational queries, and the second would offer a simple contribution to the problem of finding the relevant graphs for direct access queries (e.g. searching all graphs containing the URIs in the query as a starting point). However, on the problematic side, the first heavily relies on the availability of identity statements (which seems a very optimistic assumption on a web scale), and the second heavily relies on services for finding available URIs. Our position on this is that the two practices not only can coexist, but can support each others. They need not be thought of as mutually exclusive, like publishing HREF links on the web of documents is not incompatible with using search engines to find pages instead of just browsing. They address different needs (navigation vs. integration), and rely on different tools, and only the practical experience will say where the balance is between them on the Semantic Web.

References

1. Bao, J., Voutsadakis, G., Slutzki, G., Honavar, V.: Package-based description logics. In: Stuck-enschmidt, H., Parent, C., Spaccapietra, S. (eds.) Modular Ontologies. LNCS, vol. 5445, pp. 349–371. Springer, Heidelberg (2009)
2. Bizer, C., Cyganiak, R., Heath, T.: How to publish linked data on the web,
   http://sites.wiwiss.fu-berlin.de/suhl/bizer/pub/LinkedDataTutorial/
   20070727/
3. Carroll, J.J., Bizer, C., Hayes, P., Stickler, P.: Named graphs, provenance and trust. In: WWW 2005: Proceedings of the 14th international conference on World Wide Web, pp. 613–622. ACM, New York (2005)
4. Ding, L., Finin, T., Peng, Y., Pinheiro da Silva, P., McGuinness, D.L.: Tracking RDF Graph Provenance using RDF Molecules. Technical report, UMBC (April 2005)
5. Ghidini, C., Serafini, L.: Distributed First Order Logics. In: Gabbay, D., de Rijke, M. (eds.) Frontiers Of Combining Systems 2 (Papers presented at FroCoS 1998). Studies in Logic and Computation, pp. 121–140. Research Studies Press/Wiley (1998)
6. ter Horst, H.J.: Completeness, decidability and complexity of entailment for rdf schema and a semantic extension involving the owl vocabulary. J. Web Sem. 3(2-3), 79–115 (2005)

A Relevance-Directed Algorithm for Finding Justifications of DL Entailments

Qiu Ji[1], Guilin Qi[2,1], and Peter Haase[1]

[1] AIFB Institute, University of Karlsruhe,
D-76128 Karlsruhe, Germany
[2] School of Computer Science and Engineering, Southeast University,
211189 Nanjing, China

Abstract. Finding the justifications of an entailment, i.e. minimal sets of axioms responsible for the entailment, is an important problem in ontology engineering and thus has become a key reasoning task for Description Logic-based ontologies. Although practical techniques to find all possible justifications exist, efficiency is still a problem. Furthermore, in the worst case the number of justifications for a subsumption entailment is exponential in the size of the ontology. Therefore, it is not always desirable to compute all justifications. In this paper, we propose a novel black-box algorithm that iteratively constructs a set of justifications of an entailment using a relevance-based selection function. Each justification returned by our algorithm is attached with a weight denoting its relevance degree w.r.t. the entailment. Finally, we implement the algorithm and present evaluation results over real-life ontologies that show the benefits of the selection function.

1 Introduction

Ontologies play a central role for the formal representation of knowledge on the Semantic Web. In logic based ontology languages, such as Description Logics (DLs), finding the justifications of an entailment is an important problem that has many practical applications, such as handling inconsistency in an ontology [9] and diagnosing terminologies [13].

Several methods have been proposed to find the justifications for an entailment. Although practical techniques to find all possible justifications exist, efficiency is still a problem (see [13]). Furthermore, as shown in [4], in the worst case the number of justifications for a subsumption entailment is exponential in the size of the ontology. Therefore, it is not always desirable to find *all* justifications. As an illustration, according to the statistics given in [8], for unsatisfiability entailments in the real-life ontology Chemical ontology there exist up to 26 justifications, and a single justification contains up to 12 axioms. In such cases, understanding all the justifications is a tedious effort for a human. Instead of a guarantee to find all justifications, the user is typically only interested in a set of justifications that are *relevant* to the entailment to a certain degree, and thus intuitively easier to understand. At the same time, the restriction to only find some justifications allows for considerable improvements in efficiency.

To this end, we propose a novel algorithm for finding justifications of a DL-based entailment by using a *relevance-based selection function*. Our algorithm is based on a

A. Gómez-Pérez, Y. Yu, and Y. Ding (Eds.): ASWC 2009, LNCS 5926, pp. 306–320, 2009.
© Springer-Verlag Berlin Heidelberg 2009

black-box approach, and thus it can be implemented using any DL reasoner. We first define a relevance-based ordering on the justifications using a selection function, which allows us to associate a relevance degree with each of the justifications to facilitate the comparison among them. We then present our algorithm which incrementally selects sub-ontologies using a selection function and finds a set of justifications from these sub-ontologies for an entailment. We consider two selection functions. One is the relevance-based selection function given in [6] which can be applied to any DL. Since this selection function does not take advantage of features of specific DLs, it does not always produce very good results. Therefore, we propose a novel selection function which is designed for an important DL: $\mathcal{EL}^+$.

The advantages of our algorithm over existing algorithms for finding all justifications, especially the one in [8] (marked as All_Just_Alg), are as follows: First, our comparison shows the advantage of introducing the selection function to find justifications incrementally. Particularly, for test ontologies in $\mathcal{EL}^+$, our algorithm based on the new selection function is orders of magnitude faster than All_Just_Alg and our algorithm based on the existing syntactical selection function. Second, our algorithm allows for a strategy for computing a set of justifications that satisfy some minimality conditions. When this strategy is chosen, our algorithm outperforms the algorithms for computing all justifications. Third, our algorithm for computing all justifications can be interrupted at any time and still compute justifications which are relevant to the entailment up to some degree. In contrast, existing algorithms will compute justifications in an unordered way.

2   Preliminaries

2.1   Justification in Description Logics

We presume that the reader is familiar with Description Logics (DLs) and refer to the DL handbook [1] for more details. A DL-based ontology $O = (\mathcal{T}, \mathcal{R}, \mathcal{A})$ consists of a set $\mathcal{T}$ of concept axioms (TBox), a set $\mathcal{R}$ of role axioms (RBox), and a set $\mathcal{A}$ of assertional axioms (ABox). The TBox and RBox are used to express the *intensional level of the ontology* while the ABox is used to express the *instance level of the ontology*. In this paper, we consider only DLs which are fragments of first-order predicate logic. Thus, an ontology can be translated into a first-order knowledge base. Therefore, semantics of DLs can be defined in a standard way. An interpretation $\mathcal{I}$ is called a *model* of an ontology O, iff it satisfies each axiom in O. The main types of entailments are concept unsatisfiability (written $O \models C \sqsubseteq \bot$): C is *unsatisfiable* w.r.t. O; and concept subsumption: C *is subsumed by* D w.r.t. O (written $O \models C \sqsubseteq D$). Without loss of generality, we restrict our attention to concept subsumption in what follows.

In [8], a justification for an entailment is defined as a minimal subset of the axioms in the ontology responsible for the entailment. Formally, let O be a consistent DL-based ontology and ϕ be an axiom such that $O \models \phi$, a subset $O' \subseteq O$ is a justification for ϕ in O, if $O' \models \phi$, and $O'' \not\models \phi$ for every $O'' \subset O'$. In this paper, we only consider consistent ontologies.

2.2   Reiter's Hitting Set Tree Algorithm

Our algorithm for finding justifications is adapted from Reiter's Hitting Set Tree (HST) algorithm [11]. We follow the reformulated notions in Reiter's theory given in [8]. Given a *universal set* U, and a set $S = \{s_1, ..., s_n\}$ of subsets of U which are *conflict sets*, i.e. subsets of the system components responsible for the error, a *hitting set* T for S is a subset of U such that $s_i \cap T \neq \emptyset$ for all $1 \leq i \leq n$. A *minimal hitting set* T for S is a hitting set such that no $T' \subset T$ is a hitting set for S. A hitting set T is cardinality-minimal if there is no other hitting set T' such that $|T'| < |T|$. In the case of finding justifications, the universal set corresponds to an ontology and a conflict set corresponds to a justification, and all justifications can be found by constructing a HST.

Reiter's algorithm is used to calculate minimal hitting sets for a collection $\mathcal{C} = \{S_1, ..., S_n\}$ of sets by constructing a labeled tree, called a Hitting Set Tree (HST). Given a collection $\mathcal{C}$ of sets, a HST T is the smallest edge-labeled and node-labeled tree, such that the root is labeled by $\checkmark$ if $\mathcal{C}$ is empty. Otherwise it is labeled with any set in $\mathcal{C}$. For each node n in T, let $H(n)$ be the set of edge labels on the path in T from the root to n. The label for n is any set $S \in \mathcal{C}$ such that $S \cap H(n) = \emptyset$, if such a set exists. If n is labeled by a set S, then for each $\sigma \in S$, n has a successor, n_σ joined to n by an edge labeled by σ. For any node labeled by $\checkmark$, $H(n)$, i.e. the labels of its path from the root, is a hitting set for $\mathcal{C}$.

2.3   Selection Functions

We introduce the notion of a selection function given in [7]. Let $\mathbf{L}$ be an ontology, a selection function for $\mathbf{L}$ is a mapping $s_{\mathbf{L}}: \mathcal{P}(\mathbf{L}) \times \mathbf{L} \times \mathbb{N} \to \mathcal{P}(\mathbf{L})$ such that $s_{\mathbf{L}}(O, \phi, k) \subseteq O$, where $\mathcal{P}(\mathbf{L})$ is the power set of $\mathbf{L}$. That is, a selection function selects a subset of an ontology w.r.t. an axiom at step k.

Let ϕ be an axiom in a DL-based ontology. We use $I(\phi)$, $C(\phi)$ and $R(\phi)$ to denote the sets of individual names, concept names, and role names appearing in ϕ respectively.

We first define the direct relevance relation between two axioms.

Definition 1. *Given two axioms ϕ and ψ, ϕ is* directly relevant *to ψ iff there is a common name which appears both in ϕ and ψ, i.e., $I(\phi) \cap I(\psi) \neq \emptyset$ or $C(\phi) \cap C(\psi) \neq \emptyset$ or $R(\phi) \cap R(\psi) \neq \emptyset$.*

Based on the notion of direct relevance, we can extend it to relevance relation between an axiom and an ontology. An axiom ϕ is relevant to an ontology O iff there exists an axiom ψ in O such that ϕ and ψ are directly relevant.

We introduce the relevance-directed selection function which can be used to find all the axioms in an ontology that are relevant to an axiom to some degree.

Definition 2. *Let O be an ontology, ϕ be an axiom and k be an integer. The* signature-based selection function, *written s_{rel}^{sig}, is defined inductively as follows:*

$s_{rel}^{sig}(O, \phi, 0) = \emptyset$
$s_{rel}^{sig}(O, \phi, 1) = \{\psi \in O : \phi \text{ and } \psi \text{ are directly relevant}\}$
$s_{rel}^{sig}(O, \phi, k) = \{\psi \in O : \psi \text{ is directly relevant to } s_{rel}^{sig}(O, \phi, k - 1)\}$, *where $k > 1$.*

We call $s_{rel}^{sig}(O, \phi, k)$ the k-relevant subset of O w.r.t. ϕ. For convenience, we define $s_k^{sig}(O, \phi) = s_{rel}^{sig}(O, \phi, k) \setminus s_{rel}^{sig}(O, \phi, k - 1)$ for $k \geq 1$.

Example 1. Consider an ontology O (taken from the Proton ontology, c.f. experiments in Section 4) including the following axioms:

1:	Manager $\sqsubseteq$ Employee,	2:	Employee $\sqsubseteq$ JobPosition,
3:	Leader $\sqsubseteq$ JobPosition,	4:	JobPosition $\sqsubseteq$ Situation,
5:	Situation $\sqsubseteq$ Happening,	6:	Leader $\sqsubseteq \neg$ Patent,
7:	Happening $\sqsubseteq \neg$ Manager,	8:	JobPosition $\sqsubseteq \neg$ Employee,
9:	JobPosition(lectureship)		

Given $\phi = Manager \sqsubseteq \bot$, it is easy to check that $s_1^{sig}(O, \phi) = \{1, 7\}$, $s_2^{sig}(O, \phi) = \{2, 8, 5\}$ and $s_3^{sig}(O, \phi) = \{3, 4, 9\}$.

According to [7], the relevance set selected by the signature-based selection function will grow very fast and will become very large after a small number of iteration. This problem is due to the fact that this selection function does not take advantage of features of specific DLs. Therefore, another selection function for unfoldable $\mathcal{ALC}$ TBoxes [1] is proposed to restrict the signature-based selection function.

Definition 3. *[13] Let ϕ and ψ be two axioms in an unfoldable $\mathcal{ALC}$ TBox. $\mathcal{V}_c(C)$ denotes the set of concept names that appear in a concept C. Then ϕ is directly concept-relevant to ψ iff*

- *$\mathcal{V}_c(C_1) \cap C(\psi) \neq \emptyset$ if the axiom ϕ has the form $C_1 \sqsubseteq C_2$,*
- *$\mathcal{V}_c(C_1) \cap C(\psi) \neq \emptyset$ or $\mathcal{V}_c(C_2) \cap C(\psi) \neq \emptyset$ if the axiom ϕ has the form $C_1 = C_2$ or $disjoint(C_1, C_2)$.*

If we replace the notion of directed relevance in Definition 3 by the notion of direct concept-relevance, we get a new selection function which we call a *concept-based selection function* and we still use s_{rel}^{con} to denote it. In the following, we use s_{rel} to denote either s_{rel}^{sig} or s_{rel}^{con}.

Ontology O given in Example 1 is an unfoldable $\mathcal{ALC}$ TBox. We have $s_1^{con}(O, \phi) = \{1, 7\}$, $s_2^{con}(O, \phi) = \{5\}$. Note that axiom number 7 is a disjointness axiom.

3   The Debugging Framework for Finding Justifications

In this section, we present our debugging framework for finding justifications for an DL entailment. First of all, we introduce the relevance-based ordering on justifications which is the basic definition in our framework. Then we give an overview of our framework. Finally, the concrete algorithm to compute a set of justifications attached with relevance degrees is given.

---

[1] A TBox is called unfoldable if the left-hand sides of the axioms (the defined concepts) are atomic and unique and if the right-hand sides (the definitions) contain no direct or indirect reference to the defined concept.

3.1   Relevance-Based Ordering on Justifications

Before introducing an ordering on justifications using the relevance-directed selection function, we first need to define the relevance degree.

Definition 4. *(Relevance Degree) Let O be an ontology and $C \sqsubseteq D$ be a concept subsumption of O. For a justification J, the relevance degree w.r.t. $C \sqsubseteq D$ is defined as follows,*

$$d_{rel,C \sqsubseteq D}(J) = max\{k : J \cap s_k^{sig}(O, C \sqsubseteq D) \neq \emptyset\}$$

Definition 5. *(Relevance-based Ordering) For any two justifications J_1 and J_2 for an entailment $C \sqsubseteq D$, a relevance-based preordering on the set of all the justifications for $C \sqsubseteq D$, written $\preceq_{rel,C \sqsubseteq D}$, is defined as follows,*

$$J_1 \preceq_{rel,C \sqsubseteq D} J_2 \quad iff \quad d_{rel,C \sqsubseteq D}(J_1) \geq d_{rel,C \sqsubseteq D}(J_2).$$

That is, justification J_1 is less relevant to $C \sqsubseteq D$ than J_2 if and only if the element in J_1 which is furthest from $C \sqsubseteq D$ is less relevant to that in J_2 which is furthest from $C \sqsubseteq D$.

Example 2. (Example 1 Continued) Entailment $\phi = Manager \sqsubseteq \bot$ has the following two justifications:

$J_1 = \{$Manager$\sqsubseteq$Employee, Employee$\sqsubseteq$JobPosition, JobPosition$\sqsubseteq \neg$Employee$\}$
$J_2 = \{$Manager$\sqsubseteq$Employee, Employee$\sqsubseteq$JobPosition, JobPosition$\sqsubseteq$Situation,
Situation$\sqsubseteq$Happening, Happening$\sqsubseteq \neg$Manager$\}$.

By Example 1, we have $d_{rel,\phi}(J_1) = 2$ and $d_{rel,\phi}(J_2) = 3$. Therefore, $J_2 \preceq_{rel,\phi} J_1$. It is clear that J_1 is much easier to understand than J_2.

Definition 6. *Let O be an ontology, $C \sqsubseteq D$ be a concept subsumption of O and s_{rel} be a relevance-based selection function. Given a justification J for $C \sqsubseteq D$, suppose $k = d_{rel,C \sqsubseteq D}(J)$, then we say that J is relevance-complete if $J \cap s_j(O, C \sqsubseteq D) \neq \emptyset$ for any $0 < j \leq k$ and $J \subseteq s_{rel}(O, C \sqsubseteq D, k)$.*

According to Definition 6, a justification for a concept subsumption with relevance degree k is relevance-complete if it contains at least one axiom in each j-relevant subset of O w.r.t. the subsumption for $0 < j \leq k$ and it is contained in the k-relevant subset of O w.r.t. the subsumption.

Proposition 1. [2] *Let O be an ontology, $C \sqsubseteq D$ be a concept subsumption of O and s_{rel} be a relevance-based selection function. Suppose all the justifications for $C \sqsubseteq D$ are relevance-complete. Let J_1 and J_2 be justifications for $C \sqsubseteq D$ such that $J_1 \preceq_{rel,C \sqsubseteq D} J_2$, then $J_2 \cap s_k(O, C \sqsubseteq D) \neq \emptyset$ implies $J_1 \cap s_k(O, C \sqsubseteq D) \neq \emptyset$ for all k.*

According to Proposition 1, suppose justification J_1 is less relevant to $C \sqsubseteq D$ than J_2, if J_2 is k-relevant to $C \sqsubseteq D$, then J_1 must be k-relevant to $C \sqsubseteq D$ as well, where a

---

[2] Proofs of all propositions and theorems can be found in a technical report at http://www.aifb.uni-karlsruhe.de/WBS/gqi/ASWC-TR.pdf

Globals : $ds \leftarrow debugStrategy$, $s \leftarrow selectionFunction$
$\overrightarrow{\mathcal{J}}, O', HS \leftarrow \emptyset$
$k \leftarrow 1$
while $s_k(O, C \sqsubseteq D) \neq \emptyset$ **do**
$\quad$ $O' \leftarrow O' \cup s_k(O, C \sqsubseteq D)$
$\quad$ **if** $HS = \emptyset$ *and* $O' \models C \sqsubseteq D$ **then**
$\quad\quad$ $(\mathcal{J}, HS) \leftarrow$ Expand_HST$(C \sqsubseteq D, O')$
$\quad\quad$ $\overrightarrow{\mathcal{J}} \leftarrow \{(J, k) | J \in \mathcal{J}\}$
$\quad$ **else if** $HS \neq \emptyset$ **then**
$\quad\quad$ $HS_{tmp} \leftarrow \emptyset$
$\quad\quad$ **for** $P \in HS$ **do**
$\quad\quad\quad$ **if** $O' \setminus P \not\models C \sqsubseteq D$ **then** $HS_{tmp} \leftarrow HS_{tmp} \cup \{P\}$
$\quad\quad$ **if** $ds = $ *REL_JUSTS_ALL* or $(ds = $ *CM_JUSTS* and $HS_{tmp} = \emptyset)$ **then**
$\quad\quad\quad$ **for** $P \in (HS \setminus HS_{tmp})$ **do**
$\quad\quad\quad\quad$ $(\mathcal{J}, HS') \leftarrow$ Expand_HST$(C \sqsubseteq D, O' \setminus P)$
$\quad\quad\quad\quad$ $\overrightarrow{\mathcal{J}} \leftarrow \overrightarrow{\mathcal{J}} \cup \{(J, k) | J \in \text{newly found justifications}\}$
$\quad\quad\quad\quad$ $HS \leftarrow HS \cup \{P \cup P' | P' \in HS'\} \setminus \{P\}$
$\quad\quad$ **else**
$\quad\quad\quad$ $HS \leftarrow HS_{tmp}$
$\quad$ $k \leftarrow k + 1$
return $(\overrightarrow{\mathcal{J}}, HS)$

Fig. 1. REL_JUSTS$(C \sqsubseteq D, O, debugStrategy, selectionFunction)$

justification is k-relevant to $C \sqsubseteq D$ if and only if it has non-empty intersection with $s_k(O, C \sqsubseteq D)$. We observe that for many real life ontologies, if we arbitrarily take an entailment of the ontology, then all the justifications for it are relevance-complete when the signature-based selection function is chosen. This observation together with Proposition 1 tells us that if we want to find justifications for a subsumption that are more relevant to the subsumption, we should select those axioms in the ontology that are more relevant to the subsumption. This motivates our algorithm given in subsection 3.2.

3.2   Relevance-Directed Algorithm for Finding Justifications

Our algorithm REL_JUSTS (see Figure 1) receives an ontology O, a subsumption entailment $C \sqsubseteq D$ of O, a debugging strategy for computing justifications and a selection function as inputs, and outputs a set of weighted justifications $\overrightarrow{\mathcal{J}}$ and a set of global hitting sets HS^3.

---

[3] We refer to the HST constructed using algorithm REL_JUSTS (resp. EXPAND_HST) as a global (resp. local) HST, and the hitting sets found in this algorithm as global (resp. local) hitting sets.

$HS, HS_1 \leftarrow \emptyset$
$J \leftarrow \mathsf{SINGLE\_JUST}(C \sqsubseteq D, O)$
$\mathcal{J} \leftarrow \mathcal{J} \cup \{J\}$
for $a \in J$ **do** $HS_1 \leftarrow HS_1 \cup \{\{a\}\}$
while *true* **do**
    for $(P \in HS_1)$ **do**
        ⌊ **if** $O \setminus P \not\models C \sqsubseteq D$ **then** $HS \leftarrow HS \cup \{P\}$
    $HS_2 \leftarrow HS_1 \setminus HS$
    if $(ds = \mathsf{CM\_JUSTS}$ and $HS \neq \emptyset)$ or $(HS_2 = \emptyset)$ **then**
        ⌊ **return** $(\mathcal{J}, HS)$
    $HS_1 \leftarrow \emptyset$
    for $P \in HS_2$ **do**
        $J \leftarrow \mathsf{SINGLE\_JUST}(C \sqsubseteq D, O \setminus P)$
        $\mathcal{J} \leftarrow \mathcal{J} \cup \{J\}$
        for $a \in J$ **do** $HS_1 \leftarrow HS_1 \cup \{P \cup \{a\}\}$

Fig. 2. Expand_HST$(C \sqsubseteq D, O)$

In our algorithm, we consider two strategies when computing justifications:

1. **All_Just_Relevance**: expand all branches in the HST[4].
2. **CM_Just_Relevance**: expand those branches that are cardinality-minimal hitting sets in the local HST.

In the algorithm **REL_JUSTS** first we find the first k such that $C \sqsubseteq D$ is inferred from the k-relevant subset O' of O. Then the method Expand_HST is invoked to find a set of justifications for $C \sqsubseteq D$ in O' and a set of local hitting sets HS. Afterwards, a relevance degree will be associated to each found justification. In each following iteration, we need to find those found local hitting sets which are needed to be further expanded. For each hitting set $P \in HS_{tmp}$ to be further expanded, we compute a set of justifications in $O' \setminus P$ with a set of local hitting sets HS' if $O' \setminus P \models C \sqsubseteq D$. In such case, we then need to replace $P \in HS$ with the hitting sets (i.e. $\{P \cup P' | P' \in HS'\}$) in the newly constructed branches. The iteration will be terminated until no more relevant axioms can be selected in O. Note that unlike the algorithm given in [8], we do not check if a local hitting set is a global one. By doing this, we do not need satisfiability check over the whole ontology O but restrict to the selected subontology.

As for the algorithm Expand_HST (see Figure 2), we first calculate a single justification J which serves as the root node of a local HST. We then remove each of the axioms in J individually to create new branches of the HST. New justifications along these branches are calculated on the fly. The algorithm terminates if one of the conditions is satisfied: (1) we intend to find those cardinality-minimal justifications and some local hitting sets have been found; (2) no more branch needs to be expanded (i.e. $HS_2 = \emptyset$). To compute a single justification, we can take the methods given in [8].

---

[4] When using optimization techniques, we may not expand all branches.

Since our relevance-directed algorithm is adapted from the HST algorithm [11], we use the optimized techniques *justification reuse* and *Early path termination* in Reiter's HST algorithm.

The algorithm is an *anytime algorithm* in that it can be terminated at any point and is guaranteed to have computed all or some justifications with relevance degrees up to k, where k is the current relevance degree at the time of interruption.

We show the correctness of Algorithm REL_JUSTS for some ontologies when the strategy All_Just_Relevance is chosen and the signature-based selection function is used, i.e., our algorithm finds all the justifications for Rel_Just $(C \sqsubseteq D, O, str)$ when it terminates. We use ALL_JUST$(C \sqsubseteq D, O)$ to denote the set of all justifications in O w.r.t. $C \sqsubseteq D$.

Theorem 1. *Let O be an ontology, $C \sqsubseteq D$ be a concept subsumption of O and s_{rel}^{sig} be the signature-based selection function. Suppose all the justifications for $C \sqsubseteq D$ are relevance-complete. Let str= All_Just_Relevance in Algorithm REL_JUSTS. Suppose $\overrightarrow{\mathcal{J}}$ is the set of weighted justifications returned by Rel_Just$(C \sqsubseteq D, O, str)$, then ALL_JUST$(C \sqsubseteq D, O) = \{J | \exists k : (J, k) \in \overrightarrow{\mathcal{J}}\}$. For each $(J, k) \in \overrightarrow{\mathcal{J}}$, we have $k = d_{rel, C \sqsubseteq D}(J)$.*

The correctness of Theorem 1 is based on the assumption that all the justifications for $C \sqsubseteq D$ are relevance-complete. We do not have a general principle to check whether this condition is satisfied. However, in the next subsection, we show that there is a novel relevance-based selection function for DL $\mathcal{EL}^+$ such that for any $\mathcal{EL}^+$ ontologies and any entailment, all the justification for the entailment are relevance-complete. Furthermore, for all the test ontologies used in our experiment, our algorithm can find all the justifications for randomly selected entailments.

Next we show the correctness of Expand_HST$(C \sqsubseteq D, O)$ when the strategy CM_Just_Relevance is chosen, i.e., it computes those justifications that are nodes of the cardinality-minimal hitting sets in the local HST.

Theorem 2. *Suppose HS' is returned by Expand_HST$(C \sqsubseteq D, O)$ using strategy CM_Just_Relevance and HS is the set of all the local hitting sets in O w.r.t. $C \sqsubseteq D$. For each local hitting set $p' \in HS'$, there is no other hitting set $p \in HS \setminus HS'$ such that $|p| < |p'|$.*

Example 3. Given an ontology O including the axioms:

1: $U \sqsubseteq A$,	2: $U \sqsubseteq \neg A$,	3: $U \sqsubseteq C$,	4: $C \sqsubseteq \neg B$,	5: $A \sqsubseteq B$,
6: $U \sqsubseteq G$,	7: $G \sqsubseteq E$,	8: $U \sqsubseteq F$,	9: $F \sqsubseteq \neg E$,	10: $U \sqsubseteq D$,
11: $D \sqsubseteq E$,	12: $C \sqsubseteq K$,	13: $K \sqsubseteq \neg H$,	14: $B \sqsubseteq H$	

We consider the concept unsatisfiability $\phi = U \sqsubseteq \bot$ in O. It is easy to check that $s_1^{sig}(O, \phi) = \{1, 2, 3, 6, 8, 10\}$, $s_2^{sig}(O, \phi) = \{4, 5, 7, 9, 11, 12\}$ and $s_3^{sig}(O, \phi) = \{13, 14\}$. All the justifications for ϕ are listed as follows:

$$\mathcal{J} = \{\{1, 2\}, \{1, 3, 4, 5\}, \{6, 7, 8, 9\}, \{8, 9, 10, 11\}, \{1, 3, 5, 12, 13, 14\}\}.$$

We illustrate our algorithm using CM_JUSTS strategy in Figure 3, where each distinct node in a rectangular box represents a justification and each sub-tree outlined in a rectangular box shows the process to compute justifications for a sub-ontology by invoking Expand_HST(ϕ, O). The branches marked with '$\times$' are not expanded anymore

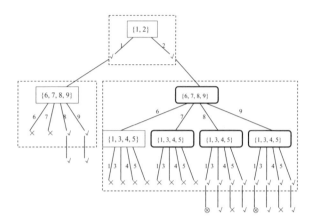

Fig. 3. Finding justifications using relevance-directed algorithm

since some cardinality-minimal hitting sets in a local hitting set tree have been found, and those marked with $\otimes$ are not expanded as well by applying the heuristic strategy of early path termination. The branches marked with '$\sqrt{}$'in a rectangle or outside a rectangle mean that we have found local hitting sets or global ones respectively. The over-bordered nodes represent reused justifications.

Iteration 1: When $O' = s_{rel}^{sig}(O, \phi, 1))$, a justification $J = \{1, 2\}$ is computed by Expand_HST(ϕ, O') and set as the root node of a HST. Then two possible branches indexed by $\{1\}$ and $\{2\}$ can be generated from J. Since ϕ can not be inferred in $O' \setminus \{1\}$ and $O' \setminus \{2\}$, we have obtained two local hitting sets.

Iteration 2: Then more axioms (i.e. $s_2^{sig}(O, \phi)$) are selected and added to O'. In this iteration, we obtain an empty set of HS_{tmp} and thus both $\{1\}$ and $\{2\}$ need to be further expanded. Take the branch $\{1\}$ as an example. We find a new justification $J' = \{6, 7, 8, 9\}$ and some new local hitting sets $\{8\}$ and $\{9\}$ in $O' \setminus \{1\}$. Then we replace the hitting set $\{1\}$ with $\{1, 8\}$ and $\{1, 9\}$. Similarly, we can find other 8 hitting sets in branch indexed by $\{2\}$.

Iteration 3: Again, more axioms $s_3^{sig}(O, \phi)$ are added to O'. The we find 6 out of 10 hitting sets which are not necessary to be expanded in the current O'.

Since no more axioms can be selected, we terminate the process by returning the found justifications and 6 hitting sets:

$$HS = \{\{1, 8\}, \{1, 9\}, \{2, 8, 3\}, \{2, 8, 5\}, \{2, 9, 3\}, \{2, 9, 5\}\}$$

and a set of weighted justifications

$$\overrightarrow{\mathcal{J}} = \{(\{1, 2\}, 1), (\{6, 7, 8, 9\}, 2), (\{1, 3, 4, 5\}, 2)\}.$$

A novel selection function. In Algorithm REL_JUSTS, the signature-based selection function is used. However, according to [7], the relevance set selected by it will grow very quickly and will become very large after a small number of iteration. The concept-based selection function performs better but it only works for unfoldable $\mathcal{ALC}$ TBoxes. Inspired by the module extraction algorithm for $\mathcal{EL}^+$ given in [5], we propose a new selection function, specifically designed for $\mathcal{EL}^+$, which is an important DL with tractable reasoning (see [2]). The idea is that an axiom $\alpha_L \sqsubseteq \alpha_R$ in an ontology O is directly relevant to a set of axioms S selected from O iff all the concept names and role names in α_L appear in the signature of S.

Definition 7. *Let O be an $\mathcal{EL}^+$ ontology and $A \sqsubseteq B$ be a concept subsumption of O, where A and B are $\mathcal{EL}^+$ concept names. We use $sig(O)$ to denote the set of all individual names, concept names and role names appearing in O. For any concept axiom or role axiom $\alpha_L \sqsubseteq \alpha_R$, we use $sig_L(\alpha_L \sqsubseteq \alpha_R)$ to denote the signature of α_L. The subsumption-based selection function, written s_{rel}^{sub}, is defined inductively as follows:*

$$s_{rel}^{sub}(O, A \sqsubseteq B, 0) = \emptyset$$
$$s_{rel}^{sub}(O, A \sqsubseteq B, 1) = \{\psi \in O : sig_L(\psi) = \{A\}\}$$
$$s_{rel}^{sub}(O, A \sqsubseteq B, k) = \{\psi \in O : sig_L(\psi) \subseteq sig(s_{rel}^{sub}(O, A \sqsubseteq B, k-1))\}, \text{ where } k > 1.$$

Similarly, we define $s_k^{sub}(O, A \sqsubseteq B)$ for $k \geq 1$.

We show a notable property of the subsumption-based selection function.

Proposition 2. *Let O be an $\mathcal{EL}^+$ ontology, $A \sqsubseteq B$ be a concept subsumption of O, where A and B are $\mathcal{EL}^+$ concept names, and s_{rel}^{sub} be the subsumption-based selection function. Then any justification for $A \sqsubseteq B$ is relevance-complete.*

Based on Proposition 2, we are able to show the following theorem.

Theorem 3. *Let O be an $\mathcal{EL}^+$ ontology, $A \sqsubseteq B$ be a concept subsumption of O, where A and B are $\mathcal{EL}^+$ concept names, and s_{rel}^{sub} be the subsumption-based selection function. Suppose all the justifications for $A \sqsubseteq B$ are relevance-complete. Let str= All_Just_Relevance in Algorithm 1. Suppose $\overrightarrow{\mathcal{J}}$ is the set of weighted justifications returned by Rel_Just($A \sqsubseteq B, O, str$), then ALL_JUST($A \sqsubseteq B, O$) = $\{J | \exists k : (J, k) \in \overrightarrow{\mathcal{J}}\}$. For each $(J, k) \in \overrightarrow{\mathcal{J}}$, we have $k = d_{rel, A \sqsubseteq B}(J)$.*

4 Experiments

In this section, we present the evaluation results for our algorithm. Our algorithm was implemented in Java using KAON2[5] as a reasoner. To fairly compare with the debugging algorithm in [8], we re-implemented the algorithm with the KAON2 API (we call it as ALL_JUSTS algorithm). The experiments were performed on a Linux 2.6.16.1 System and 2GB maximal heap space. Sun's Java 1.5.0 Update 6 was used for Java-based tools. For computing justifications of a single unsatisfiable concept or a subsumption in each run, we set a time limit of one hour.

4.1 Data Set

Table 1 shows some characteristics of the data set[6] used for our experiments – all of them are real-life ontologies from various domains. CHEM is an ontology for the

---

[5] http://kaon2.semanticweb.org/
[6] All ontologies, experimental results and binary code are available at
   http://radon.ontoware.org/downloads/data-relevance.zip

Table 1. Statistics of the ontologies in the data set. In the table, SubCl, EqCl, DisjCl, SubOp, EqOp, OpDomain and OpRange mean the number of subClassOf, equivalentClasses, disjoint-Classes, subObjectPropertyOf, equivalentObjectProperties, objectpropertyDomain and object-propertyRange axioms respectively.

Ontology	TBox	SubCl	EqCl	DisjCl	SubOp	EqOp	OpDomain	OpRange
CHEM	114	46	18	6	4	0	8	8
CONF	466	99	10	70	0	6	62	62
KM1	3140	2297	0	537	0	0	148	158
KM2	3160	2312	0	541	0	0	148	159
KM3	3180	2328	0	544	0	0	149	159
KM4	3200	2337	0	551	0	0	151	161
GALEN	4379	3238	699	0	416	0	0	0
GO	28897	28896	0	0	0	0	0	0
NCI	46940	46800	0	0	0	0	70	70

chemical domain. CONF is generated by merging two individual ontologies cmt and confOf from the conference domain and a mapping between the two individual ontologies (the mapping is generated by ontology mapping system ASMOV). Ontology KM1500 has been generated automatically from 1500 abstracts of the 'knowledge management' information space which is part of the BT Digital Library. Since KM1500 contains too many justifications (more than 100 justifications) for every unsatisfiable concept we have tested, we randomly choose an unsatisfiable concept (i.e. concept time) and construct four sub-ontologies (i.e. KMi, i=1,2,3,4) by increasing the number of the justifications w.r.t. concept time. GALEN[7], GO (Gene Ontology)[8] and NCI[9] are well-known ontologies from the biomedical domain. Both GO and NCI are formulated in the lightweight DL $\mathcal{EL}$. We used a trimmed down version (i.e. an $\mathcal{EL}^+$ fragment) of GALEN.

4.2   Evaluation Results

In this section, we show our evaluation results about computing justifications using different algorithms and computing justifications based on various selection functions.

Evaluation I.   We first evaluate the performance of our algorithm. We compute justifications for some unsatisfiability entailments using ALL_JUSTS and our relevance-directed one with both strategies (i.e. REL_ALL_JUSTS and CM_JUSTS). The unsatisfiability entailments chosen from test ontologies[10] are the following:

$$
\begin{aligned}
\text{CHEM:C1} \quad & \text{GA\_RelatedPublishedWork} \sqsubseteq \bot \\
\text{CHEM:C2} \quad & \text{GF\_RelatedPublishedWork} \sqsubseteq \bot \\
\text{CHEM:C3} \quad & \text{GF\_Precursor} \sqsubseteq \bot \\
\text{CONF:C} \quad & \text{Review} \sqsubseteq \bot \\
\text{KM1 / KM2 / KM3 / KM4:C} \quad & \text{time} \sqsubseteq \bot
\end{aligned}
$$

---

[7] http://www.openclinical.org/prj_galen.html

[8] http://www.geneontology.org

[9] http://www.mindswap.org/2003/CancerOntology/nciOntology.owl

[10] We do not show the evaluation results for other ontologies because both algorithm ALL_JUSTS and our algorithm applying strategy REL_ALL_JUSTS cannot terminate in one hour for some selected entailments (the results for our algorithm applying strategy REL_ALL_JUSTS can be found in Fig.5).

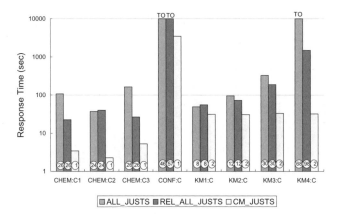

Fig. 4. The performance to compute justifications for an unsatisfiability entailment by applying different debugging algorithms or strategies. "TO" on top of the figure means time is out.

Figure 4 shows the debugging time[11] to compute justifications for each unsatisfiability entailment. The number within each circle means the number of justifications found by each corresponding debugging algorithm (within one hour).

From Figure 4, we first observe that our algorithm with the debugging strategy CM_JUSTS performs much faster than the other ones. This is because fewer justifications (e.g. one or two justifications) are generated with several global hitting sets. Secondly, comparing our algorithm using strategy REL_ALL_JUSTS with ALL_JUSTS, we can see that our algorithm outperforms ALL_JUSTS in most cases. Especially for CHEM:C1 and CHEM:C3, our algorithm is about 6 times faster. Also, for Km4:C, our algorithm can find all justifications (i.e. 98 justifications) within 1494 seconds, but ALL_JUSTS only returns 68 justifications when time is out. This shows that the advantage of incrementally increasing the size of the ontology using the selection function can be better reflected on those larger ontologies or ontologies with more justifications for an entailment.

Evaluation II. We compared the performance of algorithms instantiated from our algorithm with the strategy REL_ALL_JUSTS by using the signature-based selection function, the concept-based selection function and subsumption-based selection function, based on those $\mathcal{EL}^+$ ontologies. Note that our algorithm instantiated by the concept-based selection function may not return all the justifications for an entailment. We chose the following subsumptions:

GALEN:X1	AcuteErosionOfStomach	⊑ GastricPathology
GALEN:X2	AppendicularArtery	⊑ PhysicalStructure
GALEN:X3	UnstableKneeJoint	⊑ MirrorImagedBodyStructure
GO:X1	GO_0000024	⊑ GO_0007582
GO:X2	GO_0051803	⊑ GO_0051709
NCI:X1	APC_8024	⊑ Drugs_and_Chemicals
NCI:X2	CD97_Antigen	⊑ Protein

---

[11] Please note that the debugging time here means the total time to compute the justifications, including the time to check satisfiability, unlike the results reported in [8], which excluded the time for satisfiability checking.

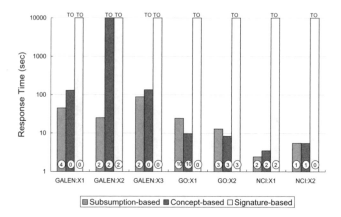

Fig. 5. Time to compute all justifications for a subsumption based on various selection functions. TO means time is out.

The results are shown in Figure 5 including the debugging time and the number of justifications found by each corresponding debugging algorithm (see the number within each circle). We use signature-based algorithm to indicate our algorithm instantiated by the signature-based selection function. Similarly, we use concept-based algorithm and subsumption-based algorithm to indicate two other algorithms.

Firstly, from Figure 5 we can see our signature-based algorithm fails to terminate the process to compute all justifications for all subsumptions for these ontologies. For example, for GO:X2, although all justifications (i.e. 3 justifications) have been found, the algorithm still yields a time out since it selects too many relevant axioms (e.g. more than 20000 axioms) to terminate the process.

Secondly, comparing the subsumption-based algorithm with the concept-based one, they have similar performance w.r.t. GO and NCI ontologies, but very different performance for ontology GALEN. One reason is that GALEN contains not only subClassOf axioms as both GO and NCI but equivalentClasses axioms, which will influence the performance of the two selection functions. Specifically, for the concept-based selection function, an equivalentClasses axiom ψ is directly relevant to axiom ϕ if and only if the concept signature (only considering concept names) of ϕ intersects with the signature of the left hand of ψ or the right hand. But for the subsumption-based selection function, ψ is directly relevant to ϕ if and only if the signature (considers concept and role names) of the left or right hand of ψ is a subset of the signature of ϕ. Therefore, for GALEN, the concept-based selection function may select many more axioms than the subsumption-based one. Take GALEN:X2 as an example. The concept-based selection function has selected 1802 axioms before the process has been terminated as the time is out. But the subsumption-based algorithm only selects 120 axioms in total, and the whole process to compute all justifications can be finished within only 25 seconds. Another reason is that some justifications contain subObjectProperty axioms or contain a subsumption whose subconcept is not atomic. Since for the concept-based selection function such kind of axioms can not be selected and thus no justification can be found for GALEN:X1 and GALEN:X3. Summarizing, this shows that a good selection function will improve the performance of our algorithm significantly.

5   Related Work

There are several methods dealing with the problem of finding all justifications of an entailment [3,4,8,9,12]. We have discussed the advantage of our algorithm over them in the introduction. Our algorithm is also related to a black-box algorithm proposed in [13] to calculate the justifications for concept unsatisfiability with the support of an external DL reasoner. Like our algorithm, their algorithm is also based on a relevance-based selection function. A shortcoming of their algorithm however is that the obtained justifications are usually not enough to resolve the unsatisfiability, i.e., removing one axiom from each of the justifications for the concept cannot make it satisfiable. Recently, methods have been proposed that use module extraction techniques to optimize the existing methods of finding all the justifications, such as [5]. We can use these methods to further improve the efficiency of our algorithm. This will be left as future work.

There are some other related methods. A heuristic method for finding justifications for concept unsatisfiability is given in [15]. The limitation of this method is that it can determine neither all the justifications nor enough justifications for explaining an unsatisfiable concept in general, because it is based on heuristics and pattern matching. There are some methods to find fine-grained justifications, such as the one in [10].

Our algorithm using the subsumption-based selection function is closely related to the modularization-based algorithm given in [14]. The main difference is that our new algorithm is an anytime algorithm and it allows different strategies when computing justifications.

6   Conclusion and Future Work

In this paper, we presented a flexible framework to find justifications for DL entailments. We first introduced a relevance-based ordering on justifications and thus provided a criterion of comparison between different justifications. Then we provided an overview of our debugging framework. After that, we proposed a concrete algorithm to find justifications for a concept entailment based on a relevance-based selection function. Specifically, our algorithm allows for two different strategies when calculating justifications: REL_ALL_JUSTS and CM_JUSTS. Using the first strategy, our algorithm finds all the justifications for an entailment. When the second strategy is chosen, our algorithm usually does not calculate all the justifications but a subset of them that satisfies some minimality condition. Since the relevance set selected by the signature-based selection function will grow very quickly and the concept-based one mainly focus on those ontologies with concept hierarchy (i.e. unfoldable $\mathcal{ALC}$), we proposed a novel selection function for more expressive ontologies (i.e. $\mathcal{EL}^+$).

Based on experimental results, we showed that our algorithm is very promising compared with ALL_JUSTS algorithm in [8] for both strategies. More specifically, our algorithm – by applying the strategy CM_JUSTS – is much more efficient than others. Although only a small subset of the justifications are returned, these justifications still provide a partially complete view of the entailment, as they correspond to some hitting sets for the set of all the justifications. Besides, our algorithm – by applying the strategy REL_ALL_JUSTS – can compute more justifications within a given time limit, and it is more efficient to find justifications. Another point is, comparing with the signature-based and concept-based selection functions, our subsumption-based algorithm performs quite well for those simple ontologies like Go and Nci. A good selection function may improve the performance of our algorithm significantly.

For the future work, we will investigate more powerful selection functions, such as a selection function based on semantic relevance.

Acknowledgments

Research reported in this paper was partially supported by the EU in the IST project NeOn (IST-2006-027595, http://www.neon-project.org/).

References

1. Baader, F., Calvanese, D., McGuinness, D., Nardi, D., Patel-Schneider, P.: The Description Logic Handbook: Theory, Implementation and Application. Cambridge University Press, Cambridge (2003)
2. Baader, F., Lutz, C., Suntisrivaraporn, B.: Efficient reasoning in $l+$. In: Proc. of DL 2006 (2006)
3. Baader, F., Peñaloza, R.: Axiom pinpointing in general tableaux. In: Olivetti, N. (ed.) TABLEAUX 2007. LNCS (LNAI), vol. 4548, pp. 11–27. Springer, Heidelberg (2007)
4. Baader, F., Peñaloza, R., Suntisrivaraporn, B.: Pinpointing in the description logic EL^+. In: Hertzberg, J., Beetz, M., Englert, R. (eds.) KI 2007. LNCS (LNAI), vol. 4667, pp. 52–67. Springer, Heidelberg (2007)
5. Baader, F., Suntisrivaraporn, B.: Debugging SNOMED CT using axiom pinpointing in the description logic EL^+. In: Proc. of KR-MED 2008 (2008)
6. Huang, Z., van Harmelen, F., ten Teije, A.: Reasoning with inconsistent ontologies. In: Proc. of IJCAI 2005, pp. 254–259 (2005)
7. Huang, Z., van Harmelen, F., ten Teije, A.: Reasoning with inconsistent ontologies: evaluation. Project Report D3.4.2, SEKT, pp. 254–259 (2006)
8. Kalyanpur, A., Parsia, B., Horridge, M., Sirin, E.: Finding all justifications of OWL DL entailments. In: Aberer, K., Choi, K.-S., Noy, N., Allemang, D., Lee, K.-I., Nixon, L.J.B., Golbeck, J., Mika, P., Maynard, D., Mizoguchi, R., Schreiber, G., Cudré-Mauroux, P. (eds.) ASWC 2007 and ISWC 2007. LNCS, vol. 4825, pp. 267–280. Springer, Heidelberg (2007)
9. Kalyanpur, A., Parsia, B., Sirin, E., Hendler, J.: Debugging unsatisfiable classes in OWL ontologies. Journal of Web Semantics 3(4), 268–293 (2005)
10. Lam, J.S.C., Sleeman, D.H., Pan, J.Z., Vasconcelos, W.W.: A fine-grained approach to resolving unsatisfiable ontologies. In: Spaccapietra, S. (ed.) Journal on Data Semantics X. LNCS, vol. 4900, pp. 62–95. Springer, Heidelberg (2008)
11. Reiter, R.: A theory of diagnosis from first principles. Artificial Intelligence 32(1), 57–95 (1987)
12. Schlobach, S., Cornet, R.: Non-standard reasoning services for the debugging of description logic terminologies. In: Proc. of IJCAI 2003, pp. 355–362 (2003)
13. Schlobach, S., Huang, Z., Cornet, R., van Harmelen, F.: Debugging incoherent terminologies. J. Autom. Reasoning 39(3), 317–349 (2007)
14. Suntisrivaraporn, B., Qi, G., Ji, Q., Haase, P.: A modularization-based approach to finding all justifications for OWL DL entailments. In: Domingue, J., Anutariya, C. (eds.) ASWC 2008. LNCS, vol. 5367, pp. 1–15. Springer, Heidelberg (2008)
15. Wang, H., Horridge, M., Rector, A., Drummond, N., Seidenberg, J.: Debugging OWL-DL ontologies: A heuristic approach. In: Gil, Y., Motta, E., Benjamins, V.R., Musen, M.A. (eds.) ISWC 2005. LNCS, vol. 3729, pp. 745–757. Springer, Heidelberg (2005)

Entropy-Based Metrics
for Evaluating Schema Reuse

Xixi Luo[1] and Joshua Shinavier[2]

[1] Computer Science Department, Beihang University,
Beijing, China 100083
xixiluo.china@gmail.com
[2] Tetherless World Constellation, Rensselaer Polytechnic Institute,
Troy, New York 12180
shinaj@rpi.edu

Abstract. Schemas, which provide a way to give structure to information, are becoming more and more important for information integration. The model described here provides concrete metrics of the momentary "health" of an application and its evolution over time, as well as a means of comparing one application with another. Building upon the basic notions of actors, concepts, and instances, the presented technique defines and measures the information entropy of a number of simple relationships among these objects. The technique itself is evaluated against data sets drawn from the Freebase collaborative database, the Swoogle search engine, and an instance of Semantic MediaWiki.

1   Introduction

Many social networking applications, such as Freebase,[1] aim to create a shared schema on the Web through collaborative contributions by online users. Collaborative tagging systems [2], such as del.icio.us,[2] employ user-contributed tags as a form of schema, while the technology of the Semantic Web is based on ontologies, which are a more sophisticated kind of schema. The use of a schema tends to facilitate information integration, but there is one prerequisite: a schema should be reused. If the schema is not frequently reused – for example, if each user creates his or her own schemas – then it may lose much of its value.

There is a fair amount research on schema reuse (for example, [1]). However, measuring the extent of schema reuse hasn't drawn much attention. In this paper, we start with the premise that dominant, widely reused schemas boost interoperability in general, and we propose a means of characterizing an application in such terms. It is apparent that naïve metrics such as the number of tags in an application, or the number of tags versus the number of instances tagged, are not sufficient for this purpose, as they don't allow us to distinguish "healthy" applications with a few dominant schemas among many, from more

---

[1] http://www.freebase.com/
[2] http://delicious.com/

A. Gómez-Pérez, Y. Yu, and Y. Ding (Eds.): ASWC 2009, LNCS 5926, pp. 321–331, 2009.

disorderly applications with the same number or proportion of schemas overall. Instead, we propose to use the entropy of tagging relationships to measure the extent of schema reuse in an application.

In the following section, we will formally define our metrics. In section 3, we will present several artificial examples to illustrate the use of the metrics in specific application scenarios, and section 4, we will discuss their use in more general ones. Finally, in section 5, we will apply the technique to several real-world applications.

2   Entropy-Based Metrics

We will first introduce the notion of information entropy, followed by the actor-concept-instance model upon which our evaluation metrics are based. We will then formally describe the evaluation of schema reuse using our metrics.

2.1   Entropy

In information theory, entropy (also known as self-information) is a measure of the uncertainty associated with a random variable [6]. The information entropy of a discrete random variable X with possible values $\{x_1, \ldots, x_n\}$ is defined by

$$H(X) = -\sum_i^n p(x_i) \log p(x_i) \tag{1}$$

where $p(x_i) = Pr(X = x_i)$ is the probability mass function of X.

For our purpose, there are two extreme cases of information entropy which are worth considering. In one extreme case, X has a single value, x, thus $p(x) = 1$ and $H(X) = 0$. In the other extreme case, all outcomes are equiprobable, making the entropy as high as possible: specifically, $H(X) = \log n$, where n is the number of possible values of variable X.

We will use the notion of entropy, as uncertainty, to measure the uncertainty of schema reuse. If concepts are frequently reused to annotate instances, then it is easier to infer the concept of an annotation given its instance, and vice versa.

Low entropy implies less uncertainty, which indicates that concepts are frequently reused. On the other hand, high entropy implies more uncertainty, which indicates that concepts are rarely reused. However, the measure of entropy itself can't fully satisfy the requirement of evaluating schema reuse. Below, we will formally introduce a more sophisticated entropy-based metric for schema reuse.

2.2   The Actor-Concept-Instance Model

Entropy-based evaluation as described in this paper can be performed on any application that can be abstracted to the *actor-concept-instance model* [5].

In this model, a *concept* (also known as a *schema*) is any tag, class, taxonomy or ontology which can be used to annotate or describe various data. The granularity of what we understand as a concept may vary widely, even within a single application. For example, in Freebase, we can consider both "types" and "domains" as concepts, depending on the requirements of the evaluation.

An *instance* is the main carrier of information in this model. An instance can be a web page, a photograph, an audio or a video file, or any other object identifiable with a URI. Instances may be annotated with arbitrary concepts by *actors*, or users.

In the following, if C is a set of concepts, I is a set of instances, and U is a set of actors (users), then let $A \subset U \times C \times I$ be the set of all possible annotations, with which actors in U associate the instances of I with concepts in C. Furthermore, let A_{c_i} be the subset of A in which the concept c_i is used to annotate an instance.

2.3   A Metric for Schema Reuse

If we were to simply use entropy to measure the uncertainty of concepts, the formula would be:

$$H(X) = -\sum_i^n p(c_i) \log p(c_i) \tag{2}$$

where $p(c_i) = Pr(X = c_i) = \frac{|A_{c_i}|}{|A|}$ with $|A_{c_i}|$ the number of annotations using concept c_i, and $|A|$ the total number of annotations. For convenience, equation 2 may be expressed in the following as $H(X) = H(p(c_1), \ldots, p(c_n))$.

For example, if there is only one concept in an application, and all instances are associated with this concept (see Figure 1(a)), then $H(X) = -1 \times \log 1 = 0$. For the example in Figure 3, $H(X) = -((\frac{3}{5} \times \log \frac{3}{5}) + (\frac{2}{5} \times \log \frac{2}{5})) = 0.67$.

However, this simple metric falls short when applied to applications with instances which are not annotated. Consider the examples in Figure 1(a) and 1(b). There are two un-annotated instances in Figure 1(b). According to Equation 2, the value of H for both examples should be 0, which is to say that all instances are annotated by the same concept. However, this is unintuitive for i_4 and i_5, which are not annotated at all.

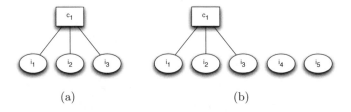

(a)   (b)

Fig. 1. Single concept

Our solution to this problem is to import a virtual concept c_v to the concept set C to form a new set C^*, and then to annotate each of the un-annotated instances "evenly" by each concept. For example, if there are 99 concepts and one un-annotated instance, we will add one virtual concept for a total of 100 concepts, then for each concept, add $1/100^{th}$ of an annotation between the concept and instance.

See Figure 2. After adding a virtual concept and distributing i_4 and i_5 to c_1 and c_v respectively, the value of H becomes 0.50.

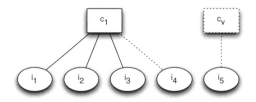

Fig. 2. Single concept with un-annotated instances

3   Examples in a Single Domain

In this section, we will discuss the application of the H metric in typical scenarios involving a single domain, with a detailed example for each scenario. A single-domain application is one in which there is only a single topic or group of concepts (for instance, an application which focuses on movies), as opposed to a multi-domain application in where there are many, possibly overlapping topics. We will discuss the scenarios for multiple domains in Section 4.

1. single domain, single concept **without** un-annotated instances
   See the example in Figure 1(a), in which $H(X) = 0$.
2. single concept **with** un-annotated instances
   See the example in Figure 2, in which $H(X) = H(\frac{4}{5}, \frac{1}{5}) = 0.50$. i_4 and i_5 are instances that are not annotated, the dotted border is the virtual concept c_v, and the dotted lines are the virtual annotations.

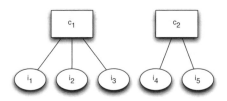

Fig. 3. Multiple concepts without un-annotated instances

3. multiple concepts **without** un-annotated instances.
   See the example in Figure 3, in which $H(X) = H(\frac{3}{5}, \frac{2}{5}) = 0.67$.
4. multiple concepts, cross-annotated instances

   In this case, the same instance is annotated with different concepts. See the example in Figure 4, in which $H(X) = H(\frac{3}{5}, \frac{2}{5}) = 0.67$. i_3 is shared by two concepts.

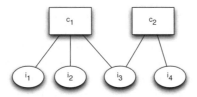

Fig. 4. Multiple concepts with cross-annotated instances

5. multiple concepts **with** un-annotated instances

   See the example in Figure 5, in which $H(X) = H(\frac{4}{8}, \frac{3}{8}, \frac{1}{8}) = 0.97$. i_6, i_7 and i_8, which are not annotated, are distributed to c_1, c_2 and virtual concept c_v, respectively.

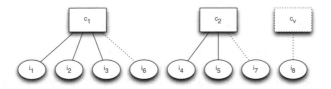

Fig. 5. Multiple concepts with un-annotated instances

4   More General Application Scenarios

In this section, we will discuss the application of the H metric in single- versus multiple-domain applications, as well as across applications.

4.1   Evaluating Schema Reuse in a Single Domain

We claim that the value of the H metric intuitively reflects the extent of schema reuse, in that a small value of H indicates a high degree of schema reuse, and conversely, a large value of H indicates relatively little schema reuse.

Case 1 above represents an extreme case of schema reuse, as there is only one concept, and all instances are associated with it. The concept is thus maximally reused, and the value of H is as small as possible: 0.

In the other extreme case of schema reuse, the value of H is maximal when instances are evenly distributed among all n concepts (with $n > 1$). Thus, H equals $\log n$. This corresponds to a scenario in which users maintain their own concepts, and never reuse other, existing concepts. In this case, the value of H increases with the total number of concepts, n.

Intermediate between these two extremes is the case in which *most*, but not all, instances are associated with a small number of concepts, in which case the value of H is low.

Finally, observe that relative to the case **without** un-annotated instances, the case **with** un-annotated instances has a higher value of H. In the case of single concepts, see case 1 vs. case 2. In the case of multiple concepts, see case 3 vs. case 5. The more un-annotated instances there are, the higher the value of H. For example, if there had been 4 rather than 3 un-annotated instances in Figure 5, then the number of instances associated with its concept would have been $3 + \frac{4}{3}$, $2 + \frac{4}{3}$ and $\frac{4}{3}$, so c_1 would have $\frac{13}{27}$ or 48% of the instances, c_2 would have $\frac{10}{27}$ or 37%, and c_v would have $\frac{4}{27}$ or 15%, with the result that $H = -(\frac{13}{27} \times \log \frac{13}{27} + \frac{10}{27} \times \log \frac{10}{27} + \frac{4}{27} \times \log \frac{4}{27}) = 1.0$, which is higher than the case with 3 un-annotated instances (0.97).

In summary, we can see that within a single domain, a relatively small number of concepts, concentrated annotations, and a relatively small number of un-annotated instances (which are typical signs of high schema reuse) lead to low values of H, while the opposite is true for large numbers of concepts and un-annotated instances, and evenly-distributed annotations.

4.2 Applying the Metric to Multiple-Domain Applications

We would also like to consider applications involving more than one domain. Within a single domain, any of cases 1 to 5 in section 3 is possible. However, we are also interested in cross-annotation over different domains. See the example in Figure 6, in which i_3 and i_4 are cross-annotated by concepts from different domains.

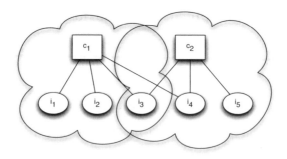

Fig. 6. Multiple concepts with un-annotated instances

The applicability of our metric to multiple-domain applications relies on the additive property of entropy.

The entropy of a system can be calculated from the entropies of its sub-systems, if the interactions between the sub-systems are known. Given a collection of n elements that are divided into k boxes (sub-systems $B_i, B_2, \ldots, B_k$) with $b_1, b_2, \ldots, b_k$ elements each, the entropy of the ensemble should be equal to the sum of the entropy of the system of boxes and the individual entropies of the boxes, each weighted with the probability of being in that particular box. For positive integers b_i where $b_1 + \cdots + b_k = n$,

$$H_n(p_1, \ldots, p_n) = H_k(\frac{b_1}{n}, \ldots, \frac{b_k}{n}) + \sum_i^k \frac{b_i}{n} H_{b_i}(p_1^{B_i}, \ldots, p_{b_i}^{B_i}) \qquad (3)$$

Therefore, the value of H over multiple domains is the entropy over those domains by size plus the weighted sum of the values of H in the individual domains.

4.3   Comparing Schema Reuse across Applications of Different Sizes

H is directly affected by the size of the set of concepts. The maximal value of H is $\log |C|$ (**without** un-annotated instances) or $\log(|C| + 1)$ (**with** un-annotated instances).

In order to compare schema reuse across applications of differing sizes, we need to normalize H with respect to size, the simplest approach being to divide by the maximal value of H, which is $\log(|C| + 1)$, so as to constrain its value to the interval $[0, 1]$.

$$H^* = H \div \log(|C| + 1) \qquad (4)$$

After normalization, we can meaningfully compare schema reuse across applications of any size.

5   Applying the Metric

Next, we will use H to evaluate three real-world applications: an instance of Semantic MediaWiki, the Swoogle search engine, and the Freebase collaborative database. Since we want to compare H among these three applications, we will use the normalized metric.

5.1   Tetherless World Wiki

The first example we will consider is the Tetherless World (TW) Wiki[3]. TW Wiki is used to share information among members of RPI's Tetherless World Constellation. It is based on the Semantic MediaWiki platform [3], and is a relatively small-scale collaborative Semantic Web application. At the time our

---

[3] http://tw.rpi.edu/wiki/Main_Page

experimental data was collected in April 2008, TW Wiki had only 29 registered users.

We adapt TW Wiki to our model by considering Semantic MediaWiki category pages as concepts and all other pages as instances. If a page uses a particular category, then we consider it to be an instance which has been annotated with a concept.

Table 1. State of the Tetherless World Wiki

Concept	Instance	H^*
327	40358	0.58

Table 1 shows the state parameters of TW Wiki at the time of data collection. We observe that a middling H value of 0.58 means that there are a variety of wiki categories which are re-used in many wiki pages. Note that this value is close to the 0.5 entropy of an 80/20 [4] arrangement of instances to concepts, in which 20% of the wiki categories account for 80% of the wiki pages.

Table 2. State of the Swoogle search engine

	Schema	Document	H^*
Ontology	148255	86410715	0.57
Class	3977509	86410715	0.27

5.2 Swoogle

Our second real-world example is Swoogle[4], a Semantic Web search engine that discovers, indexes, analyzes, and executes queries against Semantic Web documents and terms published on the Web.

There are two kinds of concepts, of different levels of granularity, in Swoogle: classes and ontologies. Classes in Swoogle are special resources which are instances of rdfs:Class, while an ontology in Swoogle is a special Semantic Web document which defines at least one class. Instances described in Swoogle's Semantic Web documents correspond to instances in our model. If an instance appears in an ontology, then we say that the ontology annotates that instance. If an instance is typed by a class, then we say that the class annotates the instance.

In Table 2 we observe that the value of H for ontologies is 0.57 (close to the 80/20 value), which may be accounted for by the fact that most instances are FOAF instances. We also observe also that the value of H for classes, 0.27, is smaller than that for ontologies, which means that classes are reused to a greater extent than ontologies. In particular, classes are reused across different ontologies: for example, *foaf:name* is commonly reused across different versions of the FOAF ontology.

---

[4] http://swoogle.umbc.edu/

5.3   Freebase

By observing changes in H within an application, we can judge whether the application is progressing in the expected direction. For example, an administrator may wish to introduce an incentive mechanism, such as auto-completion of tag labels, to stimulate schema reuse. The intended effect would be reflected by a subsequent decrease in H for schemas.

In the rest of this section, we will track the H value of the Freebase data set over the course of one year starting with its inception in 2006. To begin with, we will simply present some basic information about Freebase, most of which is taken directly from the Freebase web site.

Freebase is an open, shared database that contains structured information on millions of topics in hundreds of categories. This information is compiled from open data sets such as Wikipedia, as well as contributions from the user community.

Freebase content is made up of four basic elements. From the most general level to the most specific, these are: domains, types, properties and topics. Domains and types correspond to concepts in our model, but with different levels of granularity.

Table 3. Number of domains, classes, topics and users of Freebase from Oct. 2006 to Sep. 2007

Time	Domain	Class	Topic	User
2006-10	2	23	82	1
2006-11	4	34	134	9
2006-12	20	121	6903	36
2007-01	34	179	7318	66
2007-02	41	222	7915	87
2007-03	76	481	12522	342
2007-04	113	667	18312	1261
2007-05	188	1017	30557	3376
2007-06	273	1264	56058	5684
2007-07	410	1704	78024	10963
2007-08	542	2252	113091	12646
2007-09	598	2448	125212	13372

- A *domain* is simply a collection of types. For example, the Music Commons contains the types Musical Artist, Musical Track, and Musical Instrument, among others, which, taken together, constitute a domain.
- *types* provide a way to categorize topics in Freebase as being a particular kind of thing. For example, Arnold Schwarzenegger is typed as a Person, Film Actor, Film Producer, Politician, Pro Athlete, and Celebrity, among others. When a type is added to a topic, it is a way of saying that the topic is an example, or instance, of that type of thing.

- *properties* in Freebase are connected to types. For example, the Person type includes properties such as Birthplace, Birth Date, Parents, and Country of Nationality. If a user edits a property, we consider him or her to have contributed to the creation of the corresponding class. Likewise, if a property is used to associate a topic with a value, we consider the topic, or instance, to be annotated by the corresponding class.
- *topics* are instances, or examples, of a type, and correspond to instances in our model. For example, Robert DeNiro is an instance of the Film Actor type. Topics can be associated with many different types.

In order to analyze the Freebase data set, we first gathered behavior history for every user by crawling the web site. We will omit the details of the analysis due to space constraints, and proceed to the computed results. The numbers of domains, classes, topics and users over time are listed in Table 3.

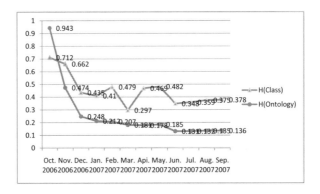

Fig. 7. Evolution of Freebase's H from October 2006 to September 2007

As we can see in the Figure 7, $H(Domain)$ is close to its highest possible value of 1.0 in October, 2006, then decreases steadily, perhaps exponentially to a stable value of around 0.1, at which point we know that most Freebase topics are annotated in only a relatively small number of dominant domains, which are the domains Freebase uses for publication of the data set. This result indicates that Freebase has successfully guided users to create and reuse a small number of very popular domains. Consequently, the heterogeneity of the relationship between topics and domains has decreased to a large extent.

$H(Type)$ is relatively high in the beginning, then decreases while oscillating between the values of about 0.3 to 0.5. Compared to $H(Domain)$, $H(Type)$ indicates that types are not reused to the same extent as domains. The greater variability of $H(Type)$ relative to $H(Domain)$ may be attributed to the fact that popular domains in Freebase are predefined and carefully maintained, which makes them relatively stable compared to types.

6   Conclusion and Future Work

Schema-based applications, such as collaborative Semantic Web applications, are new to the face of the Web. At some point, the inherent portability of Semantic Web data sets may make it possible to transfer sophisticated knowledge management techniques from one collaborative application to the next. In the meantime, the technique presented above provides a simple and efficient way to take a high-level "snapshot" of some essential features of such an application. This single technique can be used to evaluate any application which maintains basic annotation and publishing metadata. In the future, we will explore other possible metrics to describe the state of collaborative applications, such as the distribution of ontologies with respect to the classes they contain, metrics based on relationships inverse to those explored above or on compound relationships such as co-annotation or co-publishing. We would also like to supplement these global metrics with object-specific metrics relating, for instance, the ontologies a specific actor has published with the instances he or she has published.

Acknowledgements

This work has been supported by Rensselaer's Tetherless World Constellation. In particular, we would like to thank Li Ding for helpful discussions which led to the formulation of this technique, and Jim Hendler and Deborah McGuinness for their valuable feedback on various drafts of this paper.

References

1. Bontas, E.P., Mochol, M., Tolksdorf, R.: Case studies on ontology reuse. In: Proceedings of I-KNOW, Graz, Austria (June 2005)
2. Golder, S., Huberman, B.A.: The structure of collaborative tagging systems, HP Labs Technical Report (2006)
3. Krötzsch, M., Vrandečić, D., Völkel, M.: Semantic MediaWiki. In: Cruz, I., Decker, S., Allemang, D., Preist, C., Schwabe, D., Mika, P., Uschold, M., Aroyo, L.M. (eds.) ISWC 2006. LNCS, vol. 4273, pp. 935–942. Springer, Heidelberg (2006)
4. Lipovetsky, S.: Pareto 80/20 law: derivation via random partitioning. International Journal of Mathematical Education in Science and Technology (1), 1–6 (2008)
5. Mika, P.: Ontologies Are Us: A unified model of social networks and semantics. Web Semantics: Science, Services and Agents on the World Wide Web 5(1), 5–15 (2007)
6. Shannon, C.E.: A mathematical theory of communication. Bell System Technical Journal 27 (1948)

Overcoming Schema Heterogeneity between Linked Semantic Repositories to Improve Coreference Resolution

Andriy Nikolov, Victoria Uren, Enrico Motta, and Anne de Roeck

Knowledge Media Institute, The Open University, Milton Keynes, UK
{a.nikolov,v.s.uren,e.motta,a.deroeck}@open.ac.uk

Abstract. Schema heterogeneity issues often represent an obstacle for discovering coreference links between individuals in semantic data repositories. In this paper we present an approach, which performs ontology schema matching in order to improve instance coreference resolution performance. A novel feature of the approach is its use of existing instance-level coreference links defined in third-party repositories as background knowledge for schema matching techniques. In our tests of this approach we obtained encouraging results, in particular, a substantial increase in recall in comparison with existing sets of coreference links.

1 Introduction

With the emergence of the Linking Open Data initiative[1] the amount of semantic data available on the Web is constantly growing. New datasets are being published and connected to existing ones according to the Linked Data principles. Coreference links between entities (RDF individuals) in different datasets constitute a major added value: these links allow combining data about the same individuals stored in different locations. Due to large volumes of data, which have to be processed, these links cannot be produced manually, so automatic techniques are usually employed. However, discovering these links and querying data distributed over different repositories is often problematic due to the semantic heterogeneity problem: datasets often use different ontological schemas to describe data from the same domain.

In particular, ontological heterogeneity represents an obstacle for automatic coreference resolution tools, which discover links between individuals: it is not clear which classes represent overlapping sets of individuals that should be compared and which of their properties are relevant for similarity computation. For example, if a newly published dataset contains information about computer scientists, the tool needs to know which other datasets potentially contain co-referring individuals and to which classes they belong. Although the use of common schema ontologies such as FOAF, SKOS or Dublin Core is encouraged [1], existing datasets often employ their own schemas or, in other cases, terms of common ontologies do not provide full information about the data they describe

---

[1] http://esw.w3.org/topic/SweoIG/TaskForces/CommunityProjects/
LinkingOpenData

A. Gómez-Pérez, Y. Yu, and Y. Ding (Eds.): ASWC 2009, LNCS 5926, pp. 332–346, 2009.

and many important usage patterns remain implicit: e.g., in DBLP[2] a generic *foaf:Person* class in fact refers only to people related to computer science. Because of these issues, coreference resolution algorithms must be specially tuned for each pair of datasets to be linked. As a result, when a new repository is integrated into the the Linked Data cloud it is often hard for the publisher to connect it to all relevant datasets, which contain co-referring individuals, and many coreference links can remain undiscovered.

Thus, schema-level matching and instance-level coreference resolution constitute two stages of the process needed to discover coreference links between individuals stored in different repositories. Our data fusion tool KnoFuss [2] was originally developed to perform the second stage of the task. In this paper we extend its workflow and focus on resolving the first part of the problem (schema-level) in order to improve the performance at the second stage (instance-level).

Our approach focuses on inferring schema-level mappings between ontologies employed in Linked Data repositories. It implements the following novel features, which we consider our contribution:

- Use of instance-level coreference links to and from individuals defined in third-party repositories as background knowledge for schema-level ontology matching.
- Generating schema-level mappings suited for the needs of the instance coreference resolution process. In particular, our algorithm produces fuzzy mappings representing degree of overlap between classes of different ontologies rather than strict equivalence or subsumption relations.

The primary effect of the approach for the instance-level coreference resolution stage is the increase in recall because of newly discovered coreference links, which were initially missed because relevant subsets of individuals in two datasets were not directly compared. We obtained promising results in our test scenarios: coreference resolution recall increased (by 15% in the worst case and by 75% in the best case) in comparison with existing publicly available sets of links without substantial loss of precision.

The rest of the paper is organized as follows: in the section 2 we briefly discuss the most relevant existing approaches. Section 3 gives a general idea of our approach, provides illustrative example scenarios and outlines the overall workflow of the system. Sections 4 and 5 focus on the schema-level and data-level stages of the approach respectively. In the section 6 we present the results of our experiments performed with test datasets. Finally, section 7 summarizes our contribution and outlines directions for future work.

2   Related Work

This paper is related to both schema-level and data-level aspects of data integration. Both these aspects were originally studied in the database research community where many of the solutions, which were later extended and adapted in the Semantic Web domain, were originally proposed.

---

[2] http://www4.wiwiss.fu-berlin.de/dblp/

Considering the schema matching problem, two classes of techniques emerged: schema-level and instance-level approaches [3], which respectively rely on evidence defined in database schemas and in the data itself. With the emergence of ontological languages for the Semantic Web, which had different expressive capabilities from relational database schemas, specific solutions for ontology matching were developed [4]. The features of the Linked Data environment are the presence of large volumes of data and the availability of many interconnected information sources, which can be used as background knowledge. Therefore, two types of approaches are particularly relevant for this environment.

First, instance-level techniques for schema matching can be utilized. The advantage of instance-level methods is their ability to provide valuable insights into the contents and meaning of schema entities from the way they are used. This makes them suitable for the Linked Data environment for two reasons: (i) the need to capture the actual use pattern of an ontological term rather than how it was intended to be used by an ontology designer and (ii) the availability of large volumes of evidence data. In several approaches instances of classes and relations are considered as features when computing similarity between schema-level entities: e.g., CIDER [5] and RiMOM [6]. One particularly interesting approach, which uses schema alignment and coreference resolution in combination, was introduced in the ILIADS system [7]. ILIADS focuses on the traditional ontology matching scenario, where two schemas have to be integrated, and performs schema-level and instance-level matching in a loop, where newly obtained instance-level mappings are used to improve schema-level alignment and vice versa. This is similar to our approach where schema-level matching is performed to enhance the instance coreferencing process. However, unlike ILIADS, our approach was primarily motivated by the instance-level integration scenario and exploits information from many sources rather than only two.

The second relevant category of schema-matching techniques are those which utilize external sources as background knowledge. An approach proposed in [8] performs matching of two ontologies by linking them to an external third one and then using semantic relations defined in the external ontology to infer mappings between entities of two original ontologies. The SCARLET tool [9] employs a set of external ontologies, which it searches and selects using the Watson ontology search server[3]. These approaches, however, only consider schema-level evidence from third-party sources, while our approach relies on instance-level information.

Existing ontology matching tools usually produce as their output strict mappings such as equivalence and subsumption. In the Linked Data environment such mappings in many cases are impossible to derive: sometimes even strong semantic similarity between concepts does not imply strict equivalence. For instance, the concept *dbpedia:Actor* denotes professional actors (both cinema and stage), while the concept *movie:actor* in LinkedMDB refers to any person who played a role in a movie, including participants in documentaries, but excluding stage actors. Since our goal is to discover pairs of classes, which are likely to contain equivalent individuals, in our approach we produce mappings of a more

---

[3] http://watson.kmi.open.ac.uk/WatsonWUI/

generic type, which would reflect the fact that there is a significant degree of overlap between classes.

The problem of coreference resolution between instances (also called record linkage [10]) also has been studied for a long time in the database domain. In the classical model proposed in [10] by Fellegi and Sunter the decision regarding whether two individuals (records) describe the same entity was taken based on calculating an aggregated similarity between their field values. This model served as the base for the majority of algorithms developed in the domain (see [11] for a survey). Other proposed approaches exploit additional kinds of available information. For example, the algorithms described in [12] and [13] analyse links between instances and consider graphs instead of atomic data individuals, in [14] ontological schema restrictions are used as evidence and in [15] provenance information about data sources is exploited.

While originally in the Semantic Web domain the research primarily focused on the schema-level ontology matching, with the emergence of the Linked Data the problem of instance-level integration also gained importance. Popular existing tools were created and applied to discovery of the *owl:sameAs* coreference links for Linked Data. These follow the classical Fellegi-Sunter model: e.g., SILK [16] and ODDLinker (the latter used to create links from the LinkedMDB dataset [17]). However, given the decentralized character of data and the need to deal with large numbers of connections between entities in different repositories, the problem of managing and maintaining these connections received special attention in the Semantic Web community along with the problem of creating them (e.g., OKKAM [18] and CRS [19]). A particularly interesting approach is idMesh [20], where maintenance of links constitutes a complementary stage of the link discovery process: the system combines coreference links into graphs and considers their impact on each other to reason about their correctness and reliability of their sources. While these solutions focused on instance-level links, the UMBEL ontology[4] was developed to connect schemas used by Linked Data repositories. It provides an important "common denominator" for different ontologies, but, in our view, it does not always capture the actual usage patterns of ontological terms, but provides instead its own view on data, which often differs from other ontologies (e.g., the class *umbel:Dragon* contains such instances as *dbpedia:Azure_Dragon* describing the creature from the Chinese mythology and *dbpedia:Bernard_Guasch* referring to a CEO of the French rugby team "Catalans Dragons"). The VoiD ontology[5] provides meta-level descriptions of repositories including their main topic. However, these descriptions do not go into detail about the sets of individuals, which a dataset contains.

These approaches, however, focus either on the schema or the data level and do not consider them in combination, in particular, how mappings between entities at both levels influence each other. Our approach can be seen as complementary to these methods as it takes into account which schema-level mappings are needed

---

[4] http://www.umbel.org/
[5] http://rdfs.org/ns/void/html

to discover coreference links between instances and which schema-level relations are implied by available instance-level correspondences.

3   Overview: Example Scenarios and the Overall Workflow

As was said before, the focus of our approach was to produce schema-level mappings to assist instance-level coreference resolution. These mappings are primarily needed to identify, which subsets of two data repositories are likely to contain co-referring individuals, so that such subsets can be processed by a link discovery algorithm afterwards. Let us consider an example scenario, where deriving such mappings is problematic and hampers coreference resolution (Fig. 1). Both DBPedia and DBLP datasets contain individuals representing computer scientists. In many cases the same person is described in both repositories, but under different URIs. However, only a small proportion of possible coreference links between them is available[6]. More links can be discovered by performing automatic coreference resolution, but this task is complicated by two issues:

- Datasets do not contain overlapping properties for their individuals apart from personal names.
- Individuals which belong to overlapping subsets are not distinguished from others: in DBLP all paper authors belong to the *foaf:Person* class, while in DBPedia the majority of computer scientists is assigned to a generic class *dbpedia:Person* and not distinguished from other people. As a result, it becomes complicated to extract the subset of computer scientists from DBPedia which can be also represented in DBLP.

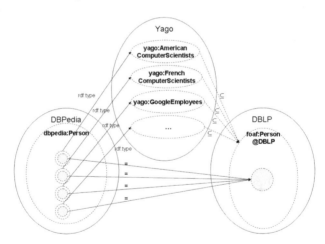

Fig. 1. DBPedia and DBLP: exploiting schema-level links with third-party datasets. Solid arrows show existing *owl:sameAs* (=) and *rdf:type* links. Dashed arrows represent discovered schema relations. The system identifies the subset of *dbpedia:Person* instances, which overlaps with DBLP *foaf:Person* instances, as a union of classes defined in YAGO.

---

[6] 196 links in total in DBPedia 3.2 on 13/06/2009

Applying name comparison for all *foaf:Person* and *dbpedia:Person* individuals is likely to produce many false positive results because of ambiguity of personal names. Before performing instance matching we need to narrow the context and exclude from comparison individuals which are unlikely to appear in both datasets. Since the actual schema ontologies used by repositories, which have to be connected, are not sufficiently detailed, then evidence data defined in other data sources should be utilized.

For the reasons outlined in section 2, instance-based ontology matching techniques are particularly suitable to infer schema-level mappings in the Linked Data environment. These techniques operate on ontologies, which share the same sets of individuals. Sometimes this scenario is present in Linked Data repositories directly: for instance, in the example shown in Fig. 1, DBPedia individuals are structured by the DBPedia own ontology, but also have *rdf:type* links to the classes defined in the YAGO[7] and Umbel ontologies. However, more often such sets can be constructed by clustering together individuals connected via existing *owl:sameAs* coreference links. Such sets are likely to be incomplete because intermediate datasets may not contain all individuals represented in their neighbour repositories or because some links on the path are not discovered, but they can still be used to derive relations between classes.

A crucial difference between the Linked Data environment and the traditional ontology matching scenario, which focuses on matching two ontologies, is the possibility of using individuals and concepts defined in other repositories and links between them as background knowledge. In our approach we exploit two types of background knowledge:

- Schema-level evidence from third-party repositories.
- Data-level evidence from third-party repositories.

In the following subsections 3.1 and 3.2 we will describe these types of evidence and briefly outline how they are used to produce schema-level mappings. Then, in the subsection 3.3 we will briefly describe the overall workflow of our KnoFuss system, which employs these mappings to discover new coreference links between individuals.

3.1   Schema-Level Evidence

In the example shown in Fig. 1 the problem is caused by insufficiently detailed classification of individuals provided by the repositories' ontologies. In this situation additional schema-level information has to be introduced from external sources.

Individuals in DBPedia are connected by *rdf:type* links to classes defined in the YAGO repository. The YAGO ontology is based on Wikipedia categories and provides a more detailed hierarchy of classes than the DBPedia ontology. Our algorithm uses this external ontology to identify the subset of DBPedia which overlaps with the DBLP repository. The procedure involves the following steps:

---

[7] http://www.mpi-inf.mpg.de/yago-naga/yago/

1. Construct clusters of identical individuals from DBPedia and DBLP using existing *owl:sameAs* mappings. In this scenario each cluster corresponds to one *owl:sameAs* link and contains two individuals: one from DBLP and one from DBPedia.
2. Connect these clusters to classes in the YAGO and DBLP ontologies respectively. In the later case only the class *foaf:Person* is involved. For example, the cluster containing the individual *dbpedia:Andrew_Herbert* is connected to several YAGO classes (e.g., *yago:MicrosoftEmployees*, *yago: BritishComputerScientists* and *yago:LivingPeople*) and to *foaf:Person*.
3. Infer mappings between YAGO classes and the *foaf:Person* class used in DBLP using instance-based matching (see section 4). A set of overlapping YAGO classes is produced as a result: e.g., mappings between *foaf:Person* and *yago:MicrosoftEmployees* and between *foaf:Person* and *yago: BritishComputerScientists*.
4. Run instance-level coreference resolution for individuals belonging to mapped classes to discover more coreference resolution links. For example, at this stage we discover the link between the individual *dbpedia:Charles_P._Thacker* belonging to the class *yago:MicrosoftEmployees* and its DBLP counterpart, which did not exist in the original link set.

3.2   Data-Level Evidence

Data-level evidence includes individuals defined in third-party repositories and coreference links to and from them. The scenario shown in Fig. 2 illustrates the use of this type of evidence. The LinkedMDB repository[8] contains data about movies structured using a special Movie ontology. Many of its individuals are also mentioned in DBPedia under different URIs. Some of these coreferent individuals, in particular, those belonging to classes *movie:film* and *movie:actor*, are explicitly linked to their counterparts in DBPedia by automatically produced *owl:sameAs* relations. However, for individuals of some classes, direct links are not available. For instance, there are no direct links between individuals of the class *movie:music_contributor* representing composers, whose music was used in movies, and corresponding DBPedia resources. Then, there are relations of the type *movie:relatedBook* from movies to related books in RDF Book Mashup but not to books mentioned in DBPedia. Partially, such mappings can be obtained by computing a transitive closure for individuals connected by coreference links. However, many links are missed in this way because of the omission of an intermediate link in a chain (e.g., 32% of *movie:music_contributor* instances were not connected to corresponding DBPedia instances). Again, such links can be discovered by comparing corresponding subsets of LinkedMDB and DBPedia directly. To discover these subsets our approach computes a transitive closure over existing mappings and combines co-referring individuals into clusters. These clusters are used as evidence for the schema matching procedure to derive schema-level mappings: in our example, we derive the correspondence between

---

[8] http://data.linkedmdb.org/

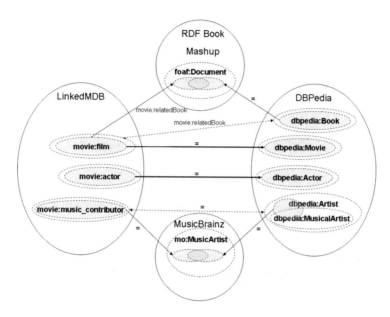

Fig. 2. LinkedMDB and DBPedia: exploiting instance-level coreference links with third-party datasets. Solid arrows show existing *owl:sameAs* (=) and *movie:relatedBook* links. Dashed arrows connect sets containing potentially omitted links.

movie:music_contributor and *dbpedia:Artist* and the *rdfs:range* relation between the property *movie:relatedBook* and the class *dbpedia:Book*. These mappings are used afterwards to perform coreference resolution over related subsets.

3.3   KnoFuss Architecture and the Fusion Workflow

Performing integration at the data level constitutes the main focus of our data fusion tool called KnoFuss [2], originally implemented to merge datasets (knowledge bases) structured according to the same ontology. The KnoFuss architecture implements a modular framework for semantic data fusion. The fusion process is divided into subtasks as shown in the Fig. 3 and the original architecture focuses on its second stage: *knowledge base integration*. The first subtask is *coreference resolution*: finding potentially coreferent instances based on their attributes. The next stage, *knowledge base updating*, refines coreferencing results taking into account ontological constraints, data conflicts and links between individuals (algorithms of this stage were not employed in the tests described in this paper). The tool uses SPARQL queries both to select subsets of data for processing and to select the most appropriate processing techniques depending on the type of data. In order to make the tool applicable to datasets, which use different ontologies, it was extended with the *ontology integration* stage. This stage consists of two subtasks: *ontology matching*, which produces mappings between schema-level concepts, and *instance transformation*, which uses these mappings to translate

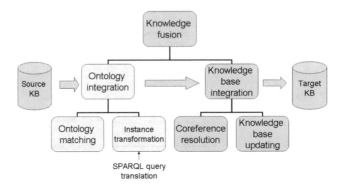

Fig. 3. Fusion task decomposition incorporating schema matching

SPARQL queries into different ontologies, so that the following stages can operate in the same way as in the single-ontology scenario. In the workflow described in this paper the *ontology integration* stage is performed in a "bottom-up" way exploiting links defined at the data level while the *knowledge base integration* stage uses them in a "top-down" way. In sections 4 and 5 we will describe these two main stages of the workflow in more detail.

4   Inferring Schema Mappings: The "Bottom-Up" Stage

The process of inferring schema mappings starts by composing clusters of individuals from different repositories. At this stage pairs of connected individuals belonging to different datasets are retrieved. Then the system forms clusters of coreferent individuals by computing transitive closures over available links.

These clusters represent the basic evidence, which we use to infer schema-level mappings. For each individual in a cluster we extract its class assertions. We consider that a cluster belongs to a certain class if at least one individual from this cluster belongs to a class. At this stage classes which are used in different datasets are always treated as different classes, even if they have the same URI. For instance, in our example, the Movie ontology used in LinkedMDB and the Music ontology used in Musicbrainz both extend the standard FOAF ontology. But we treat the class foaf:Person in both these ontologies as two distinct classes: *foaf:Person@Movie* and *foaf:Person@Music*. This is done in order to discover the actual usage pattern for each class, which may implicitly extend its ontological definition, as was pointed out before.

At the next step we construct mappings between classes. As we said before, instead of equivalence and subsumption the algorithm produces a special type of relation, which we called *#overlapsWith*. Formally this relation is similar to the *umbel:isAligned* property[9] and states that two classes share a subset of

---

[9] http://www.umbel.org/technical_documentation.html

their individuals. However, in our case, a quantitative assessment of the relation is necessary to distinguish between strongly correlated classes (like *dbpedia:Actor* and *movie:actor*) and merely non-disjoint ones (like *movie:actor* and *dbpedia:FootballPlayer*, which share several instances such as "Vinnie Jones"). This relation has a quantitative measure varying between 0 (meaning the same as *owl:disjointWith*) and 1 (meaning that there is a *rdfs:subClassOf* relation in one direction or both). We calculate similarities between classes based on the sets of clusters assigned to them. Two criteria are used to produce the output set of *#overlapsWith* relations between classes:

1. The value of the overlap coefficient compared to a threshold.

$$sim(A, B) = overlap(c(A), c(B)) = \frac{|c(A) \cap c(B)|}{min(c(A), c(B))} \geq t_{overlap},$$

   where c(A) and c(B) are sets of instance clusters assigned to classes A and B respectively. The overlap coefficient was chosen as a similarity metric to reduce the impact of dataset population sizes. If the first dataset is populated to a lesser degree than the second one, then for most classes $|c(A)| << |c(B)|$. In this case relatively small changes in $|c(B)|$ would have a big impact on such distance metrics as Jaccard score or Dice coefficient, while different values of $|c(A)|$ would not change the value significantly.

2. Choosing the "best match" mapping among several options. It is possible that for the same class, A, several relations are produced, which connect it to classes at different levels of the hierarchy. For instance, we can have both *overlapsWith(A, B)* and *overlapsWith(A, C)*, where $B \sqsubseteq C$. In our scenario the relation with a more generic class will always override the more specific one: it will mean that individuals of A will have to be compared to individuals of C. Only one such relation should be chosen, and the original overlap coefficient value cannot be used as a criterion: if $|c(A)| \leq |c(B)|$, then the relation $sim(A, B) \leq sim(A, C)$ always holds. Selecting the relation with the more generic class will mean that possibly more coreference resolution links will be discovered between individuals of A and $C \setminus B$. On the other hand, if the overlap between A and $C \setminus B$ is small and $|C \setminus B|$ is big, then more erroneous mappings can be produced and the damage to results quality due to the loss of precision will be higher than a possible gain from recall increase. To make this decision we use the following criterion:

$$\frac{(|A \cap C| - |A \cap B|)/|A \cap C|}{(|C| - |B|)/|C|} \geq \lambda,$$

   where λ reflects both the expected ratio of errors for the instance coreference resolution algorithm and relative importance of precision comparing to recall. If the inequality holds, then *overlapsWith(A, C)* is chosen, otherwise *overlapsWith(A, B)* is preferred.

In our tests we used an additional restriction: pairs of classes (A, B) where either $|A| = 1$, $|B| = 1$ or $|A \cap B| = 1$ were ignored. This was done to filter out weak overlap mappings such as the one between *foaf:Person@DBLP* and *yago:PeopleFromRuralAlberta*, which led to noise at the instance-level matching stage.

In general, schema-level mappings obtained by the system in this way can be saved and used on its own in any scenario where individuals stored in several datasets have to be queried by class. As was said before, in our approach we focus on one specific use case scenario where these mappings are reused at the "top-down" stage of the KnoFuss workflow to produce coreference resolution links between individuals and improve the recall in comparison with existing relations.

5   Exploiting Inferred Schema Mappings for Coreference Resolution: The "Top-Down" Stage

The schema-level mappings obtained at the previous stage are used to identify sets of individuals in different repositories, which are likely to contain equivalent individuals not discovered before. These sets of relevant schema-level mappings are provided as input to the *instance transformation* stage of the KnoFuss tool (Fig. 3). It uses schema mappings to translate SPARQL queries, which select sets of individuals to be compared, from the vocabulary of one ontology into the terms of another one. It is possible that a class in one ontology is found to be connected to several classes in another ontology (not related via a *rdfs:subClassOf* relation). Such mappings are aggregated into a single *ClassUnion* mapping. For instance, in our DBLP example to select individuals from the DBLP dataset we use the following query:

SELECT *?uri* **WHERE**
{ *?uri rdf:type foaf:Person*}

To select potentially comparable individuals from the DBPedia repository this query is translated into:

SELECT *?uri* **WHERE**
{ { *?uri rdf:type yago:AmericanComputerScientists*}
UNION { *?uri rdf:type yago:GermanComputerScientists*}
UNION { *?uri rdf:type yago:GoogleEmployees*}
UNION... }

Using these translated queries individuals from both repositories are processed in the same way as if they shared the same schema. The system can employ several basic matching techniques, which can be selected and configured depending on the type of data as described in [2].

To avoid redundancy and potential errors, individuals, which were already connected either directly or indirectly via a third-party dataset, are excluded from analysis. The final set of instance-level mappings produced by the tool then can be added to existing ones.

A decision, which has to be taken at this stage, concerns the choice, whether two subsets of individuals should be compared at all: it is possible that all relevant links between individuals were already discovered and the algorithm can only produce errors. A naive way to decide it could be to use an upper threshold on the overlap degree between classes assuming that many existing links between individuals can mean that all possible ones were already discovered. However, this misses important information regarding how existing mappings were produced. Using a coreference resolution system in combination with a link maintenance and trust computation system such as idMesh [20] can be interesting.

6 Evaluation

For our initial experiments we used three scenarios mentioned before:

1. Finding equivalence links between individuals representing people in DBPedia and DBLP (auxiliary dataset: YAGO, gold standard size 1229).
2. Finding equivalence links between *movie:music_contributor* individuals in LinkedMDB and corresponding individuals in DBPedia (auxiliary dataset: Musicbrainz, gold standard size 942).
3. Finding *movie:relatedBook* links between *movie:film* individuals in Linked-MDB and books mentioned in DBPedia (auxiliary dataset: RDF Book Mashup, gold standard size 419).

Our goal was to check the applicability of our approach in general and, in particular, the possibility to improve coreference resolution recall in comparison with already existing links. Thus, all three scenarios were of relatively small scale so that both precision and recall could be checked manually and the actual coreference resolution was performed using simple label-based similarity (Jaro metric[10]). Test results (precision, recall and F1 measure) are given in the Table 1. For each scenario three sets of results are provided:

- Baseline, which involves computing the transitive closure of already existing links.
- Results obtained by the algorithm when applied to all comparable individuals.
- Combined set of existing results and new results obtained by the algorithm.

As was expected, in all cases applying instance-level coreference resolution using automatically produced class-level mappings led to improvement in recall due to the discovery of previously missed mappings, and to a better overall performance, as measured by the F1-measure. In all cases, the best performance and F1-measure was achieved by combining newly produced mappings with existing ones. It means that the algorithms, which produced these sets of links, could generate complementary results and no set of links was redundant.

Obviously, the precision of the combined set of links was lower than the precision of the best algorithm in all three tests. In our tests this decrease was relatively small and was compensated by the increase in recall. However, in cases where the same data were already processed by algorithms of higher quality, the

---

[10] In the first two scenarios the metric was adapted for personal names, e.g., to match complete name with initials.

Table 1. Test results

Dataset	Test	Precision	Recall	F1
	Baseline	0.90	0.14	0.25
DBPedia vs DBLP	All individuals	0.95	0.88	0.91
	Combined set	0.93	0.89	0.91
LinkedMDB vs DBPedia	Baseline	0.99	0.68	0.81
(music contributors)	All individuals	0.98	0.91	0.94
	Combined set	0.98	0.97	0.98
LinkedMDB vs DBPedia	Baseline	0.97	0.82	0.89
(books)	All individuals	0.98	0.90	0.93
	Combined set	0.96	0.97	0.96

situation can be different. It makes the issue of tracing provenance of existing links important, as mentioned in section 5.

Considering the schema matching stage we found two factors which were potential causes of errors. The first factor was insufficient evidence. When only a small number of existing coreference links are available as evidence, distinguishing between "weakly overlapped" and "strongly overlapped" classes is problematic. For example, in the DBPedia-DBLP scenario the class *yago: FellowsOfWolfsonCollege,Cambridge* received a higher overlap score with *foaf: Person@DBLP* than the class *yago:IsraeliComputerScientists*, which in fact was strongly overlapped. This happened because for both of these classes there were only 2 evidence links available and the class *yago: FellowsOfWolfsonCollege,Cambridge* contained fewer instances. At the coreference resolution stage instances of such weakly overlapped classes caused the majority of false positive mappings because of name ambiguity.

The second factor concerned the quality of the ontologies themselves and of the class assertion statements. For instance, in the DBPedia dataset many musicians were not assigned to an appropriate class *dbpedia:MusicalArtist* but instead were assigned to more general classes *dbpedia:Artist* or even *dbpedia:Person*. As a result the "best fit" mappings produced by the algorithm did not correspond to the originally intended meaning of classes, because this originally intended meaning was not followed in the dataset (e.g., based on instance data the class *movie:music_contributor* was mapped to the class *dbpedia:Artist* instead of *dbpedia:MusicalArtist*). More serious issues involved instances being assigned to classes, which were actually disjoint (e.g., the individual *dbpedia:Jesse_Ventura* was classified as both a *dbpedia:Person* and a *dbpedia:TelevisionShow*). While in our scenarios spurious schema mappings caused by these errors were filtered out by the threshold, in other cases their impact can be significant. Explicit specification of ontological constraints can help to deal with such situations.

7   Conclusion and Future Work

In this paper we presented an approach which uses existing links between individuals stored in different repositories as evidence to generate schema-level mappings between classes. A distinctive feature of our approach is the use of information defined in third-party datasets as background knowledge to enhance instance-based ontology matching techniques. These mappings are then used to discover new coreference links between individuals, which were missed before. Our initial experiments have shown an improvement in resulting coreference resolution performance in comparison with existing sets of links. However, there are still issues, which have to be resolved in the future work.

First, we plan to continue our experiments with instance-based schema alignment algorithms over different public datasets in order to evaluate their capabilities when applied to large networks of connected datasets and determine factors and conditions in real-world datasets, which influence their performance, in order to improve the reusability of the approach. In particular, one such factor is the presence of subsets of individuals, for which there are few or no connections available to infer any schema-level patterns.

In the test scenarios described in this paper there was no need for matching properties in addition to classes: only data described by standard attributes such as *rdfs:label*, *foaf:name* and *dc:title* was available. In the future inferring schema mappings between properties and reuse of axioms should be elaborated.

Considering the data-level integration stage, as mentioned in the section 5, there is an issue of automatic assessment of datasets and distinguishing the cases when new coreference links can be discovered with a sufficient precision.

Finally, there are infrastructural issues, which have to be taken into account to make the approach reusable. In particular, this concerns storing, publishing and maintaining both schema-level and instance-level links. There are several interesting directions to follow, such as applying the coreference bundles approach [19] instead of maintaining sets of pairwise *owl:sameAs* links and integrating with the *idMesh* approach [20], which reasons about sets of coreference links and their reliability.

Acknowledgements

This work was funded by the X-Media project (www.x-media-project.org) sponsored by the European Commission as part of the Information Society Technologies (IST) programme under EC grant number IST-FP6-026978.

References

1. Bizer, C., Heath, T., Berners-Lee, T.: Linked data - the story so far. International Journal on Semantic Web and Information Systems (IJSWIS) (to appear)
2. Nikolov, A., Uren, V., Motta, E., de Roeck, A.: Integration of semantically annotated data by the KnoFuss architecture. In: Gangemi, A., Euzenat, J. (eds.) EKAW 2008. LNCS (LNAI), vol. 5268, pp. 265–274. Springer, Heidelberg (2008)
3. Rahm, E., Bernstein, P.A.: A survey of approaches to automatic schema matching. The VLDB Journal 10(4), 334–350 (2001)
4. Euzenat, J., Shvaiko, P.: Ontology matching. Springer, Heidelberg (2007)

5. Gracia, J., Mena, E.: Matching with CIDER: Evaluation report for the OAEI 2008. In: 3rd Ontology Matching Workshop (OM 2008), Karlsruhe, Germany (2008)
6. Zhang, X., Zhong, Q., Li, J., Tang, J., Xie, G., Li, H.: RiMOM results for OAEI 2008. In: 3rd Ontology Matching Workshop (OM 2008), Karlsruhe, Germany (2008)
7. Udrea, O., Getoor, L., Miller, R.J.: Leveraging data and structure in ontology integration. In: SIGMOD 2007, Beijing, China, pp. 449–460 (2007)
8. Aleksovski, Z., Klein, M.C.A., ten Kate, W., van Harmelen, F.: Matching unstructured vocabularies using a background ontology. In: Staab, S., Svátek, V. (eds.) EKAW 2006. LNCS (LNAI), vol. 4248, pp. 182–197. Springer, Heidelberg (2006)
9. Sabou, M., d'Aquin, M., Motta, E.: Exploring the Semantic Web as background knowledge for ontology matching. In: Spaccapietra, S., Pan, J.Z., Thiran, P., Halpin, T., Staab, S., Svatek, V., Shvaiko, P., Roddick, J. (eds.) Journal on Data Semantics XI. LNCS, vol. 5383, pp. 156–190. Springer, Heidelberg (2008)
10. Fellegi, I.P., Sunter, A.B.: A theory for record linkage. Journal of American Statistical Association 64(328), 1183–1210 (1969)
11. Elmagarmid, A.K., Ipeirotis, P.G., Verykios, V.S.: Duplicate record detection: A survey. IEEE Transactions on Knowledge and Data Engineering 19(1), 1–16 (2007)
12. Dong, X., Halevy, A., Madhavan, J.: Reference reconciliation in complex information spaces. In: SIGMOD 2005, pp. 85–96. ACM, New York (2005)
13. Kalashnikov, D.V., Mehrotra, S.: Domain-independent data cleaning via analysis of entity-relationship graph. ACM Transactions on Database Systems 31(2), 716–767 (2006)
14. Saïs, F., Pernelle, N., Rousset, M.C.: L2R: a logical method for reference reconciliation. In: AAAI 2007, Vancouver, BC, Canada, pp. 329–334 (2007)
15. Shen, W., DeRose, P., Vu, L., Doan, A., Ramakrishnan, R.: Source-aware entity matching: A compositional approach. In: ICDE 2007, Istanbul, Turkey (2007)
16. Volz, J., Bizer, C., Gaedke, M., Kobilarov, G.: Silk - a link discovery framework for the web of data. In: Workshop on Linked Data on the Web (LDOW 2009), Madrid, Spain (2009)
17. Hassanzadeh, O., Consens, M.: Linked movie data base. In: Workshop on Linked Data on the Web (LDOW 2009), Madrid, Spain (2009)
18. Bouquet, P., Stoermer, H., Bazzanella, B.: An Entity Name System (ENS) for the Semantic Web. In: Bechhofer, S., Hauswirth, M., Hoffmann, J., Koubarakis, M. (eds.) ESWC 2008. LNCS, vol. 5021, pp. 258–272. Springer, Heidelberg (2008)
19. Glaser, H., Jaffri, A., Millard, I.: Managing co-reference on the semantic web. In: Workshop on Linked Data on the Web (LDOW 2009), Madrid, Spain (2009)
20. Cudré-Mauroux, P., Haghani, P., Jost, M., Aberer, K., de Meer, H.: idMesh: Graph-based disambiguation of linked data. In: WWW 2009, Madrid, Spain, pp. 591–600. ACM, New York (2009)

Social Semantic Rule Sharing and Querying in Wellness Communities

Harold Boley, Taylor Michael Osmun, and Benjamin Larry Craig

Institute for Information Technology,
National Research Council of Canada,
Fredericton, NB, E3B 9W4, Canada

Abstract. In this paper we describe the Web 3.0 case study WellnessRules, where ontology-structured rules (including facts) about wellness opportunities are created by participants in rule languages such as Prolog and N3, and translated for interchange within a wellness community using RuleML/XML. The wellness rules are centered around participants, as profiles, encoding knowledge about their activities, nutrition, etc. conditional on the season, the time-of-day, the weather, etc. This distributed knowledge base extends fact-only FOAF profiles with a vocabulary and rules about wellness group networking. The communication between participants is organized through Rule Responder, permitting translator-based reuse of wellness profiles and their distributed querying across engines. WellnessRules interoperates between rules and queries in the relational (Datalog) paradigm of the pure-Prolog subset of POSL and in the frame (F-logic) paradigm of N3. These derivation rule languages are implemented in the engines OO jDREW and Euler, and connected via Rule Responder to support wellness communities. An evaluation of Rule Responder instantiated for WellnessRules found acceptable Web response times.

1 Introduction

Medicine 2.0 [Eys08], Health 2.0 [MO09], and Wellness 2.0 have emerged as interconnected application areas of Web 2.0 (Social Web) techniques. Web 2.0 combined with Semantic Web techniques is currently leading to Web 3.0 (Social Semantic Web) techniques, which can also be applied in these areas. As part of NRC-IIT's Health & Wellness and Learning & Training efforts, we are exploring Wellness 3.0, employing Social Semantic Web rules plus ontologies to plan wellness-oriented activities and nutrition.

We focus here on WellnessRules[1], a system supporting the management of wellness practices within a community based on rules plus ontologies. The idea is the following. As in Friend of a Friend (FOAF)[2], people can choose a (community-unique) nickname and create semantic profiles about themselves, here about their wellness practices, for their own planning and to network with other people supported by a system that 'understands' those profiles. As in FindXpRT [LBBM06],

---

[1] http://ruleml.org/WellnessRules/
[2] http://www.foaf-project.org/

A. Gómez-Pérez, Y. Yu, and Y. Ding (Eds.): ASWC 2009, LNCS 5926, pp. 347–361, 2009.

such FOAF-like fact-only profiles are extended with rules to capture conditional person-centered knowledge such as each person's wellness activity depending on the season, the time-of-day, the weather, etc. People can use rules of various refinement levels and rule languages ranging from pure Prolog to N3, which will be interoperated through RuleML/XML [Bol07]. Like our (RuleML-20xy) SymposiumPlanner [CB08] (and unlike FindXpRT), WellnessRules is based on Rule Responder [PBKC07, CB08], which is itself based on the Mule Enterprise Service Bus (ESB).

We will discuss an example where John (p0001) advertises Prolog-style rules on his wellness community profile, including a refinement of the following: p0001 may do outdoor running if it is summer and not raining. Hence, Peter and Paul can find p0001 via Prolog or N3 queries to Rule Responder expressing their own preferences, so that an initial group might be formed. Interoperating with translators, WellnessRules thus frees participants from using any single rule language. In particular, it bridges between Prolog as the main Logic Programming rule paradigm and N3 as the main Semantic Web rule paradigm.

The distributed nature of Rule Responder profiles, each queried by its own (copy of an) engine, permits scalable knowledge representation and processing. Since participants of a wellness community are supposed to meet in overlapping groups for real-world events such as skating, this kind of community (unlike a virtual-only community) has a maximal effective size (which we estimate to be less than 1000 participants). Beyond that size, it can be split into two or more subcommunities based on preferred wellness practices, personal compatibility, geographic proximity, etc. Rule Responder support thus needs to extend only to that maximal size, but can be cloned as subcommunities emerge.

The rest of the paper is organized as follows. Section 2 details the design goals with a focus on Rule Responder. Section 3 discusses the hybrid global knowledge bases of WellnessRules. Section 4 explains its local knowledge bases distributed via Rule Responder. Section 5 focusses on the interoperation between Prolog and N3. Section 6 explains and evaluates Rule Responder querying of WellnessRules knowledge. Section 7 concludes the paper. Appendix A contains two rule signatures of WellnessRules. Appendix B shows its RDFS taxonomy.

2   Design Goals and Rule Responder Instantiation

A range of design goals is being pursued with the WellnessRules prototype:

1. Identify a language of appropriate expressiveness for wellness rules in the layered family of RuleML [Bol07], RIF [BK09], and N3 [BLCK+08] such as unary/binary vs. n-ary Datalog and rules with or without slots, objects, negation as failure (Naf), etc.
2. Identify a language for wellness ontologies in the layered family of OWL, OWL 2, and RDFS such as subClassOf taxonomies, description logics from ALC to SHOIN, etc. The ontology language should be combined with the rule language.

3. Permit rule plus ontology authoring in different human-oriented syntaxes, while translating the rulebases plus ontologies to and from XML for interchange.
4. Permit different rule- (e.g., OO jDREW, Prova, and Euler) plus ontology-oriented (e.g., Racer, Pellet, and HermiT) engines to execute local rulebases plus ontologies, while transporting XML rulebases, ontologies, queries, and answers over an ESB.
5. Create profile examples and templates of typical wellness practitioners from which new participants can glean ideas and copy & edit for their own profiles.
6. Create an infrastructure to support wellness communities in the management of their existing practices, the adoption of new practices from participants, and the formation of ad hoc groups through profile querying. This also prepares the social network analysis of evolving WellnessRules communities.
7. Explore the appropriateness of the rule plus ontology languages and engines for expressing and translating the knowledge required in wellness profiles.
8. Explore the stand-alone efficiency and network scalability of the distributed WellnessRules architecture, including its ESB and engines, w.r.t. the different languages employed.

The current version of WellnessRules focusses goal 1 on Datalog with Naf and on N3 [BLCK$^+$08] with scoped Naf, interoperating between these rule paradigms through RuleML/XML (a Naf Core is not currently planned for RIF [BK09]). It focusses goal 2 on light-weight ontologies, namely a sorted Datalog and an N3 with RDFS subClassOf taxonomies, for both using sorted (or typed) variables in RuleML/XML. Accordingly, we focuss goal 3 on POSL [Bol04] and N3, and goal 4 on their engines OO jDREW[3] and Euler[4], respecively. Goal 5 has been a joint effort with the wellness community at NRC-IIT Fredericton.

Goal 6 instantiates the Rule Responder multi-agent architecture as follows: Rule Responder's virtual organization is instantiated to a wellness community. An organizational agent (OA) becomes an assistant for an entire wellness community. Each personal agent (PA) becomes an assistant for one participant. Newcomers and participants can assume the role of an external agent (EA), (indirectly) querying participants' profiles.

Rule Responder uses the following sequence of steps: An EA asks queries to an OA. The OA maps and delegates each query to the PA(s) most knowledgeable about it. Each PA poses the query to its local rulebase plus ontology, sending the derived answer(s) back to the OA. The OA integrates relevant answers and gives the overall answer(s) to the EA, by default not revealing the coordinates of the answering PA(s).

In this way, the OA acts as a mediator that protects the privacy of profiles of participants in a wellness community. Participants within the same community can of course later decide to reveal their real name and open up their wellness profiles for (direct) querying by selected other participants.

---

[3] http://www.jdrew.org/oojdrew/
[4] http://eulersharp.sourceforge.net/

The above Rule Responder steps have been instantiated earlier, including to the SymposiumPlanner system [CB08]. The new instantiation for the Wellness-Rules system relates to this in interesting ways:

Like SymposiumPlanner's OA stands for a specific symposium such as RuleML-2009, WellnessRules' OA stands for a specific wellness community such as the one at NRC-IIT Fredericton. Like SymposiumPlanner's PAs formalize knowledge to assist (publicity, panel, etc.) chairs within a symposium, Wellness-Rules' PAs formalize knowledge to assist (running, skating, etc.) coaches within a wellness community.

Unlike in SymposiumPlanner, where, besides the chairing participants, regular (symposium) participants *are not* supported by PAs, in WellnessRules, besides the coaching participants, regular (wellness) participants *are* supported by PAs: Each WellnessRules participant publishes a more or less specific profile used by his/her PA to respond to queries about his/her wellness preferences, constraints, etc.

Goal 7, the expressiveness exploration of WellnessRules, so far found that both the Prolog paradigm of top-down n-ary relational rules and the N3 paradigm of bottom-up binary frame rules can express the required Web 3.0 knowledge, although more directly and compactly in pure Prolog and more closely to Semantic Web standards in N3. Goal 8, the efficiency exploration, so far found that Rule Responder instantiated for WellnessRules has acceptable Web efficiency and provides a good balance between knowledge centralization and distribution.

3   Hybrid Global Knowledge Bases in WellnessRules

WellnessRules employs a hybrid combination [Bol07] of ontologies and rules. While the entire ontology and a portion of the rulebase is globally shared by all participants (agents), the other portion of the rulebase is locally distributed over the participants (agents). The global knowledge base (re)uses an ontology and defines rules about activities, locations, forecasts, etc.

As its (light-weight) ontology component, WellnessRules employs `subClassOf` taxonomies. We reuse parts of the Nuadu ontology collection [SLKL07], especially from the Activity and Nutrition ontologies. WellnessRules currently employs an `Activity` taxonomy using Nuadu classes (based on activity codes [AHW+00]) `Running`, `Walking`, `WaterSports` subsuming `SwimmingCalm`, `WinterSports` subsuming `IceSkating`, and `Sports` subsuming a WellnessRules class `Baseball`, as well as WellnessRules classes `Hiking`, and `Yoga`. Appendix B contains the corresponding RuleML- and N3-readable RDFS `subClassOf` statements.

As its rule component, WellnessRules employs Naf Datalog POSL and N3 with scoped Naf. We restrict the use of Naf Datalog POSL to atoms with positional arguments, leaving F-logic-like frames with property-value slots to N3, thus demonstrating the range of our approach through complementary rule styles. For that reason, the POSL syntax corresponds to pure-Prolog syntax except that POSL variables are prefixed by a question mark while Prolog variables are upper-cased.

This Datalog POSL sublanguage uses (positional) n-ary relations (or, predicates) as its central modeling paradigm. N3 instead uses (unordered) sets of binary relations (or, properties) centered around object identifiers (OIDs, in the role of 'subjects' in N3). For example, in POSL we use the 4-ary predicate `meetup` with a positional signature: `meetup(?MapID,?Activity,?Ambience,?Location)`. In N3 this becomes a slotted signature with subject `_:meetup`, an `rdf:type` of `:Meetup`, and the 4 arguments as the remaining slots:

```
_:meetup
    rdf:type      :Meetup;
    :mapID        ?MapID;
    :activity     ?Activity;
    :inOut        ?Ambience;
    :location     ?Location.
```

Generally, as a foundation for our interchange study, the signatures of all WellnessRules relations are described in a Datalog-N3-generalizing manner, using the same mnemonic names for predicate arguments in Datalog and properties in N3. The WellnessRules signatures needed here are given in appendix A.

Shared rules (including underlying facts) of the WellnessRules system were extracted from participants and collected for the rulebase of the OA. They formalize global knowledge of the WellnessRules system.

For example, this is a global `meetup` fact according to the above positional signature: `meetup(m0001,run,out,conniesStation)`. Similarly, this is its slotted counterpart:

```
:meetup_1
    rdf:type      :Meetup;
    :mapID        :m0001;
    :activity     :run;
    :inOut        :out;
    :location     :conniesStation.
```

Both express that one `meetup` for `run` activities of the supported wellness community is `conniesStation` as found on map `m0001`.

An example of a global POSL rule defines a `participation` as follows:

```
participation(?ProfileID,?Activity,?Ambience,?MinRSVP,?MaxRSVP) :-
    groupSize(?ProfileID,?Activity,?Ambience,?Min,?Max),
    greaterThanOrEqual(?MinRSVP,?Min),
    lessThanOrEqual(?MaxRSVP,?Max).
```

As in FindXpRT, the first argument of a WellnessRules conclusion predicate always is the person the rule is about. Similar to Prolog, the rule succeeds for its five positional arguments if the acceptable `groupSize` of the participant with ?ProfileID, for an ?Activity in an ?Ambience, is between ?Min and ?Max, and the emerging group has size between $?MinRSVP \geq ?Min$ and $?MaxRSVP \leq$?Max, where `greaterThanOrEqual` and `lessThanOrEqual` are SWRL built-ins as implemented in OO jDREW 0.961.

The corresponding global N3 rule represents this in frame form as follows:

```
{
 ?rsvpQuery
     rdf:type      :RSVPQuery;
     :profileID    :p0001;
     :minRSVP      ?MinRSVP;
     :maxRSVP      ?MaxRSVP.

 ?groupSize
     rdf:type      :GroupSize;
     :profileID    ?ProfileID;
     :activity     ?Activity;
     :inOut        ?Ambience;
     :min          ?Min;
     :max          ?Max.

 ?MinRSVP math:notLessThan ?Min.

 ?MaxRSVP math:notGreaterThan ?Max.
}
=>
{
 _:participation
     rdf:type      :Participation;
     :profileID    :p0001;
     :activity     ?Activity;
     :inOut        ?Ambience;
     :min          ?MinRSVP;
     :max          ?MaxRSVP.
}.
```

Here, the first premise passes the input arguments ?MinRSVP and ?MaxRSVP into the rule (cf. its use in section 6). The remaining premises correspond to those in the POSL version, where math:notLessThan and math:notGreaterThan are N3 built-ins as implemented in Euler.

The global OA rulebase is being maintained in both languages at http://ruleml.org/WellnessRules/WR-Global.posl and *.n3.

4   Locally Distributed Knowledge Bases in WellnessRules

Each PA has its own local rules, which were selected from profiles created by participants of the NRC-IIT Fredericton wellness community.

This is an example of a local POSL rule from the PA rulebase of a participant p0001, defining the main predicate myActivity for running:

```
myActivity(p0001,?:Running,out,?MinRSVP,?MaxRSVP,?StartTime,?EndTime,
                                    ?Place,?Duration,?Level) :-
    calendar(p0001,?Calendar),
    event(?Calendar,?:Running,possible,?StartTime,?EndTime),
    participation(p0001,run,out,?MinRSVP,?MaxRSVP),
```

```
season(?StartTime,summer),
forecast(?StartTime,sky,?Weather),
notEqual(?Weather,raining),
map(p0001,?Map),
meetup(?Map,run,out,?Place),
level(p0001,run,out,?Place,?Duration,?Level),
fitness(p0001,?StartTime,?ExpectedFitness),
greaterThanOrEqual(?ExpectedFitness,?Level),
goodDuration(?Duration,?StartTime,?EndTime).
```

The rule conclusion starts with the person's profile ID, p0001, followed by the kind of activity, ?:Running (the anonymous variable "?" has type Running), and its ambience, outdoors, followed by variables for the group limits ?MinRSVP and ?MaxRSVP, the earliest ?StartTime and ?EndTime, its actual ?Duration and its ?Level. The rule premises query p0001's ?Calendar, an event of a possible (or tentative) ?:Running, the participation predicate (for simplicity and N3 compatibility called with a constant run rather than the typed variable ?:Running), an appropriate season and forecast, p0001's ?Map, a meetup ?Place, the required level less than the expected fitness, and a goodDuration.

The corresponding local N3 rule is given abridged below (complete, online at http://ruleml.org/WellnessRules/PA/p0001.n3):

```
{
 ...

?forecast
    rdf:type     :Forecast;
    :startTime   ?StartTime;
    :aspect      :sky;
    :value       ?Weather.

?Weather log:notEqualTo :raining.

 ...
}
=>
{
 _:myActivity
    rdf:type        :MyActivity;
    :profileID      :p0001;
    :activity       :Running;
    :inOut          :out;
    :minRSVP        ?MinRSVP;
    :maxRSVP        ?MaxRSVP;
    :startTime      ?StartTime;
    :endTime        ?EndTime;
    :location       ?Place;
    :duration       ?Duration;
    :fitnessLevel   ?FitnessLevel.
}.
```

The online version of the above POSL rule employs the premise `naf(forecast(?StartTime,sky,raining))` instead of separate `forecast` and `notEqual` premises. For the N3 online version, the above `log:notEqualTo` built-in call is more convenient. An irreducible Naf used in POSL's online version adds the following premises in the `myActivity` rule (after the current `event` premise):

```
yesterday(?StartTime,?StartTimeYtrday,?EndTime,?EndTimeYtrday),
naf(event(?Calendar,?:Running,past,?StartTimeYtrday,?EndTimeYtrday)),
```

It makes sure that p0001's calendar does not contain a running event on the day before. The counterpart in N3 could use `log:notIncludes`, which in Euler, as in our online version, is replaced with `e:findall`, checking that the result is the empty list, '`()`'.

The resulting PA rulebases, which require Datalog with Naf and N3 with '`()`'-`e:findall`, are being maintained in both of these languages at `http://ruleml.org/WellnessRules/PA`, e.g. those for p0001 at `http://ruleml.org/WellnessRules/PA/p0001.posl` and `*.n3`.

5   Cross-Paradigm Rulebase Alignment and Translation

The WellnessRules case study includes a testbed for the interoperation (i.e., alignment and translation) of rulebases in the main two rule paradigms: Prolog-style (positional) relations and N3-style (slotted) frames. In our interoperation methodology, we make iterative use of alignment and translation: An initial alignment permits the translation of parts of a hybrid knowledge base. This then leads to more precise alignments, which in turn leads to better translations, etc. Using this methodology for WellnessRules, we are maintaining a relational as well as a frame version of the rules, both accessing the same, independently maintained, RDFS ontology.

For rulebase alignment, the signatures of WellnessRules relations and frames are aligned in a shared signature document (cf. appendix A), discussed in section 3, which includes the slot names for frames.

For rulebase translation, we first use a pair of online translators (`http://ruleml.org/posl/converter.jnlp`) between the human-oriented POSL syntax and its XML serialization in OO RuleML. Based on the RDF-RIF combinations in [dB09], similar translators are being developed between N3, RIF, and RuleML.

The interoperation between WellnessRules PAs that use different rule paradigms is then enabled by RuleML, which has sublanguages for both the relational and the frame paradigms, so that the cross-paradigm translations can use the common XML syntax of RuleML.

The alignment of sample relations and frames in sections 3 and 4 suggests translations between both paradigms. We consider here translations that are 'static' or 'at compile-time' in that they take an entire rulebase as input and return its entire transformed version. We thus make the 'closed-arguments' assumption of fixed signatures for relations and frames. In particular, the arity of relations cannot change at run-time and no slots can be dynamically added or removed from a frame. The translations work in both directions:

Objectify: Mapping from an n-ary relation `rel` to a frame, this constructs a new frame with a generated fresh OID `rel_j`, where $j > 0$ is the first integer making `rel_j` a unique name, and with the argument positions `p_1, p_2, ..., p_n` as slot (or property) names.

Positionalize: Mapping from a frame to a relation, this constructs a new relationship with the first argument taking the frame OID and the remaining arguments taking the slot values of the sorted slot names from all frames of OID's class (null values for those properties not used in the current frame).

Formally, positionalizing is specified as follows, using POSL's frame notation with slots arrows (->) as in F-logic and an OID separated from its slots by a hat (^).

1. Unite all slot names from all frames whose OID is an instance of a class cl into a finite set SN_{cl} of n_{cl} elements.

2. Introduce $(SN_{cl}, <)$ as a total order '<' over SN_{cl}, where '<' usually is the lexicographic order. Assume without loss of generality that the elements of SN_{cl} are $prop_1 < ... < prop_{n_{cl}}$.

3. For each frame frel = $cl(oid\hat{\ }prop_{k_1}$->$TERM_{k_1};...;prop_{k_m}$->$TERM_{k_m})$ assume without loss of generality that $prop_{k_1} < ... < prop_{k_m}$. Replace frel by a relation frel' = $cl(oid, TERMCOMP_1,...,TERMCOMP_{n_{cl}})$, where for $1 \leq i \leq n_{cl}$ and $1 \leq j \leq m$ we have $TERMCOMP_i = TERM_{k_j}$ if $i = k_j$ and $TERMCOMP_i = \bot$ otherwise ('$\bot$' is the null value formalized as the bottom element of the taxonomy, e.g. owl:NOTHING, which is equal only to itself, not to any other sort, constant, or variable).

Step 3 can be thought of as 'replenishing' the lexicographically sorted slots of a frame frel with slots $prop_x$->$\bot$ for all slot names $prop_x$ 'missing' for their class cl, and then making cl the relation name, inserting the oid, and omitting all slot names (keeping only their slot values).

An XSLT implementation of such a translator is available online (`http://ruleml.org/ooruleml-xslt/oo2prml.html`).

For example, the POSL `meetup` fact of section 3 is serialized in positional RuleML thus:

```
<Atom>
  <Rel>meetup</Rel>
  <Ind>m0001</Ind>
  <Ind>run</Ind>
  <Ind>out</Ind>
  <Ind>conniesStation</Ind>
</Atom>
```

Since RuleML allows intermediate forms, a translation between such a relation and a frame need not insert the relationship OID as a new first positional argument but can directly introduce it as a role tag:

```
<Atom>
  <oid><Ind iri=":meetup_1"/></oid>
  <Rel>meetup</Rel>
  . . .
</Atom>
```

Extending the mappings at `http://ruleml.org/indoo/n3ruleml.html`, this is an intermediate step towards the following frame in OO RuleML, where the RuleML `Rel` represents the `rdf:type` and literals are written as `Data`:

```
<Atom>
  <oid><Ind iri=":meetup_1"/></oid>
  <Rel iri=":Meetup"/>
  <slot>
    <Ind iri=":mapID"/>
    <Ind iri=":m0001"/>
  </slot>
  <slot>
    <Ind iri=":activity"/>
    <Data>run</Data>
  </slot>
  <slot>
    <Ind iri=":inOut"/>
    <Data>out</Data>
  </slot>
  <slot>
    <Ind iri=":location"/>
    <Data>conniesStation</Data>
  </slot>
</Atom>
```

Practical objectifying looks up the slot names in the shared signature document (cf. appendix A).

For the translation of a rule, the above relation-frame translation is applied to the relation (frame) in the conclusion and to all the relations (frames) in the premises. For a rulebase, the translation then applies to all of its rules.

With the above-discussed human-oriented syntax translators, rulebases containing rules like the `myActivity` rule in section 4 can thus be translated via Prolog, POSL, RuleML (relations, frames), and N3. These translators permit rule, query, and answer interoperation, via RuleML/XML, for the Rule Responder infrastructure of WellnessRules communities.

6   Distributed Rule Responder Querying of WellnessRules

On the basis of Rule Responder (cf. section 2) the knowledge bases of sections 3 and 4 can be queried, using the translators of section 5 for interoperation. The implemented Rule Responder for WellnessRules is available for online use at `http://ruleml.org/WellnessRules/RuleResponder/`.

For example, this is a POSL query regarding p0001's wellness profile, for execution by a top-down engine such as OO jDREW TD:

```
myActivity(p0001,?:Running,out,1:Integer,20:Integer,"2009-06-10T10:15:00",
                 "2009-06-10T11:15:00",?Place,?Duration,?Level)
```

It uses the rule from section 4 to check whether p0001 will possibly be ?:Running, outdoors, in a group of 1:Integer to 20:Integer, between start time "2009-06-10T10:15:00" and end time "2009-06-10T11:15:00", filling three named variables. Using further rules and facts from p0001's profile (http://ruleml.org/WellnessRules/PA/p0001.posl), it produces multiple solutions, binding the meetup ?Place, the ?Duration, and the required fitness ?Level.

The corresponding N3 query for execution by a bottom-up engine such as EulerSharp EYE uses a temporary fact to pass the input arguments:

```
:rsvpQuery
    rdf:type      :RSVPQuery;
    :profileID    :p0001;
    :minRSVP      1;
    :maxRSVP      20.
```

The N3 query itself then is as follows:

```
@prefix : <wellness_profiles#>.
@prefix rdf: <http://www.w3.org/1999/02/22-rdf-syntax-ns#>.

_:myActivity
    rdf:type         :MyActivity;
    :profileID       :p0001;
    :activity        :Running;
    :inOut           :out;
    :minRSVP         ?MinRSVP;
    :maxRSVP         ?MaxRSVP;
    :startTime       "2009-06-10T10:15:00";
    :endTime         "2009-06-10T11:15:00";
    :location        ?Place;
    :duration        ?Duration;
    :fitnessLevel    ?FitnessLevel.
```

After declaring two prefixes, it builds an existential '(_)' node, _:myActivity, using slots for the fixed parameters and the fact-provided ?MinRSVP and ?MaxRSVP bindings to fill slots with the ?Place, ?Duration, and ?Level solutions.

An evaluation of the response times of the Mule infrastructure and the Rule Responder engines (OO jDREW, Euler, and Prova) instantiated for Wellness-Rules has been conducted using the previously discussed scenario. We found that this Rule Responder instantiation operates with acceptable Web response times.

Specifically, the execution times for the above myActivity query in Euler (N3), OO jDREW (POSL), and Rule Responder on average were 157ms, 1483ms, and 5053ms, respectively, measured as the Java system time, running in Java JRE6, Windows XP Professional SP3, on an Intel Core 2 Duo 2.80GHz processor.

For this and similar WellnessRules queries, the major contribution to the overall execution time has come from the ESB (Mule), which is not our focus. Rule Responder operates using a 'hub and spoke' connection architecture, where the OA dispatches network traffic to and from the most appropriate PA. A separate study in distributed querying has done research on minimizing this communication overhead.[5]

The above query could be specialized to produce exactly one solution, e.g. by changing the parameter `outdoors` to `indoors`. It would fail for `?MaxRSVP` greater than 20. Using such queries, WellnessRules participants can check out profiles of other participants to see if they can join, or should create, an activity group.

7   Conclusion

The WellnessRules case study demonstrates FOAF-extending Web 3.0 profile interoperation between a pure Prolog subset (Datalog with Naf) and N3 through RuleML/XML. With all of its source documents available, it has become a major use case for exploring Web AI technology, including scalability of (distributed) knowledge on the (Social Semantic) Web, starting with derivation rules and light-weight ontologies. While WellnessRules so far has probed the OO jDREW, Euler, and Prova engines, its open Rule Responder architecture will make it easy to bring in new engines.

WellnessRules currently emphasizes rulebase translation and querying. These constitute basic services that we intent to extend by superimposed update services, e.g. for changing calendar entries for activities from status `possible` to `planned`; this will require production rules. The next extension will be relevant for `performing` wellness events, which will call for event-condition-action rules. Both of these extended rule types are covered by Reaction RuleML [PKB07].

This case study will also provide challenges for RIF [BK09]: WellnessRules' current derivation rules, for RIF-BLD implementations; its envisioned production rules, for RIF-PRD implementations; and its interest in reaction rules, for a potential RIF Reaction Rule Dialect.

Acknowledgements

We thank the wellness community at NRC-IIT Fredericton for their advice & enthusiasm. Thanks also go to Jos de Roo for his help with the Euler engine. NSERC is thanked for its support through a Discovery Grant for Harold Boley.

References

[AHW+00]   Ainsworth, B., Haskell, W., Whitt, M., Irwin, M., Swartz, A., Strath, S., O'Brien, W., Bassett Jr., D., Schmitz, K., Emplaincourt, P., Jacobs Jr., D., Leon, A.: Compendium of Physical Activities: an update of activity codes and MET intensities (2000)

[BK09]   Boley, H., Kifer, M.: RIF Basic Logic Dialect. W3C Working Draft (July 2009), http://www.w3.org/2005/rules/wiki/BLD

---

[5] http://ruleml.org/papers/EvalArchiRule.pdf

[BLCK⁺08]   Berners-Lee, T., Connolly, D., Kagal, L., Scharf, Y., Hendler, J.:
            N3Logic: A Logical Framework For the World Wide Web. Theory and
            Practice of Logic Programming (TPLP) 8(3) (May 2008)
[Bol04]   Boley, H.: POSL: An Integrated Positional-Slotted Language for Se-
            mantic Web Knowledge (May 2004),
            http://www.ruleml.org/submission/ruleml-shortation.html
[Bol07]   Boley, H.: Are Your Rules Online? Four Web Rule Essentials. In:
            Paschke, A., Biletskiy, Y. (eds.) RuleML 2007. LNCS, vol. 4824, pp.
            7–24. Springer, Heidelberg (2007)
[CB08]   Craig, B.L., Boley, H.: Personal Agents in the Rule Responder Archi-
            tecture. In: Bassiliades, N., Governatori, G., Paschke, A. (eds.) RuleML
            2008. LNCS, vol. 5321, pp. 150–165. Springer, Heidelberg (2008)
[dB09]   de Bruijn, J.: RIF RDF and OWL Compatibility. W3C Working Draft
            (July 2009), http://www.w3.org/2005/rules/wiki/SWC
[Eys08]   Eysenbach, G.: Medicine 2.0: Social Networking, Collaboration, Par-
            ticipation, Apomediation, and Openness. J. Med. Internet Res. 10(3)
            (July-September 2008)
[LBBM06]   Li, J., Boley, H., Bhavsar, V.C., Mei, J.: Expert Finding for eCollab-
            oration Using FOAF with RuleML Rules. In: Montreal Conference of
            eTechnologies 2006, pp. 53–65 (2006)
[MO09]   Molyneaux, H., O'Donnell, S.: Patient Portals 2.0: The Potential for
            Online Videos. In: COACH Conference: e-Health: Leadership in Action
            (May/June 2009)
[PBKC07]   Paschke, A., Boley, H., Kozlenkov, A., Craig, B.: Rule Responder:
            RuleML-Based Agents for Distributed Collaboration on the Pragmatic
            Web. In: 2nd ACM Pragmatic Web Conference 2007, ACM, New York
            (2007)
[PKB07]   Paschke, A., Kozlenkov, A., Boley, H.: A Homogenous Reaction Rule
            Language for Complex Event Processing. In: Proc. 2nd International
            Workshop on Event Drive Architecture and Event Processing Systems
            (EDA-PS 2007), Vienna, Austria (September 2007)
[SLKL07]   Sachinopoulou, A., Leppänen, J., Kaijanranta, H., Lähteenmäki, J.:
            Ontology-Based Approach for Managing Personal Health and Wellness
            Information. In: Engineering in Medicine and Biology Society (EMBS
            2007). 29th Annual Int'l. Conference of the IEEE (August 2007)

A   Signatures of WellnessRules

We give exemplary signatures, for Activity and Meetup. The complete Wellness
signatures are at http://ruleml.org/WellnessRules/Signatures/.

Activity

The objective of the wellness rules is to suggest a wellness activity to the user
according to their profile preferences. Each activity's requirements can be described by the user in an incremental manner.

POSL Syntax

```
myActivity(?ProfileID,?Activity,?Ambience,?MinRSVP,?MaxRSVP,
          ?StartTime,?EndTime,?Location,?Duration,?FitnessLevel).
```

N3 Syntax

```
_: myActivity
    rdf:type        :SuggestedActivity;
    :profileID      ?ProfileID;
    :activity       ?Activity;
    :inOut          ?Ambience;
    :minRSVP        ?MinRSVP;
    :maxRSVP        ?MaxRSVP;
    :startTime      ?StartTime;
    :endTime        ?EndTime;
    :location       ?Location;
    :duration       ?Duration;
    :fitnessLevel   ?FitnessLevel.
```

Usage

ProfileID = The profile identification corresponding to the suggested activity.
Activity = The kind of suggested activity determined.
Indoors/Outdoors = Whether the suggested activity is indoors or outdoors.
MinRSVP = Submitted minimum number of participants.
MaxRSVP = Submitted maximum number of participants.
StartTime = The start time of the suggested activity.
EndTime = The end time of the suggested activity.
Location = The meet up location for the suggested activity.
Duration = The duration of the suggested activity.
FitnessLevel = The required fitness level of the suggested activity.

Meetup

This defines the possible meetup locations for activities of participants. Each location is tied to its map and activity type.

POSL Syntax

```
meetup(?MapID,?Activity,?Ambience,?Location).
```

N3 Syntax

```
_:meetup
    rdf:type    :Meetup;
    :mapID      ?MapID;
    :activity   ?Activity;
    :inOut      ?Ambience;
    :location   ?Location.
```

Usage

MapID = The map that the meetup location is assigned to. In order for profiles to share meetup locations, their MapID's must be the same.

Activity = The kind of activity being performed.

Indoors/Outdoors = Whether the activity is being performed inside or outside.

Location = The name of the meetup location.

B   RDFS Taxonomy of WellnessRules

These RDFS type definitions are being maintained at `http://ruleml.org/WellnessRules/WR-Taxonomy.rdf`.

```
<rdf:RDF
  xmlns:rdf="http://www.w3.org/1999/02/22-rdf-syntax-ns#"
  xmlns:rdfs="http://www.w3.org/2000/01/rdf-schema#"
  xml:base="http://example.org/schemas/vehicles">
    <rdf:Description rdf:ID="Wellness">
      <rdf:type rdf:resource="http://www.w3.org/2000/01/rdf-schema#Class"/>
    </rdf:Description>
    <rdf:Description rdf:ID="Activity">
      <rdf:type rdf:resource="http://www.w3.org/2000/01/rdf-schema#Class"/>
      <rdfs:subClassOf rdf:resource="#Wellness"/>
    </rdf:Description>
    <rdf:Description rdf:ID="Walking">
      <rdf:type rdf:resource="http://www.w3.org/2000/01/rdf-schema#Class"/>
      <rdfs:subClassOf rdf:resource="#Activity"/>
    </rdf:Description>
    <rdf:Description rdf:ID="Running">
      <rdf:type rdf:resource="http://www.w3.org/2000/01/rdf-schema#Class"/>
      <rdfs:subClassOf rdf:resource="#Activity"/>
    </rdf:Description>
    <rdf:Description rdf:ID="Swimming">
      <rdf:type rdf:resource="http://www.w3.org/2000/01/rdf-schema#Class"/>
      <rdfs:subClassOf rdf:resource="#Activity"/>
    </rdf:Description>
    <rdf:Description rdf:ID="Skating">
      <rdf:type rdf:resource="http://www.w3.org/2000/01/rdf-schema#Class"/>
      <rdfs:subClassOf rdf:resource="#Activity"/>
    </rdf:Description>
    <rdf:Description rdf:ID="Yoga">
      <rdf:type rdf:resource="http://www.w3.org/2000/01/rdf-schema#Class"/>
      <rdfs:subClassOf rdf:resource="#Activity"/>
    </rdf:Description>
    <rdf:Description rdf:ID="Hiking">
      <rdf:type rdf:resource="http://www.w3.org/2000/01/rdf-schema#Class"/>
      <rdfs:subClassOf rdf:resource="#Activity"/>
    </rdf:Description>
    <rdf:Description rdf:ID="Baseball">
      <rdf:type rdf:resource="http://www.w3.org/2000/01/rdf-schema#Class"/>
      <rdfs:subClassOf rdf:resource="#Activity"/>
    </rdf:Description>
</rdf:RDF>
```

Semantic Rules on Drug Discovery Data

Sashikiran Challa, David Wild, Ying Ding, and Qian Zhu

Indiana University, 901 E 10th Street,
Bloomington, Indiana, USA
{schalla,djwild,dingying,qianzhu}@indiana.edu
http://www.informatics.indiana.edu

Abstract. Aggregating and presenting a wide variety of information pertinent to the biological and pharmacological effects of chemical compounds will be a critical part of 21st century drug discovery. However there is currently a lack of tools for effectively integrating and aggregating information about chemical compound. In this paper we tackle this problem using Semantic Web Technologies, particularly OWL ontologies,compound-centric RDF networks, and RDF inference to detect relationships between compounds and biological affects, genes,and diseases, and to present information to a user clustered by disease area.

Keywords: Ontology, RDF data model, Jena, reasoner.

1 Introduction

In the field of drug discovery there are many repositories of information of many different types.Many are publicly available. At Indiana University, we recently developed an infrastructure to make a wide variety of these sources available as web services [4], including database services (including PubChem Compound, Pubchem BioAssay, and Drug Bank) computation services (particularly predictive models that predict biological properties of compounds), and literature searches which can identify the compounds and ontological terms that are contained in scholarly journal articles. Aggregating information about a compound from several different web services is being achieved using a tool called WENDI (Web Engine for Non-obvious Drug Information). We have now taken the next step which is to not only integrate information but also make some new inferences and establish relationships among the data aggregated using OWL, RDF and reasoning engines.

Drug discovery data used here was collected from the WENDI aggregate web service. WENDI takes a single compound as a query and then aggregates comprehensive information about a compound from web services that represent several diverse sources (including predictive models,chemical compound databases, and journal articles). The aggregate information was obtained as an XML document from which active compounds, journal articles information was extracted using minidom package in Python Scripting language. This information was converted into RDF triples[2] using Python according to Wendi Ontology (created

A. Gómez-Pérez, Y. Yu, and Y. Ding (Eds.): ASWC 2009, LNCS 5926, pp. 362–364, 2009.

in house using Protege4.0). The Ontology has Chemical Compound, BioAssay, Journal Article as Classes;isSimilarTo, isAssociatedWith, isContainedIn as the Object Properties; hasPubMedID,hasTitle as Data Type Properties.Here are some of the triples in Turtle format generated based on the Wendi Ontology. These triples mean that a compound with ID 15940175 is a type of Chemical Compound and that it is associated with a Bio Assay with an ID 1004.

```
WO:cid15940175 rdf:type owl:ChemicalCompound;
               WO:isAssociatedWith WO:aid1004.
```

2 Framing the Rules

RDF triples thus generated based on Wendi Ontology were loaded into Ont Model class in Jena, a java framework for building semantic web applications. Then rules were written as shown below.

```
[rule1:(?querycmpd WO:isSimilarTo ?cid)
(?cid  WO:isContainedIn ?journal)
->(?qerycmpd  WO:mightBeContainedIn ?journal)];
[rule2:(?querycmpd WO:isSimilarTo ?cid)
(?cid WO:isAssociatedWith ?aid)
->(?querycmpd WO:mightBeAssociatedWith ?aid)];
```

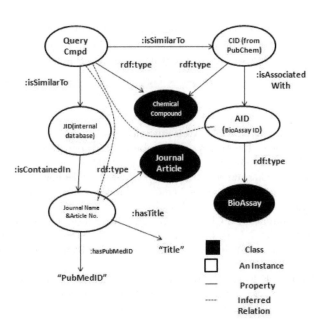

Fig. 1. RDF graph generated on Drug Discovery Data

The rule1 means that if there is a triple with 'isSimilarTo' as property, query compound as subject and say a compound C as object and if there is a triple with 'isContainedIn' as property, compound C as subject, JournalId as object, then infer a relationship between query compound and JournalId by creating a triple with 'mightBeContainedIn' as property, query compound as subject, JournalId as object. Similar is rule2 which infers new property 'mightBeAssociatedWith' between query compound and Bio Assay ID. Rules can be extended to include diseases, gene names. The rules were parsed using the Generic rule reasoner belonging to Reasoner class in Jena. On these additionally generated triples, SPARQL queries were written to output the information about the query compound, Assay ids, Journal article titles, and their Pubmed ids.

An example of the kinds of inference we can produce: If a compound A was to found to be similar to compound B and if compound B was found to be active in a Bio Assay C, then an inference that Compound A might be active in Bio Assay C is one such inference. And say if a compound A was found to be similar to compound B and if compound B was found to be contained in a Journal C, then an inference that compound A's relevant information could be contained in Journal C is another inference achieved.

3   Conclusion

We are now able to infer relationships between compounds, genes and diseases based on RDF chains which are interpretable by a medicinal chemist. Thus we can not only associate a compound with a disease, say, but also present the evidence for this association. Thus the work described here helps us to realize the potential of having the data as RDF data model triples based on an Ontology and the potential of rules in data integration and knowldege discovery.

References

1. Stephens, S.: Enabling Semantic Web Inferencing with Oracle Technology: Applications in Life Sciences. In: Adi, A., Stoutenburg, S., Tabet, S. (eds.) RuleML 2005. LNCS, vol. 3791, pp. 8–16. Springer, Heidelberg (2005)
2. Wang, X., Robert, G., Almeida, J.S.: From XML to RDF: how semantic web technologies will change the design of 'omic' standards. Nature Biotechnology 23(9), 1099–1103 (2005)
3. Hugo, Y.K.L., Marenco, L., Clark, T., Gao, Y., Kinoshita, J., Shepherd, G., Miller, P., Wu, E., Wong, G.T., Liu, N., Crasto, C., Morse, T., Stepehen, S., Cheung, K.-H.: AlzPharm: integration of neurodegeneration data using RDF. BMC Bioinformatics 8 (2007)
4. Dong, X., Gilbert, K.E., Guha, R., Heiland, R., Kim, J., Pierce, M.E., Fox, G.C., Wild, D.J.: Web service infrastructure for chemoinformatics. Journal of Chemical Information and Modeling 47(4), 1303–1307 (2007)

RepOSE: An Environment for Repairing Missing Ontological Structure

Patrick Lambrix, Qiang Liu, and He Tan

Department of Computer and Information Science,
Linköpings universitet, 581 83 Linköping, Sweden

Developing ontologies is not an easy task and often the resulting ontologies are not consistent or complete. Such ontologies, although often useful, lead to problems when used in semantically-enabled applications. Wrong conclusions may be derived or valid conclusions may be missed. Defects in ontologies can take different forms. Syntactic defects are usually easy to find and to resolve. Defects regarding style include such things as unintended redundancy. More severe defects are the modeling defects which require domain knowledge to detect and resolve, and semantic defects such as unsatisfiable concepts and inconsistent ontologies.

In this paper we present a system, RepOSE (*Rep*air of *O*ntological *S*tructure *E*nvironment), that tackles a special case of the problem of repairing modeling defects, i.e. the repairing of missing is-a relations, and to our knowledge this system is the first in its kind. In the given setting it is known that a number of intended is-a relations are not present in the source ontology. The problem is then to find is-a relations (called a *structural repair*) such that when these are added to the ontology, all missing is-a relations can be derived from the extended ontology. For formal definitions we refer to [1]. Although the missing is-a relations themselves constitute a structural repair, this may not be the most interesting solution for the domain expert. For instance, in a real case based on the Anatomy track from the Ontology Alignment Evaluation Initiative 2008, we know that an is-a relation between *wrist joint* and *joint* is missing and could be added to the ontology. However, knowing that there is an is-a relation between *wrist joint* and *limb joint*, a domain expert will most likely prefer to add an is-a relation between *limb joint* and *joint* instead. This is more informative and would lead to the fact that the missing is-a relation can be derived. In this particular case, it would also lead to the repairing of 6 other missing is-a relations (e.g between *elbow joint* and *joint*).

We have developed a tool that generates and recommends possible ways to repair the structure of the ontology (based on named concepts and subsumption axioms) and that allows a domain expert to repair the structure of an ontology in a semi-automatic way. As input our system takes an ontology in OWL format as well as a list of missing is-a relations. We use a framework and reasoner provided by Jena (version 2.5.7). The domain knowledge that we use is WordNet and the Unified Medical Language System. The ontology and missing is-a relations can be imported using the *Load/Derive Missing IS-A Relations* button. The user can see the list of missing is-a relations under the *Missing IS-A Relations* menu (see figure 1). In our example case there are 7 missing is-a relations. Clicking on the *Compute Repairing Actions* button, results in the computation of possible ways to repair. The user can select which missing is-a relation to repair first. The missing is-a relations are ranked with respect to the number of possible

A. Gómez-Pérez, Y. Yu, and Y. Ding (Eds.): ASWC 2009, LNCS 5926, pp. 365–366, 2009.

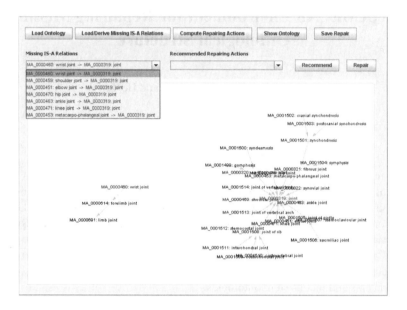

Fig. 1. Screenshot

ways to repair, and the first in the list may therefore be a good starting point. When the user chooses a missing is-a relation, the possible sources and targets for is-a relations that can be chosen for repairing are shown in the panels on the left and the right, respectively. The concepts in the missing is-a relation are highlighted in red. The figure illustrates this for the missing is-a relation between *wrist joint* and *joint*. The user can also ask for recommendations by clicking the *Recommend* button. In our case, the system recommends to add an is-a relation between *limb joint* and *joint*. In general, the system presents a list of recommendations. By selecting an element in the list, the concepts in the repairing action are highlighted in the panels. The user can repair a missing is-a relation by selecting a concept in the left panel and a concept in the right panel and clicking on the *Repair* button. The is-a relation is then added to the ontology, and may lead to updates for other missing is-a relations. At all times during the process the user can inspect the ontology by clicking the *Show Ontology* button. Newly added is-a relations will be highlighted. After adding the is-a relation between *limb joint* and *joint*, not only (*wrist joint,joint*) is repaired, but all other missing is-a relations as well, as they can be derived in the extended ontology. The list of missing is-a relations is therefore updated to be empty. After completing the repair of all missing is-a relations, the repaired ontology can be exported into an OWL file by clicking the *Save Repair* button.

Reference

1. Lambrix, P., Liu, Q., Tan, H.: Repairing the missing is-a structure of ontologies. In: Gómez-Pérez, A., Yu, Y., Ding, Y. (eds.) ASWC 2009. LNCS, vol. 5926, pp. 76–90. Springer, Heidelberg (2009)

SHARE: A Semantic Web Query Engine for Bioinformatics

Ben P. Vandervalk, E. Luke McCarthy, and Mark D. Wilkinson

The Providence Heart + Lung Research Institute at St. Paul's Hospital,
University of British Columbia, Department of Medical Genetics,
Vancouver, BC, Canada
ben.vvalk@gmail.com

Abstract. Driven by the goal of automating data analyses in the field of bioinformatics, SHARE (Semantic Health and Research Environment) is a specialized SPARQL engine that resolves queries against Web Services and SPARQL endpoints. Developed in conjunction with SHARE, SADI (Semantic Automated Discovery and Integration) is a standard for native-RDF services that facilitates the automated assembly of services into workflows, thereby eliminating the need for ad hoc scripting in the construction of a bioinformatics analysis pipeline.

The SHARE Query System

The task of coordinating datasets and software packages from different research laboratories is a recurring practical problem in bioinformatics [1]. While the size, quantity, and diversity of bioinformatics resources continues to grow rapidly, there are still no widely adopted standards for encoding data or designing software interfaces. As a result, the orchestration of bioinformatics analyses typically requires a large time investment for studying database schemas and for implementing ad hoc scripts that connect incompatible data formats and software.

To address this problem, the authors have developed the SADI service standard and the SHARE query engine. SADI is a straightforward set of recommendations for exposing bioinformatics resources as RDF-native Web Services, while SHARE is a specialized SPARQL query engine that enables the flexible rewiring of SADI services into complex workflows. In more detail, the two data sources queried by SHARE are:

SADI services: SADI [2] services are stateless, native-RDF Web Services that are invoked by means of an HTTP POST request. Both the request and the response are single RDF documents, and the input/output URIs within these documents are instances of a provider-specified input/output OWL class. The principal rule regarding input/output OWL instances is that they must be rooted at the same URI (i.e. the input URI). It follows from this rule that a service must report its results by attaching new predicates to the input URI, and thus *the precise relationship between the input data and the output data is explicitly stated.* The SADI service registry identifies the predicates that a service attaches to its

A. Gómez-Pérez, Y. Yu, and Y. Ding (Eds.): ASWC 2009, LNCS 5926, pp. 367–369, 2009.

input URIs by comparison of its input and output OWL classes, and services may be retrieved from the registry using these predicates as a key.

SPARQL Endpoints: Any standard SPARQL endpoint may be registered, indexed and queried by the SHARE query engine. As in the DARQ [3] system, SPARQL endpoints are indexed by the set of predicates that occur within their datasets.

The syntax of a SHARE query is identical to that of a standard SPARQL query. SHARE answers a SPARQL query by successively resolving each triple pattern in the query against a matching set of SADI services/SPARQL endpoints. A service/endpoint matches a triple pattern with predicate p if that service/endpoint attaches p (or the OWL inverse of p) to its input. The output triples generated by services/endpoints are accumulated in a local triple store, and the subject/object values of output triples are recorded as bindings (values) for any unbound variables in the current pattern. Bindings are carried forward into subsequent triple patterns, and thus the resolution of a particular pattern may require multiple invocations of the same service/endpoint. After all triple patterns in the query have been resolved, the solutions are displayed.

The SHARE demonstration site [4] allows users to issue SPARQL queries against a sample set of SADI services and SPARQL endpoints. By accessing several well known databases in bioinformatics, the provided services generate predicates that connect proteins (UniProt), genes (KEGG/Entrez), molecular structures (PDB), biological pathways (KEGG), and several other types of biological entities. The relationships between the service-generated predicates (and thus the services themselves) are depicted schematically at [5]. Fig. 1 shows an example query which asks: "What motifs are present in enzymes of the human caffeine metabolism pathway?". (Motifs are contiguous, evolutionarily-conserved subsequences of proteins that often have a known chemical function.)

```
PREFIX pred: <http://sadiframework.org/ontologies/predicates.owl#>
PREFIX kegg: <http://biordf.net/moby/KEGG_PATHWAY/>
SELECT ?gene ?protein ?motif
WHERE {
    kegg:hsa00232 pred:hasPathwayGene ?gene .
    ?gene pred:encodes ?protein .
    ?protein pred:hasMotif ?motif .
}
```

Fig. 1. A sample SHARE query which asks: "What motifs are present in enzymes of the human caffeine metabolism pathway?"

SHARE answers the example query in Figure 1 by resolving each of the triple patterns in the WHERE clause sequentially. In the first step, SHARE identifies the set of services/endpoints that attach the predicate **pred:hasPathwayGene** to their input URIs. In this case, the SADI registry contains one such service called *getKEGGGenesByPathway*. The URI for the human caffeine metabolism pathway (**kegg:hsa00232**) is sent as input to *getKEGGGenesByPathway*, and in response, the service generates a set of triples with **kegg:hsa00232** as the subject, **pred:hasPathwayGene** as the predicate, and KEGG genes as the objects. The object URIs (genes) of these triples then become bindings for the variable **?gene**. In the second step, SHARE identifies services/endpoints that attach the predicate **pred:encodes** to their input URIs, and finds a service called *KeggToUniProt*. Each binding of **?gene** is sent as an input to this service, and in response, *KeggToUniProt* returns a set of triples that connect genes to proteins via the predicate **pred:encodes**. In the third and final step, the query engine resolves **pred:hasMotif** to a service called *FindMotifById*, in order to connect the bindings of **?protein** to the URIs of corresponding motifs.

While the first two services invoked by the sample query perform simple database retrievals, the third service (*FindMotifById*) invokes a search algorithm for discovering known patterns in sequence data. This demonstrates that SHARE is suitable for coordinating both data retrieval and arbitrary software analysis within a single query, which is a highly desirable charactertistic for our bioinformatics user base.

The main innovation of the SHARE system is its ability to resolve SPARQL queries across Web Services in a transparent manner. In this sense, it can be viewed as an extension of the ideas underlying the SemWIQ [6] and DARQ [3] projects that enable querying over SPARQL endpoints. The mapping from SPARQL queries to Web Services is facilitated by the SADI service standard, which allows services to be discovered via the relationships that connect their inputs and outputs. In addition, SADI offers a service model that is straightforward in comparison to existing Semantic Web Service standards such as OWL-S [7] and WSMO [8], provided only stateless and non-world-altering services are required.

References

1. Stein, L.: Creating a Bioinformatics Nation. Nature 417, 119–120 (2002)
2. The SADI Web Service Framework, http://sadiframework.org
3. Quilitz, B., Leser, U.: Querying Distributed RDF Data Sources with SPARQL. In: Bechhofer, S., Hauswirth, M., Hoffmann, J., Koubarakis, M. (eds.) ESWC 2008. LNCS, vol. 5021, pp. 524–538. Springer, Heidelberg (2008)
4. SHARE demonstration, http://biordf.net/cardioSHARE/query
5. SHARE predicates page, http://biordf.net/cardioSHARE/predicates.html
6. Langegger, A., Wöß, W., Blöchl, M.: A Semantic Web Middleware for Virtual Data Integration on the Web. In: Bechhofer, S., Hauswirth, M., Hoffmann, J., Koubarakis, M. (eds.) ESWC 2008. LNCS, vol. 5021, pp. 493–507. Springer, Heidelberg (2008)
7. Martin, D., Paolucci, M., McIlraith, S., et al.: Bringing Semantics to Web Services: The OWL-S Approach. In: Cardoso, J., Sheth, A.P. (eds.) SWSWPC 2004. LNCS, vol. 3387, pp. 26–42. Springer, Heidelberg (2005)
8. Roman, D., Keller, U., Lausen, H., et al.: Web Service Modeling Ontology. Applied Ontology 1(1), 77–106 (2005)

IP-Explorer: A Semantic Web Based Intellectual Property Knowledge Base and Trading Platform

Lei Zhang[1], Hugh Glaser[2], Hai-Tao Zheng[1], and Ian Millard[2]

[1] Tsinghua-Southampton Web Science Laboratory at Shenzhen,
Graduate School at Shenzhen, Tsinghua University
{zhanglei, zheng.haitao}@sz.tsinghua.edu.cn
[2] School of Electronics and Computer Science,
University of Southampton
{hg, icm}@ecs.soton.ac.uk

Abstract. In this paper, we demonstrate IP-Explorer, a semantic web based IP knowledge base and trading platform with the following characteristics: First, it is based on the semantic web technology; Second, it is built by mining the China High-Tech Fair(CHTF) database and the State Intellectual Property Office of PRC (SIPO) database; Third, "Smart Pushing" proactively pushes useful information based on user profiles, browsing and searching records, and activities of peers; Fourth, "IP family" collects IP's from different industry sectors and integrate them as a "family", helping the user to broaden his horizon; Fifth, "Technology Roadmap" provides a visualization tool for decision makers to observe technology history and trends.

Keywords: Semantic Web, Intellectual Property, Data Mining.

1 Introduction

Intellectual Property (IP) is one of the key incentives that help to keep industry innovations active by granting exclusive rights to the owners. Acquisition of new IP's is considered to be a feasible way to survive the financial crisis for Small and Medium size Enterprises (SMEs). There have been a number of IP trading platforms supporting browsing and searching of IP's. However, most of them require good background knowledge when selecting searching keywords in order to get acceptable results.

We are facing a dilemma now. On the one hand, there have been a huge number of IP's available. On the other hand, owners of SME's find it difficult to use current IP systems. Therefore, we need to build a new IP knowledge base and trading platform to bridge IP providers and consumers. The employment of the Semantic Web technology makes it possible.

2 The System

The platform builds a semantic web database for IP's from the High-Tech Fair [1] and the SIPO [3][1].

---

[1] To access the data, visit http://www.weblab.sz.tsinghua.edu.cn/

A. Gómez-Pérez, Y. Yu, and Y. Ding (Eds.): ASWC 2009, LNCS 5926, pp. 370–373, 2009.

The China High-Tech Fair is the largest and most influential exhibition on science and technology in China, regarded as the No.1 Chinese exhibition in this industry. Each year, CHTF would receive more than 500,000 people, and accomplish a total of 13 billion US dollars in terms of product and technical transactions. The database of CHTF contains information about 16726 projects along with their kernel technology IP numbers.

SIPO is affiliated to the State Council of China with the responsibility to organize, administrate, and coordinate IPR protection work nationwide. The database of SIPO contains more than 7'000'000 entries of granted IP's classified with the International Patent Classification (IPC)[2].

Our IP-Explorer system is based on the following two steps of data processing:

1. Semantic Web representation of IP-related information including IP number, author, affiliation, IPC category, abstract, and full-text. This is implemented using 3store [5] RDF repository and the core RKBExplorer [4] as user interface.

2. An ontology design for the IPC tree structure and a cross-link discovery algorithm to establish links among semantic-related categories. For original tree links, the ontology is simply "isA" relation. For cross-links connecting semantic-related IPC categories, they often take forms of "useMaterial" and "useDesign". The cross-links are discovered by abstracting keywords for IP descriptions, clustering co-authors, and then mapping back to the IPC structure.

Based on the semantic web technology and data mining, IP-Explorer is different from traditional IP platforms in the following three phases of transaction:

- **Pre-sale Smart Pushing:** Instead of asking the consumers to "pull" information by searching, the new platform should proactively "push" related IP's based on the user profile, browsing and searching history, and peer activities. This feature helps users without computer skills to get exposed to many useful IP's quickly and is supported by client-end multiple-panel browser and server-end mining and pushing engines, as shown in Fig. 1 and Fig. 2.

- **In-sale IP Family Integration and Technology Roadmap:** The user may start with one query on chips without realizing the importance of other devices, packaging, design, or materials. By integrating IP's from different industry sectors, the system would help the user to broaden his target area. Technology Roadmap provides a visualization tool for decision makers by observing technology development over time and indicating future trends. Analysis of IP location distribution (see Fig. 3) is provided to evaluate industry-chain maturity.

---

[2] The International Patent Classification (IPC), established by the Strasbourg Agreement 1971, provides for a hierarchical system of language independent symbols for the classification of patents and utility models according to the different areas of technology to which they pertain.

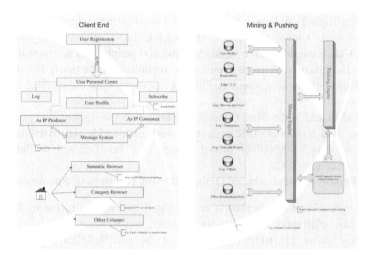

Fig. 1. Client-side Module   **Fig. 2.** Mining and Pushing Engines

Fig. 3. Geographic proximity analysis of China photovoltaic industry technology roadmap

■ **After-sale Service and Support:** The system helps to identify people with matching expertise to provide after-sale service and support. The feature is similar to that of ExpertFinder [1], but the module is seamlessly integrated as part of our platform instead of another separate system.

3   Conclusion

IP-Explorer, to the best of our knowledge, is the first Semantic Web based IP knowledge base and trading platform. Ontology design and main features are original and beneficial.

References

1. ExpertFinder Initiative, http://expertfinder.info/
2. The High-Tech Fair, http://www.chtf.com/
3. State Intellectual Property Office of China, http://www.sipo.gov.cn/
4. Glaser, H., Millard, I., Carr, L.: RKBExplorer: Repositories, Linked Data and Research Support. Eprints User Group, Open Repositories 2009, Atlanta, GA, USA (May 20, 2009)
5. Harris, S., Gibbins, N.: 3store: Efficient Bulk RDF Storage, http://eprints.ecs.soton.ac.uk/7970/

Author Index

Printing: Mercedes-Druck, Berlin
Binding: Stein+Lehmann, Berlin